Biomass Burning in South and Southeast Asia

Biomass Burning in South and Southeast Asia

Mapping and Monitoring, Volume One

Edited by
Krishna Prasad Vadrevu,
Toshimasa Ohara, and Christopher Justice

CRC Press
Taylor & Francis Group
Boca Raton London New York

CRC Press is an imprint of the
Taylor & Francis Group, an **Informa** business

First edition published 2021
by CRC Press
6000 Broken Sound Parkway NW, Suite 300, Boca Raton, FL 33487-2742

and by CRC Press
2 Park Square, Milton Park, Abingdon, Oxon, OX14 4RN

© 2021 Taylor & Francis Group, LLC

CRC Press is an imprint of Taylor & Francis Group, LLC.

ISBN: 978-0-367-07681-8 (hbk)
ISBN: 978-1-032-01351-0 (pbk)
ISBN: 978-0-429-02225-8 (ebk)

Typeset in Times
by codeMantra

Contents

Introduction

SECTION I   Mapping and Monitoring of Fires, including Burned Areas

SECTION II   Land Use, Forests, and Biomass Burning

SECTION III   Climate Drivers and Biomass Burning

Contents

Foreword

It is my pleasure to write a foreword to the two-volume book on the topic of vegetation fires and biomass burning edited by Dr. Krishna Prasad Vadrevu, Dr. Toshimasa Ohara, and Prof. Christopher Justice. I had the privilege to write a foreword to their earlier 2018 book on Land Atmospheric Interactions in South and Southeast Asia, which attracted broad attention. While society is facing more envi-

ronmental problems with disasters, such as fires, floods, droughts, and landslides, scientists work hard on their detection, spatiotemporal variability, and impacts and quantify the uncertainties. Remote sensing and geospatial technologies offer unique opportunities in mapping and monitoring natural disasters and their impacts and can provide useful information for land management.

The current two-volume compilation of articles focuses on biomass burning processes, the tools for its mapping and monitoring, and quantifying impacts on atmospheric composition including air quality. The contributions in this book are a result of several international workshops organized in Asia under the NASA South/Southeast Asia Research Initiative (SARI, http://sari.umd.edu), funded under the Land-Cover/Land-Use Change Program (http://lcluc.umd.edu), which I manage at NASA Headquarters. Prior to SARI, the NASA LCLUC program contributed to other regional initiatives in Africa, northern Eurasia, and South America by soliciting, selecting, and funding research projects and supporting NASA scientists to attend regional science workshops. Recognizing the pervasive land-use changes driven by rapid population growth and economic development in Asia, the NASA LCLUC program launched the SARI, with Dr. Krishna Prasad Vadrevu as the lead, a few years ago. Since then, SARI workshops have been organized annually, with more than 150 scientists in attendance, and where important regional science issues have been addressed and research needs and priorities identified. One of the topics requiring immediate attention is biomass burning and the associated pollution widespread in the region.

As Land-Cover/Land-Use Change Program Manager, during my 21-year tenure at NASA Headquarters, I have funded multiple projects that involved land–atmospheric interactions and included analysis of drivers and impacts. The NASA LCLUC program is a truly interdisciplinary science program within the Science Mission Directorate. The program is distinct from other NASA discipline programs as it integrates both physical and socioeconomic sciences with the use of remote sensing observations in an interdisciplinary framework to address environmental and societal issues. The LCLUC program aims to develop and use NASA space- and airborne remote sensing technologies, relying on US and non-US satellite data sources, to improve our understanding of human interactions with the environment. One of the critical questions addressed in the program is: "What are the causes and consequences of LCLUC?" Several decades of research in the SARI region have revealed the drivers of extensive modifications of the natural environment and environmental impacts of LCLUC, ranging from changes in atmospheric composition and air quality, biomass burning being a significant contributor.

My scientific interests in biomass burning extend over 30 years back when I worked on my research with remote sensing applications over land and in the atmosphere. During the early 1990s, while working at NOAA, I collaborated with Israeli scientists on remote sensing of atmospheric processes with Dr. U. Dayan on Dust Intrusions Events into the Mediterranean Basin (*J. Appl. Met.*, 1991, 30, 1185–1199) and with Prof. D. Rosenfeld on Retrieving Microphysical Properties Near the Tops of Potential Rain Clouds by Multispectral Analysis of AVHRR Data (*Atmos. Research J.*, 1994, 34, 259–283). However, most of my research was on detecting/screening clouds to study land surface. I have used remote sensing to address atmosphere–vegetation feedbacks across diverse geographical regions and at multiple scales, including land–climate interactions using long-term observations from satellites. I have always promoted the use of satellite data in climatology even though 30 years ago the length of time series might not have been sufficient but hoped that continuous observations would one day provide a decent observational base for analyzing climate-related land-cover anomalies. For example, I specifically published my paper of 2000 on 1997–1998 fires in Indonesia (*Bulletin Amer. Met. Soc.* 2000, 81, 1189–1205) in a meteorological journal in an attempt to attract the attention of meteorologists and climatologists to the richness of the Advanced Very High-Resolution Radiometer (AVHRR) observations for land–climate applications. In that paper, I demonstrated how many land and atmospheric products can be derived from AVHRR alone and highlighted the compound effect of fires from slash-and-burn agriculture and one of the strongest El Niño events on the atmosphere (smoke and cloud microphysics) and land (surface reflectance, temperature, and greenness).

This two-volume book is a collection of papers from several SARI workshops organized in South/Southeast Asian countries. More than 85 authors contributed to the book chapters. Scientists from different disciplines gathered systematic information and knowledge guided by robust theories and models useful for fire management and analyzing atmospheric impacts. Each chapter is unique, highlighting the drivers of fire regime to study its characteristics and biomass burning impacts in the region, integrating remote sensing technologies with ground-based measurements. The material in this book will provide information pertinent to fire mitigation, preparedness, response, and monitoring vegetation recovery in various ecosystems. Several chapters also highlight the impacts of biomass burning pollution on the environment in South/Southeast Asian countries. Overall, the scientific results will be of interest to the research community and useful to policymakers, emergency managers, and other practicing professionals in the field. With contributions from several international scholars and professionals, and edited by three eminent scientists, this two-volume book is a major addition to the research and applications literature on fires and biomass burning in South/Southeast Asia (S/SEA). I am impressed with the quality of cutting-edge science and contributions from regional scientists. The book is a timely contribution to the NASA LCLUC SARI science and reflects the current status in this research field.

The Editors of this book are highly active and reputed researchers. I commend all the team members, particularly the Chief Editor, Dr. Krishna Prasad Vadrevu, on accomplishing the exhaustive task of creating this comprehensive book, rich with material and the latest references. I am confident that it will motivate researchers on this pertinent topic and trigger innovative ideas. I welcome more such contributions from the US and regional scientists and wish you all an exciting and informative reading.

Dr. Garik Gutman
Land-Cover/Land-Use Change Program Manager
NASA Headquarters, Washington DC., USA

Preface

In the past few decades, there has been an increasing intensity and frequency of natural disasters worldwide with severe socioeconomic impacts, especially in South/Southeast Asia (S/SEA). Of the different disasters, vegetation fires are one of the most critical. Fires can destroy the vegetation cover, thereby impacting the biodiversity, nutrient cycles, and hydrology. The vegetation combustion releases large volumes of radiatively active gases, aerosols, and other chemically active species at the regional and global scales, significantly influencing the radiative budget and atmospheric chemistry. The smoke pollutants released during the burning can impact air quality and pose risks to human health. Thus, mapping and monitoring of fires and biomass burning events and quantifying their impacts on the environment gain significance. Addressing fire characteristics and related impacts is quite challenging as they show considerable spatial and temporal variability due to fuel loads, moisture content, temperature, topography, humidity, wind speed, etc. The challenge facing scientists worldwide is to effectively integrate various datasets, approaches, and technologies to help inform those that manage and mitigate fires in different ecosystems.

This two-volume book (Volume 1: "Biomass Burning in South/Southeast Asia – Mapping and Monitoring"; and Volume 2: "Biomass Burning in South/Southeast Asia – Impacts on the Biosphere") is a collection of papers from several South/Southeast Asia Research Initiative (SARI; sari.umd.edu) workshops organized in Asia since 2015. SARI is a NASA Land-Cover/Land-Use Change Program (lcluc.umd.edu)-funded research activity. SARI's goal is to develop innovative regional research, education, and capacity-building programs involving state-of-the-art remote sensing, natural sciences, engineering, and social sciences to enrich LCLUC science in S/SEA. To address LCLUC science, SARI has been utilizing a systems approach to problem-solving that examines both biophysical and socioeconomic aspects of land systems, including the interactions between land use and climate and the interrelationships among policy, governance, and land use. During the last few decades, fire events in S/SEA have attracted international attention due to the significant transboundary pollution they have been causing, impacting air quality and human health. Several SARI meetings identified fires and biomass burning as a vital issue in the region, requiring immediate attention. This two-volume book series on biomass burning was planned to meet community research and application needs. All three editors of the current volumes have published significant papers on fires, biomass burning, and emissions.

Volume 1 is divided into three sections: A) Mapping and Monitoring of Fires, including Burned Areas; B) Land Use, Forests, and Biomass Burning; and C) Climate Drivers and Biomass Burning. These sections are preceded by an introductory chapter in which we provide details on fire regime characteristics in S/SEA, including spatial and temporal patterns, intensity, and burned area statistics. Also, we highlight some of the critical research needs and priorities on fire mapping and monitoring, drivers, emissions, and health impacts including international policy and capacity-building efforts needed to address biomass burning pollution in S/SEA.

Section 1, "Mapping and Monitoring of Fires, including Burned Areas," includes seven different chapters, all focusing on mapping, monitoring, and spatial patterns of fires using remote sensing and geospatial technologies. Chapter 2 focuses on using nighttime Visible Infrared Imaging Radiometer Suite (VIIRS) satellite data for detecting smoldering fires in Indonesia. Detecting smoldering fires in peatlands is essential as they are a significant source of transboundary smoke and greenhouse gas emissions in Indonesia. The authors highlight the use of a spectral un-mixing procedure for identifying the low-intensity fires. Their results indicate that it is possible to detect and characterize combustion ranging from high-temperature natural gas flaring and flaming, glowing embers, and low-temperature smoldering biomass burning events. In Chapter 3, the authors use the Sea and Land Surface Temperature Radiometer (SLSTR) to detect active fires over forests, peatlands, and croplands. They highlight the importance of regional fine-tuning of fire detection algorithms for effective mapping and monitoring. In Chapter 4, the authors use the Sentinel-1A&1B Synthetic Aperture Radar (SAR) data to detect burnt areas in Laos. They highlight significant differences in burnt areas compared to other land-cover classes such as forests, urban, croplands, and barren, using VV and VH SAR data. In Chapter 5, the authors apply differential interferometry SAR (DInSAR) analysis using ALOS-2 PALSAR-2 data to monitor peatlands before and after a fire in Central Kalimantan, Indonesia. The authors demonstrate the potential of DInSAR for delineating peatland surface height difference due to fire incidents. In Chapter 6, the authors use very high-resolution PlanetScope data for burnt area delineation in India's Uttarakhand State. Using the daily PlanetScope data, the authors capture burnt area progression more robustly than with Landsat data for the study area. Chapter 7 focuses on spatial distribution and intensity of agriculture and forest fires in north India. Results suggest that Punjab, Haryana, and the foothill plains and southern slopes of Himalayan Uttarakhand collectively contribute to about 85% of the total number of fires in north India. In Chapter 8, the authors use various spatial statistical techniques to quantify spatial patterns of fire events in Laos and Cambodia. Their results on scaling and clustering of fires can be useful for fire management and mitigation efforts in these countries.

Section 2, "Land Use, Forests, and Biomass Burning," includes three different chapters. Chapter 9 focuses on both the historical and current fire situations in Bhutan. The authors provide details on the local geography, vegetation types, land use/cover, protected areas, and satellite-derived fire analysis. For effective fire prevention and mitigation, the authors call for building robust spatial data infrastructure and implementing viable management practices involving local stakeholders. In Chapter 10, the authors highlight important biomass burning sources and impacts in Malaysia. The authors infer that air quality in Malaysia is highly impacted due to transboundary pollution from neighboring Indonesia, mainly attributed to the conversion of forests to oil palm plantations and related biomass burning events. In Chapter 11, the authors describe the swidden agriculture and biomass burning pollution in the Philippines. The authors report a significant decline in swidden agriculture in the past few decades.

Section 3, "Climate Drivers and Biomass Burning," comprises five different chapters. Chapter 12 provides an exhaustive account of fire danger indices, methods, and rating systems. The author also highlights the usefulness of satellite data for retrieving the meteorological, fire, vegetation, and topographic parameters useful for fire mapping and monitoring studies. In Chapter 13, the authors report the air pollution situation at Palangkaraya, Kalimantan, Indonesia, and attribute drought conditions during El Niño as an important driver of fires and emissions. Similar conclusions were presented in Chapter 14, focusing on Sumatera and Kalimantan in Indonesia using satellite remote sensing, aerosols, and meteorological information. Both chapters highlight the role of climatology in fire occurrence in Indonesia. Chapter 15 shows how daily weather data are related to fire events in Nilgiris, Western Ghats, India. The author identifies the potential evapotranspiration as a significant predictor of the anomalous fire activity at landscape scales in the study area. Finally, in Chapter 16, the last chapter, the authors summarize the current understanding of biomass burning signals on Tibetan glaciers. The authors infer that levoglucosan accumulation on glaciers can be used as a proxy for detecting regional biomass burning changes. The authors attribute the increase in wildfires at the beginning of the 21st century to climate change events.

These contributions cover various topics on fire, biomass burning, mapping, and monitoring. The book will serve as a valuable source of information for remote sensing scientists, geographers, ecologists, atmospheric scientists, environmental scientists, and all who wish to advance their knowledge on fires and biomass burning in S/SEA. As Editors, we would like to thank the authors for their contributions and cooperation in bringing out this volume. We also acknowledge the tremendous effort of the reviewers who have spent their valuable time working on this book's contents. We are immensely grateful to Irma Britton and Rebecca Pringle, CRC Press, and codemantra team for their constant help with this publication. We wish you all an exciting read.

Editors
Krishna Prasad Vadrevu, Huntsville, Alabama, USA
Toshimasa Ohara, Tsukuba, Japan
Christopher Justice, College Park, Maryland, USA

Editors

Dr. Krishna Prasad Vadrevu is a remote sensing scientist at NASA Marshall Space Flight Center, Huntsville, Alabama, USA. His research focuses on land-cover and land-use change (LCLUC) studies, fires, and biomass burning emissions. He has 20 years of research experience in satellite remote sensing. He is currently serving as the Deputy Program Manager for the NASA LCLUC Program (lcluc.umd.edu) and leading the South/Southeast Asia Research Initiative (www.sari.umd.edu).

Dr. Toshimasa Ohara is a scientist at the National Institute of Environmental Studies, Tsukuba, Japan. He has 32 years of research experience in air quality modeling, emission inventories, and pollution research. He is a lead developer for Regional Emission Inventory in Asia (REAS) and one of the highly cited researchers on the emissions. He is currently working on linking top-down and bottom-up approaches for emission quantification from different sectors in Asia.

Dr. Christopher Justice is Chair of the Department of Geographical Sciences, University of Maryland, College Park, USA. He has 40 years of research experience in remote sensing. His current research is on land-cover and land-use change and global agricultural monitoring using remote sensing. He is an authority on satellite remote sensing of fires. He serves as Project Scientist for the NASA LCLUC Program and the Land Discipline Lead for the NASA MODIS and the Suomi-NPP VIIRS Science Team. He is the Co-Chair of the GEO Global Agricultural Monitoring (GEOGLAM) Initiative, Chief Scientist for NASA HARVEST, and Chair of the international Global Observations of Forest and Land Use Dynamics (GOFC-GOLD) program.

Dedication

We dedicate the book to Dr. Badarinath Venkata Srinivasa Kandalam, an internationally renowned physicist for his significant research contributions in remote sensing, vegetation fires, aerosols, and climate change studies. He was popularly known as KVSB among the research community.

Born on the 18th May 1959, Badarinath received his bachelor's degree in Physics, Chemistry and Mathematics (1977), Master's degree in Physics with Electronics specialization from Andhra University, Visakhapatnam (1979), and a Ph.D. degree in Experimental Solid State Physics from Indian Institute of Technology, Madras (1984).

From 1986 till May 2011, he worked as a scientist at different levels at the National Remote Sensing Agency (Department of Space-Government of India), Hyderabad, India. He focused on remote sensing research applications to environmental problems. As a part of the same, he led the International Geosphere-Biosphere Program Section and the Atmospheric Science section at the NRSA. During his career, he received numerous awards. He received the International START visiting scientist award to work at National Resources Ecology Laboratory, Colorado State University, Colorado, U.S.A., during 1996–1997. He also worked as a visiting Professor at the Centre for Climate System Research headed by Prof. Nakajima, University of Tokyo, Japan, from July 2010 to October 2010. For several years, he also served as an expert on satellite remote sensing of vegetation fires from India and contributed to the international GOFC-GOLD Fire Implementation team.

Dr. Badarinath supervised 10 Ph.D. students and more than 50 M.Tech and B.Tech project studies on Digital Image Processing, Atmospheric sciences, Land surface temperature estimation studies, Environmental Monitoring and modeling, Mesoscale Model studies for dust prediction, cyclone track prediction, and heatwave conditions. He had an excellent publication record, International Journals: 134; National Journals: 73; Total Publications: 207. He served as an editorial member for five various journals.

Dr. K.V.S. Badarinath (1959–2012)

Several students received various awards under his supervision, including IGBP-START Young Scientist Award, Indian Science Congress Young Scientist Award, ISRS-OPTOMECH Award, Kalpana Chawla Memorial Award, and others. Dr. Krishna Vadrevu, the main editor of this book, is Dr. Badarinath's first Ph.D. student.

Dr. Badarinath's contributions to Indian science were exemplary. Several of his publications are replete with examples of outstanding contributions to remote sensing and atmospheric science. His research focused on integrating bottom-up ground-based measurements with top-down remote sensing methodologies to address scientific questions in atmospheric science. His work produced huge amounts of aerosol and greenhouse gas emissions data over the Indian region that was not previously available; for example, Multiwavelength Radiometer for Aerosol Optical Depth, MFRSR, UV-MFRSR, MICROTOPS-II, QCM/GRIMM Aerosol particle analyzer, PREDE Skyradiometer, Nephelometer, LIDAR, and various gas analyzers-$CO/CO_2/NO_x/O_3/SO_2$. He was the pioneer in designing and implementing a satellite-based fire alert system over the Indian region. He collaborated with several international researchers from countries such as U.K., U.S.A., Italy, Canada, Germany, Japan, Greece, and others.

He was extremely creative, unusually prolific, and highly elegant in his choice of research topics. He was fully committed to science. As a mentor, he made things simple with great enthusiasm and caring empathy. He was sincere in his approach to solving student issues. He was gifted with the ability to recognize and clarify the key concepts that are ambiguous. He was also in constant demand as an invited lecturer and journal article reviewer. Most of his students recognized for undergoing rigorous training in fundamentals under his able guidance have been sought after by reputed labs. They are pursuing successful careers in academia and industry.

He was a devoted teacher, husband, father, and a good scientist. His students, colleagues, and well-wishers will remember him as a humble, down to the earth, helping, and kind person. He will continue to inspire us. We miss him sorely.

Contributors

Edvin Aldrian
Agency for Assessment
and Implementation of Technology
(BPPT), Kawasan PUSPIPTEK
Serpong, Indonesia

Yessy Arvelyna
Remote Sensing Technology Center
of Japan
Tokyo, Japan

Kimberly Baugh
Earth Observation Group
Payne Institute, Colorado School
of Mines
Golden, Colorado, USA

Sumalika Biswas
Smithsonian Conservation Biology
Institute
Front Royal, Virginia, USA

Mylene G. Cayetano
Institute of Environmental Sciences
and Meteorology
University of the Philippines Diliman
Quezon City, Philippines

Josefino C. Comiso
Institute of Environmental Sciences
and Meteorology
University of the Philippines Diliman
Quezon City, Philippines
and
Institute of Environmental Science
and Meteorology
NASA Goddard Space Flight Center
Greenbelt, Maryland, USA

Naito Daisuke
Center for Southeast Asian Studies
Kyoto University
Kyoto, Japan

Aditya Eaturu
Department of Atmospheric and Earth
Science
University of Alabama Huntsville
Huntsville, Alabama, USA

Christopher D. Elvidge
Earth Observation Group
Payne Institute, Colorado School
of Mines
Golden, Colorado, USA

Hiroshi Hayasaka
Arctic Research Center,
Hokkaido University
Sapporo, Japan

Franky Herman
Faculty of Science and Natural
Resources
Universiti Malaysia Sabah
Kota Kinabalu, Malaysia

Feng-Chi Hsu
Cooperative Institute for Research
in the Environmental Sciences
University of Colorado
Boulder, Colorado, USA

Lies Indrayanti
Forestry Product Department
University of Palangka Raya
Palangka Raya, Indonesia

Christopher Justice
Department of Geographical Sciences
University of Maryland
College Park, Maryland, USA

Hirose Kazuyo
Japan Space Systems
Minato, Japan

Narendran Kodandapani
Center for Advanced Spatial and
  Environmental Research (CASER)
Bangalore, India

Ashish Kumar
Aryabhatta Research Institute
  of Observational Sciences (ARIES)
Nainital, India

Roni Kurniawan
Indonesia Agency for Meteorology
  Climatology and Geophysics
  (BMKG)
Jakarta, Indonesia

Sheila Dewi Ayu Kusumaningtyas
Indonesia Agency for Meteorology
  Climatology and Geophysics
  (BMKG)
Jakarta, Indonesia

Retno Maryani
Ministry of Forestry
Jakarta, Indonesia

Mohd Sharul Mohd Nadzir
School of Environmental and Natural
  Resources
National University of Malaysia
Bangi, Malaysia

Toshimasa Ohara
National Institute for Environmental
  Studies
Tsukuba, Japan

Mitsuru Osaki
Graduate school of Agriculture
Hokkaido University
Sapporo, Japan

Gay J. Perez
Institute of Environmental Sciences and
  Meteorology
University of the Philippines Diliman
Quezon City, Philippines

Chittana Phompila
Faculty of Forest Sciences (FFS)
National University of Laos (NUoL)
Vientiane Capital, Laos

Carolyn Melissa Payus
Faculty of Science and Natural
  Resources
Universiti Malaysia Sabah
Kota Kinabalu, Malaysia

Justin Sentian
Faculty of Science and Natural
  Resources
Universiti Malaysia Sabah
Kota Kinabalu, Malaysia

Alpon Sepriando
Climatology and Geophysics (BMKG),
  Indonesian Agency for Meteorology
Jakarta, Indonesia

Narendra Singh
Aryabhatta Research Institute
  of Observational Sciences (ARIES)
Nainital, India

Sunaryo Sunaryo
Indonesia Agency for Meteorology
  Climatology and Geophysics
  (BMKG)
Jakarta, Indonesia

Hidenori Takahashi
Hokkaido Institute of Hydro-climate
Sapporo, Japan

Pankaj Thapa
Sherubtse College
Royal University of Bhutan
Kanglung, Bhutan

Aswin Usup
Department of Forestry
Palangka Raya University
Palangkaraya, Indonesia

Krishna Prasad Vadrevu
Earth Science Office
NASA Marshall Space Flight Center
Huntsville, Alabama, USA

Vivian Kong Wan Yee
Faculty of Science and Natural
   Resources
Universiti Malaysia Sabah
Kota Kinabalu, Malaysia

Chao Xu
Institute of Atmospheric Physics,
   Chinese Academy of Sciences, China

Chao You
College of Environment and Ecology,
   Chongqing University, China
Institute of Tibetan Plateau Research,
   Chinese Academy of Sciences, China

Mikhail Zhizhin
Earth Observation Group
Payne Institute, Colorado School
   of Mines
Golden, Colorado, USA
and
Department of Space Dynamics
   and Data Analysis
Russian Space Research Institute
Moscow, Russian Federation, Russia

1 Biomass Burning in South/Southeast Asia – Needs and Priorities

Krishna Prasad Vadrevu
NASA Marshall Space Flight Center, Alabama, USA

Toshimasa Ohara
National Institute for Environmental Studies, Japan

Christopher Justice
University of Maryland College Park, USA

CONTENTS

INTRODUCTION

Vegetation fires are common in several regions of the world, including the South/ Southeast Asian countries (Figure 1.1). Fire occurrence and spread are governed by various factors such as fuel type, topography, climate, weather, and lightning. In addition to the natural factors, most of the fires in South/Southeast Asia (S/SEA) are human initiated (Vadrevu et al., 2019). For example, fire is often used as a land-clearing tool in slash-and-burn agriculture in several regions of S/SEA such as in Dhading and Chitwan districts of Nepal (Mukul and Byg, 2020); the Eastern Ghats (Prasad et al., 2001) and northeast India (Toky and Ramakrishnan, 1983); Chittagong Hill tracts of Bangladesh (Borggaard et al., 2003); Bago Mountains and northern Myanmar (Suzuki et al., 2009; Biswas et al., 2015a, b); Sarawak in Malaysia (de Neergaard et al., 2008); Caraballo Mountain in Carranglan and Mount Mingan, Philippines (Gabriel et al., 2020); Jambi Province, Sumatra, and others in Indonesia (Ketterings et al., 1999); northern Thailand (Grandstaff, 1980); northern Laos (Inoue, 2018); Cambodia (Scheidel and Work, 2016); and northern Vietnam (Nguyen et al., 2008). Fires are also extensively used for clearing land for rubber and oil palm expansion in Indonesia (Ketterings et al., 2002; Dhandapani and Evers, 2020). In addition to slash and burn, most of the countries in S/SEA are agrarian where farmers use fire for the burning of agricultural residues to clear the land for the next crop (Vadrevu and Lasko, 2018; Oanh et al.,

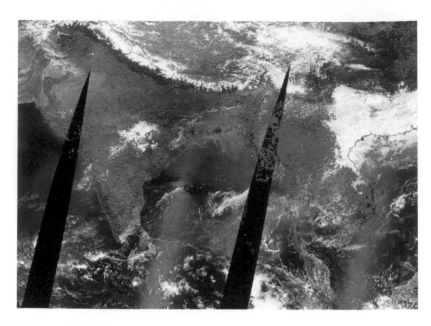

FIGURE 1.1   VIIRS 375 m day and night fires over South and Southeast Asia, March 18, 2020.

2018). All these burning practices alter landscape structure and function at various spatial scales. For example, clearing of forests through slash and burn results in the loss of biodiversity including the disruption of soil microbial processes during vegetation combustion and altering the biogeochemical cycles (Bruun et al., 2009). The resulting secondary vegetation after slash and burn can create a mosaic of mixed landscape patches leaving complex heterogeneous patterns across the landscape (Baker et al., 2008). In the case of cropland fires, incorporating residues into the soil has more benefit than open burning (Badarinath et al., 2006). Vegetation fires can release large greenhouse gas and air pollutant emissions such as CO_2, CO, NO_x, CH_4, non-methane hydrocarbons, and other chemical species including aerosols impacting radiative budget, air quality, and health at both local and regional scales (Gupta et al., 2001; Ramachandran, 2018). In addition, the biomass burning pollutants can be transported over long distances impacting regional climate. Thus, it is important to characterize fires and biomass burning in different regions of the world including in South/Southeast Asian countries. In this study, we briefly describe some of the fire regime characteristics, i.e., occurrence, seasonality, extent, and intensity, using some of the satellite fire products. We then highlight some of the important needs and priorities to address biomass burning issues in the region.

Broadly, satellite-derived fire datasets can be categorized into three different types: (1) active fire products that rely on middle-infrared and thermal-infrared (usually around 3.7–11 mm) satellite bands (Dozier, 1981) for detecting fires (Giglio et al. 2009; Schroeder et al., 2014); (2) burned areas (BAs) that rely on the visible and short-wave infrared wavelengths for detecting land surface changes after burning (Roy et al., 2005; Giglio et al., 2009); and (3) fire radiative energy products: instantaneous fire radiative power (FRP) measured from middle-infrared satellite bands to characterize the intensity and which can be further used to estimate the amount of biomass burned (Wooster et al., 2005). We use these products to describe fire regime characteristics in S/SEA.

The temporal variations of fires obtained from the VIIRS 375-m-resolution active fire product for South Asia from 2012 to 2019 are shown in Figure 1.2 and the averaged

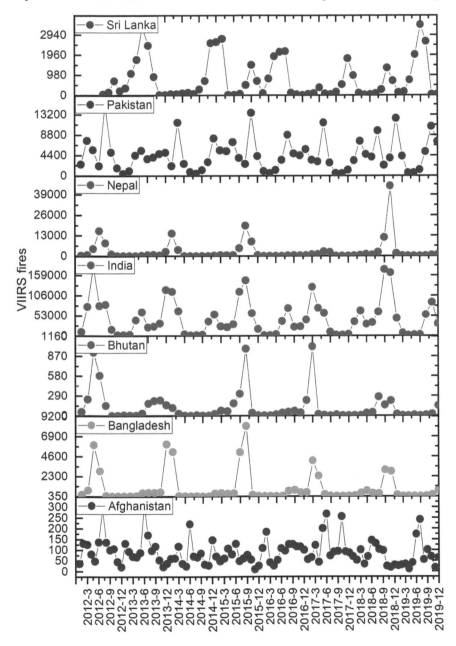

FIGURE 1.2 VIIRS (375 m) active fires over South Asian countries (2012–2019).

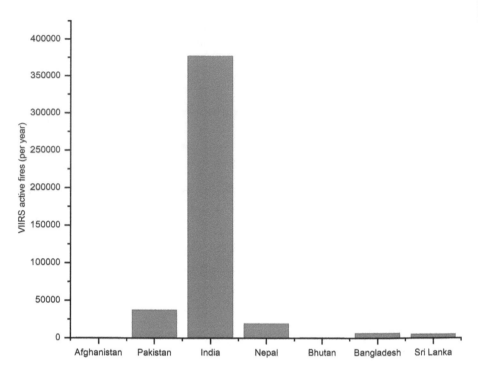

FIGURE 1.3   VIIRS (375 m) averaged annual fires (2012–2019) over South Asian countries.

annual fire counts in Figure 1.3. The VIIRS algorithm builds on the well-established MODIS Fire and Thermal Anomalies product, using a contextual approach to detect fires (Schroeder et al., 2014). Due to its higher spatial resolution, the VIIRS 375 m active fire product captures more fire pixels than the MODIS MCDML 1 km product. In South Asia, in this period, India had the highest number of fires with more than 350,000 fire counts per year, followed by Pakistan (37513), Nepal (19259), etc., with the least occurring in Afghanistan (714) (Figure 1.3). Figure 1.4 depicts the temporal variations in VIIRS retrieved active fires for Southeast Asian countries. In the case of Southeast Asia, Indonesia (230243) had the most fires, followed by Myanmar (230156), Cambodia (121212), Thailand (120782), etc., with the least in East Timor (3627) (Figures 1.4 and 1.5).

The vegetation seasonality for different countries is shown in Table 1.1. The peak fire season for the South Asian countries of India, Nepal, Bhutan, and Bangladesh runs from March till May, whereas in Pakistan it is from April till June, in Afghanistan from June to August, and in Sri Lanka from July to September. In Southeast Asia, Myanmar, Laos, Vietnam, Thailand, Cambodia, and the Philippines had a peak fire season from February till April, whereas in Malaysia the fire season is from July to September, in Indonesia from August to October, and in East Timor from September to November.

The burnt area temporal variations derived from the MODIS (MCD64A1) product for South and Southeast Asian countries are shown in Figures 1.6 and 1.7. The burnt area detection is based on an automated method using 500-m MODIS imagery and 1-km MODIS active fire observations. A hybrid approach is used in several steps for

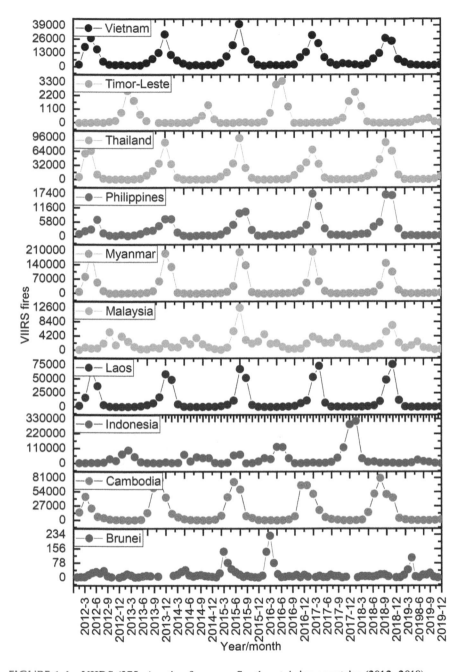

FIGURE 1.4 VIIRS (375 m) active fires over Southeast Asian countries (2012–2019).

mapping post-fire BAs. A time series of daily band 1 (0.620–0.670 μm), 5 (1.230–1.250 μm), and 7 (2.105–2.155 μm) reflectances for each pixel is used to compute a daily burn-sensitive vegetation index (VI = (B5 − B7)/(B5 + B7)), to detect a decrease in the Vegetation Index signal to detect BAs. A mask with active fires is also used to

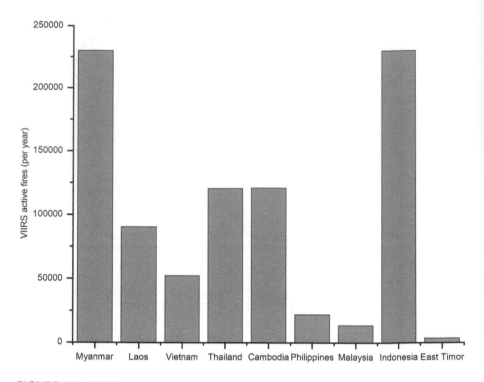

FIGURE 1.5   VIIRS (375 m) averaged annual fires (2012–2019) over Southeast Asian countries.

sort burned/unburned pixels. Results suggest India with the highest burnt areas of 3.7 million ha per year (Mha/yr), followed by Pakistan (0.17 Mha/yr), Nepal (0.16 Mha/yr), Bangladesh (55,000 Mha/yr), etc. In Southeast Asia, Myanmar had the highest burnt areas (3.09 Mha/yr), followed by Cambodia (2.49 Mha/yr), Thailand (1.37 Mha/yr), Indonesia (1.1 Mha/yr), Laos (0.66 Mha/yr), etc., with the least for Timor-Leste (0.01 Mha/yr).

FRP, which is a measure of intensity derived from the VIIRS data (VNP14IMG), is shown in Figure 1.8. The FRP is calculated using a combination of both VIIRS 375 m and 750 m data. The former is used to identify fire-affected, cloud, water, and valid background pixels; then, colocated M13 channel radiance data at 750 m coinciding with fire pixels and valid background pixels are used in the FRP calculation. More details about the product can be found in Schroeder et al. (2014). The results suggest FRP varying from ~60 MW to greater than 1000 MW for the individual fire events. We used the kernel density interpolation algorithm to depict FRP density. A clear spatial variation can be seen in different countries with relatively higher FRP in northeast India, Myanmar, northern Thailand, and parts of Indonesia, with the highest in Laos.

Important drivers and impacts of these fire regime characteristics are poorly understood. There is a need to integrate both top-down approaches such as on using remotely sensed data and airborne measurements and bottom-up approaches such

TABLE 1.1

Peak Biomass Burning Months in South/Southeast Asian Countries

Region/ Country	Jan	Feb	Mar	Apr	May	Jun	Jul	Aug	Sep	Oct	Nov	Dec
South Asia												
Afghanistan					▓	▓	▓	▓				
Pakistan				▓	▓	▓						
India			▓	▓	▓	▓						
Nepal			▓	▓	▓	▓						
Bhutan			▓	▓	▓	▓						
Bangladesh			▓	▓	▓	▓						
Sri Lanka							▓	▓	▓	▓		
Southeast Asia												
Myanmar		▓	▓	▓	▓							
Laos		▓	▓	▓	▓							
Vietnam		▓	▓	▓	▓							
Thailand		▓	▓	▓	▓							
Cambodia		▓	▓	▓	▓							
Philippines		▓	▓	▓	▓							
Malaysia						▓	▓	▓				
Indonesia							▓	▓	▓	▓		
East Timor								▓	▓	▓	▓	

as field measurements to address biomass burning issues in S/SEA. In the following section, we highlight some of the important needs and priorities.

1. Impacts of low-intensity fires on biodiversity in S/SEA are poorly understood

Most of the forest ecosystems in Asia harbor high levels of species richness and endemism. In South and Southeast Asia, most of the fires are ground fires caused due to human interference (Stott et al., 1990). How these fires influence the mosaic of evergreen and deciduous forests, and the rich biodiversity they support, is poorly understood (Baker et al., 2008). There is a need to integrate biodiversity survey information with the climate and anthropogenic drivers to evaluate fire impacts on a long-term scale. The fire effects need to be compared in different forest types to assess mortality, recovery, and growth. It is also unclear how fires create gaps in different forests. Some of the forest tree species have resilience to fires, and their growth patterns needs to be documented. It is not clear how frequent small-scale fires affect biodiversity and how they influence the landscape mosaic of the seasonal tropical forests in S/SEA, which thus needs attention.

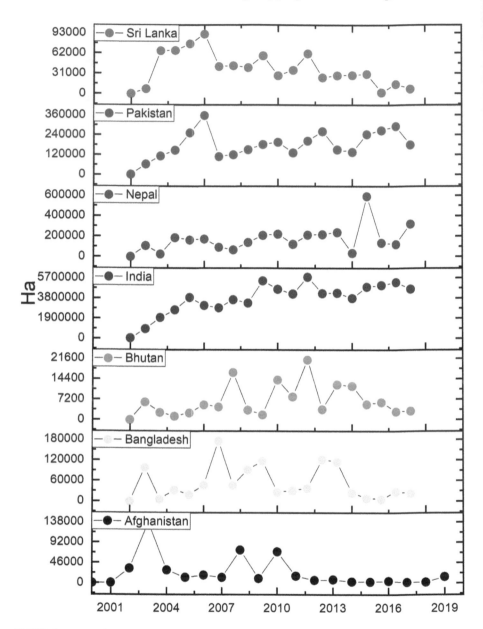

FIGURE 1.6 MODIS burnt areas over South Asian countries (2001–2019).

2. Harmonized multi-sensor fire datasets that integrate polar and geostationary data are needed to address fire regimes

For characterizing the fire diurnal cycle, the geostationary satellites can provide repeated observations on a sub-hourly basis, making it possible to detect fires not detectable by polar-orbiting satellites with lower temporal frequency (Schroeder et al., 2008). Despite their enormous potential,

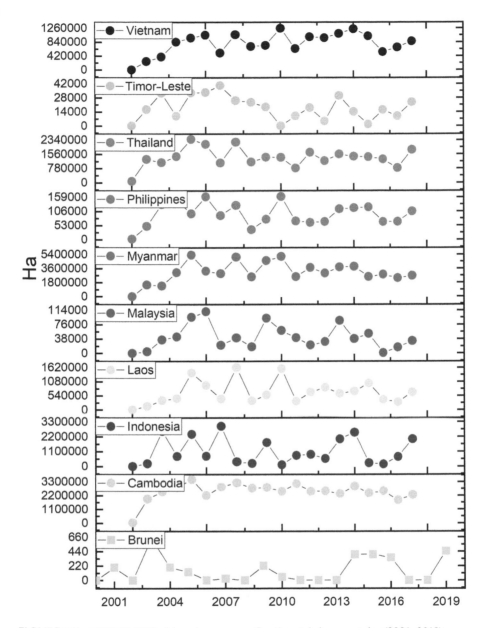

FIGURE 1.7 MODIS (500 m) burnt areas over Southeast Asian countries (2001–2019).

geostationary-based satellite fire products have remained underutilized, largely due to their coarser spatial resolution. Multiple factors have contributed to this underutilization, including the lack of well-established programs to support algorithm development and validation, the intrinsic quality limitations of previous-generation geostationary sensor datasets, and poor product documentation and/or access. The recent deployment of the next generation

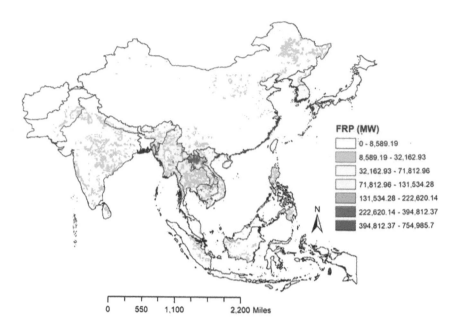

FIGURE 1.8 VIIRS fire radiative power (FRP) density in the South/Southeast Asian countries. The FRP for individual fire events varied from 66 to 1250 MW. The kernel density technique is used to depict FRP magnitude per area. FRP indicates the strength of the fires and has been shown to the amount of biomass consumed on the ground by various authors.

of geostationary weather satellites such as the Advanced Himawari Imager (AHI) series launched by the Japan Meteorological Agency (JMA) has greatly improved spatial, temporal, and radiometric qualities useful for fire detection over the Asian region (Jang et al., 2019). We strongly recognize the need to develop harmonized multi-sensor fire products that integrate both polar and geostationary fire datasets to quantify fire regimes and provide information useful for fire management. We also stress the need for the establishment of a robust global network of geostationary fire data that can greatly complement existing polar-orbiting satellite fire products in line with the goals set by the Fire Implementation Team of the Global Observations of Forest and Land Use Dynamics (GOFC-GOLD, www.gofc-gold.org).

3. Global satellite fire products need strong validation

Accurate information on fire regime characteristics such as location, frequency, extent, and seasonality is necessary for fire strategic management. The global fire products that are routinely used in Asia and other regions of the world need calibration and validation based on local conditions, i.e., vegetation type, topography, and climate, to assess the accuracy and provide omission and commission errors. Most of the fire products require validation using an independent source of data to assess the product quality and determine the level of accuracy. The comparison should be done at global representative sites (Justice et al., 2002) involving regional scientists, covering various landscapes. An inter-comparison exercise with similar products should be done to assess their robustness. Most importantly, the field-based validation activities are needed to assess the accuracy of products in addition to the implementation of robust calibration and validation protocols in different countries. The conventional approaches to fire product validation include integrating of ground, unmanned aerial vehicles (UAVs), and aircraft measurements of prescribed burns over wildfires. However, such methods can be costly; thus, simultaneous acquisition of high-resolution, coincident data that can aid in fire validation measures of both active fires and burnt areas should be explored. Specific to FRP products, integrated lab and ground-based measurements should be planned for validation. Collaborations with the international programs, such as through the Global Observations of Forest Cover and Land Use Dynamics (GOFC/GOLD) and the regional network of scientists they support, following the protocols set by the Land Product Validation (LPV) subgroup of the Calibration and Validation Working Group of the Committee on Earth Observation Satellites (CEOS) (https://lpvs.gsfc.nasa.gov/), should be part of any new fire product development, as these international programs have established standards for new product development involving experts from the science community.

4. Mapping and monitoring of agriculture fires and related emissions need more attention

Recent data from United Nations Food and Agriculture (FAO, 2020) suggest that sectoral emissions from Agriculture, Forestry, and Other Land Use (AFOLU) are a net source of carbon in S/SEA and that biomass burning from agriculture is an important source. Of the different databases, the Global Fire Emissions Database (GFED) is most widely by researchers to account for fire-related emissions from vegetation. The database uses the Seiler and Crutzen (1980) approach that multiplies mapped BA by modeled fuel consumption and species-specific gas and aerosol emission factors to estimate fire emissions at a 0.25° spatial resolution (Van der Werf et al., 2010). The BA maps used by GFED are provided by NASA's 500-m-spatial-resolution MODIS BA product (MCD64A1), which classifies pixels as burned using a spectral reflectance-based change detection technique (Giglio et al., 2013). The BA signature of landscape fires typically lasts for several days to several months post-fire depending on biome in forested landscapes. However, in agricultural fires, the signal is mostly ephemeral and can last from a few hours to a couple of days and the burnt area patches are mostly small and non-contiguous, particularly in Asian countries. The burnt area products typically require a substantial part (approximately $\geq 20\%$) of a pixel to be burned before a BA algorithm can identify it as "fire-affected" (Giglio et al., 2006; 2009; Zhang et al., 2018). The MCD64A1 product can have low detection of fires in the agricultural areas, mostly due to the coarser spatial resolution (500 m), which means small fires of <120 ha in size are not detected well

(Giglio et al., 2009). In contrast, the active fires filling only 0.01%–0.1% of a pixel are relatively easily identified due to their high thermal contrast relative to the surrounding ambient background (Wooster et al., 2005; Schroeder et al., 2014). Although GFED has been shown to represent emissions where "large-fires" dominate accurately, its performance in regions dominated by "small fires," more specifically agricultural fires, remains questionable (Randerson et al., 2012; Zhang et al., 2018). Due to the smaller field size of agricultural fields and the ephemeral nature of fires, both active fire detection and burnt area mapping productions from coarse-resolution systems miss many fires, and thus, the emissions are highly underestimated in the agriculture sector. Hence, we strongly recommend the need to focus efforts on new approaches utilizing higher temporal and spatial frequency data to mapping and monitoring agricultural fires in the Asian region and elsewhere.

5. Peatland fires are unique and need more attention

Of the different countries in S/SEA, Indonesia contains more than half of all known peatlands in the tropics, with an area ranging from 16 to 27 Mha (Page et al., 1999), accounting for ~82–92 gigatons of carbon (Immirzi and Maltby, 1992). In recent years, Indonesia's peat swamp forests have been severely degraded due to industrial and illegal logging, plantation activities, and other failed development projects such as the Mega Rice Project (Hoscilo et al., 2011). Fire is often utilized as a cheap, effective method to clear the peatlands for agricultural and plantation development in Indonesia. Since most fires in peatlands occur underground and release huge amounts of smoke, fire detection from satellites is often challenging. Novel algorithms are needed to account for smoke as well as low-intensity fire detection from satellites. Also, the smoke released from the peatland fires has significant smoldering-stage chemical compounds that vary based on the fire intensity, peatland depth, water content, etc. Ground-based instrumentation is needed to characterize the emission characteristics. Further, the water table in peatlands is influenced by drought conditions coinciding with climatological events such as El Niño–Southern Oscillation (ENSO); thus, addressing the linkages between peatland fires and climate drivers is also essential (Usup et al., 2004) to arrive at fire management and mitigation options.

6. Dust and biomass burning pollution are intermixed and thus need more robust algorithms to separate these components from the satellite data

Both South and Southeast Asian countries are influenced by severe dust events (Figure 1.9). In South Asia, primarily the Indo-Gangetic plains and the northern Himalayas, India, are influenced by the dust from the Arabian Peninsula and the Thar Desert located in northwestern India. In East Asia, the Taklamakan Desert in western China and Gobi Desert in Mongolia and northern China are a major source. Asian countries are affected by pollution from various sources such as dust, biomass burning, industrial and vehicular pollution, and other anthropogenic emissions throughout the year. The pollutants from these sources impact the air quality and health of millions of people in the region. Information on the distribution

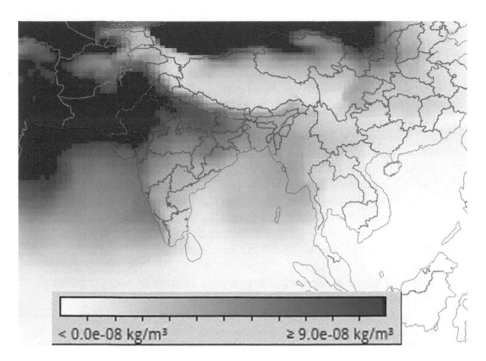

FIGURE 1.9 Dust surface mass concentration during April 2019 derived from MERRA reanalysis datasets for South/Southeast Asia. The dust often gets intermixed with biomass burning pollutants in the region causing significant haze.

of dust sources and the quantity of dust produced can help address climate change-related questions, and quantifying the source strengths can help address desertification issues (Arimoto et al., 2006). The dust, biomass burning smoke, and other pollutants are often intermixed; thus, it is not easy to separate individual constituents. From the satellites, UV Aerosol Index (UVAI) is widely used to detect elevated absorbing aerosols such as airborne dust and biomass burning smoke (Torres et al., 2007). Although the UVAI is sensitive to absorbing aerosol height, the index is not sensitive to the cloudy weather condition and terrestrial surface types. One of the alternatives is the Cloud-Aerosol Lidar and Infrared Pathfinder Satellite Observations (CALIPSO), which can provide height-resolved aerosol and cloud information at 532 and 1064 nm and a horizontal resolution of 330 m. However, the data are coarse, and much higher Lidar data are needed to resolve the differences between dust and other types of aerosols, such as smoke and industrial pollutants. More robust algorithms and ground-based studies are needed to resolve differences between the dust versus biomass burning smoke plumes in Asia and elsewhere.

7. **Addressing climate–fire relationships is important to understand inter-annual variations and related emissions**

Fires are highly influenced by vegetation type and pattern, topography, climate, weather, and anthropogenic factors. The fuel amount, type,

structure, and moisture govern fire occurrence and spread. Fuel moisture is a critical parameter that influences flammability and is a function of the weather. At a regional scale, the temperature, precipitation,
wind, and atmospheric moisture are the other weather-driven parameters
that govern the fire activity. As a result of climate change, more extreme
weather events are predicted, meaning more fires. For example, as per
the 2020 IPCC Climate Change and Land report (IPCC, 2020), anthropogenic warming has resulted in shifts of climate zones, primarily as
an increase in dry climates and a decrease in polar climates (high confidence). Ongoing warming is projected to result in new, hot climates in
tropical regions (Hoegh-Guldberg et al., 2018). However, many climate
models show higher uncertainties over the tropics concerning precipitation and seasonality which can have a significant impact on emissions.
Thus, more research is needed to understand climate–fire relationships,
in particular in South/Southeast Asian countries.

8. Regional biomass campaigns are needed in Asia to address pollution and climate impacts

There is a significant variation in the fire regime characteristics in Asia
as they vary based on the ecosystem type, local conditions, and meteorology. Thus, there is a strong need to collect ground-based data through
field-based campaigns to include fuel loads, fuel moisture content, combustion completeness, emission factors, etc. Ground-based instrumentation
and airborne studies are needed to refine biomass burning emission factors
and aerosol optical properties. Satellite data can help map and monitor
fires, burnt areas, and pollutants at varied spatial scales. Lessons can be
learned from the earlier international biomass burning campaign studies
such as The Biomass Burning Experiment (BIBEX)-The Southern Tropical
Atlantic Region Experiment (STARE) field campaign in the southern
Tropical Atlantic Ocean during 1992 (Andreae et al., 1994), Smoke, Clouds,
and Radiation-Brazil (SCAR-B) in Brazil during 1995 (Kaufman et al.,
1998), Zambian International Biomass Burning Emissions Experiment in
Zambia (ZIBBEE) during 1997 (Eck et al., 2001), The Southern African
Regional Science Initiative (SAFARI) during 1992 (Justice et al., 1996) and
2000 (Swap et al., 2002), and South American Biomass Burning Analysis
(SAMBBA) in South America during 2014 (Brito et al., 2014). In contrast
to these international campaigns, very few studies have focused on biomass
burning in Asia, such as Biomass burning Aerosols in South-East Asia:
Smoke Impact Assessment (BASE-ASIA) during 2006 (Tsay et al., 2013)
and the Seven South-East Asian Studies (7-SEAS) during 2010 and subsequent years in northeast Asia (Reid et al., 2013; Lin et al., 2013). There is a
need for more such biomass burning campaigns in South/Southeast Asian
countries to characterize the chemical, physical, optical, and radiative
properties of biomass burning pollutants and aerosols to investigate the
sources, sinks, atmospheric transport, and the chemical processes through
integrating satellite, aircraft, and ground-based measurements.

9. Health impacts of biomass burning pollution are poorly understood

The adverse health effects of biomass burning pollutants are poorly understood in South/Southeast Asian countries. The smoke that emanates during biomass burning is composed of hundreds of toxic chemicals harmful to humans. In particular, the biomass burning smoke particles with the particulate matter (PM) with aerodynamic diameter $\leq 2.5\,\mu m$ ($PM_{2.5}$) and $1\,\mu m$ (PM_1) primarily consist of organic carbon and black carbon components, along with smaller contributions from inorganic species (Pope and Dockery, 2006; Johnston et al., 2012). PM has been associated with a wide range of adverse health effects such as respiratory illness, cardiorespiratory mortality, and oxidative stress. (Naeher et al., 2007). The adverse health effects of PM derived from burning biomass are not investigated well, and very few toxicological studies exist in South/Southeast Asian countries. Some studies have focused on biomass burning and smoke exposure and mortality in Malaysia (Sastry, 2002) and Indonesia (Jayachandran, 2009; Uda et al., 2019; Zaini et al., 2020), whereas such studies specific to South Asian countries are almost lacking. There is a strong need to understand the health impacts of biomass burning pollutants more rigorously and in a systematic framework, to aid policymakers and local people on the urgency of tackling biomass burning fires to aid in healthy air quality measures.

10. Modeling studies are needed to address climate and health impacts of biomass burning pollutants

Biomass burning pollution can have significant influences on regional climate and air quality. The pollutants can affect the dispersion and transport of other chemical constituents in the atmosphere, impacting the regional climate. To address the impacts of biomass burning pollution on climate, weather, air quality, and atmosphere, systematic modeling studies are needed. Air quality models such as The Community Multiscale Air Quality Modeling System (CMAQ, 2020) and Weather Research and Forecasting/Chemistry model (WRF/Chem, UCAR, 2020) that combine emission inventory information with meteorology and atmospheric fate and transport mechanisms are one option for addressing the pollution impacts. Also, there is a need to explore the potential of ensemble modeling systems that can integrate various ground-based emissions data, airborne with the satellite to quantify the pollutant exposure estimates to populations and aid in health risk assessments. Such studies in S/SEA are very meager and thus need attention.

11. Addressing transboundary pollution requires strong implementation of multilateral agreements among countries

Several European countries during the 1970s recognized that addressing air pollution requires action at all policy levels from local government to international institutions (Wettestad, 2018). More specifically, they realized that cooperation at a broader international level is necessary to curb air pollution effectively. A Convention on Long-Range Transboundary Air Pollution (CLRTAP), also abbreviated as Air Convention, to curb transboundary air pollution was formulated and signed by several European countries on November 13, 1979, and the CLRTAP entered into force on March 16, 1983. The CLRTAP is implemented by the European Monitoring and Evaluation Program (EMEP) and directed by the United Nations Economic Commission for Europe (UNEC). The Convention has 51 parties, which also include North America. The Convention addresses multiple pollutant controls through national emission ceilings and regulatory commitments. CLRTAP is quite effective in fighting acidification of the environment and reducing ozone and photochemical smog, persistent organic pollutants, and heavy metals. Specific to Asia, to curb emissions from the biomass burning and fires, and more specifically from Indonesia, the ASEAN Agreement on Transboundary Haze Pollution came into effect in November 2003, involving the member states in Southeast Asia (Jones, 2006). The Agreement recognizes that transboundary haze pollution resulting from land and/or forest fires should be mitigated through concerted national efforts and international cooperation. As of 2014, all ten ASEAN countries have ratified the haze agreement. However, concerns remain on the stringent implementation plan. For example, the treaty was not effective in preventing fires and haze pollution from Indonesia between 2004 and 2010, and again in 2013, 2014 and 2015, 2018, and 2019. The ASEAN treaty is accused of being vague and lacking enforcement mechanisms or tools for dispute resolution (Varkkey, 2014). Further, no such treaties exist among South Asian countries. For addressing pollution in general and biomass burning in particular, more stringent multilateral agreements are needed through the cooperation of all countries in S/SEA.

12. Regional capacity-building activities are a must to train scientists and address the pollution problems

We strongly recommend that capacity-building activities are needed to train the younger generation of scientists on the new tools and techniques, instrumentation, and satellite data analysis, including modeling related to air pollution and biomass burning studies. There is a need for strengthening capacity-building activities through effective partnerships between public, private, and governmental organizations. There is a need to build and sustain regional research initiatives such as the South/Southeast Asia Regional Initiative involved in research, education, and capacity-building activities in different Asian countries to benefit locals and build a broad scientific community (Vadrevu et al., 2017). Most scientific organizations and national programs in Asia are understaffed and lack sufficient resources to tackle the problem. Thus, developing adequate infrastructures and programs in conjunction with training and education is vital for solving environmental and pollution problems in these countries.

This two-volume book provides several insights on some of the needs and priorities highlighted above. The chapters cover various topics on mapping and monitoring fires, including biomass burning impacts at varied spatial scales and contributions from the regional scientists. We infer that more such collaborative works involving regional country researchers are needed to strengthen both research and international collaborations to address the biomass burning pollution in S/SEA.

ACKNOWLEDGMENTS

We are grateful to several authors from the USA and South/Southeast Asia who contributed to the two-volume book. This work is a part of the South/Southeast Asia Research Initiative (SARI) funded by the NASA Land Cover/Land Use Change Program.

REFERENCES

Andreae, M.O., Fishman, J., Garstang, M., Goldammer, J.G., Justice, C.O., Levine, J.S., Scholes, R.J., Stocks, B.J., Thompson, A.M. and Van Wilgen, B. 1994. Biomass burning in the global environment: First results from the IGAC/BIBEX field campaign STARE/TRACE-A/SAFARI-92. In *Global Atmospheric-Biospheric Chemistry*. Prinn, R. (Ed.). Springer, Boston, MA. pp. 83–101.

Arimoto, R., Kim, Y.J., Kim, Y.P., Quinn, P.K., Bates, T.S., Anderson, T.L., Gong, S., Uno, I., Chin, M., Huebert, B.J. and Clarke, A.D. 2006. Characterization of Asian dust during ACE-Asia. *Global and Planetary Change*. 52(1–4), 23–56.

Badarinath, K.V.S., Chand, T.K. and Prasad, V.K. 2006. Agriculture crop residue burning in the Indo-Gangetic Plains–A study using IRS-P6 AWiFS satellite data. *Current Science*. 1085–1089.

Baker, P.J., Bunyavejchewin, S. and Robinson, A.R. 2008. The impacts of large-scale, low-intensity fires on the forests of continental South-east Asia. *International Journal of Wildland Fire*. 17, 782–792.

Biswas, S., Lasko, K.D. and Vadrevu, K.P. 2015a. Fire disturbance in tropical forests of Myanmar—Analysis using MODIS satellite datasets. *IEEE Journal of Selected Topics in Applied Earth Observations and Remote Sensing*. 8(5), 2273–2281.

Biswas, S., Vadrevu, K.P., Lwin, Z.M., Lasko, K. and Justice, C.O. 2015b. Factors controlling vegetation fires in protected and non-protected areas of Myanmar. *PLoS One*. 10(4), e0124346.

Borggaard, O.K., Gafur, A. and Petersen, L. 2003. Sustainability appraisal of shifting cultivation in the Chittagong Hill Tracts of Bangladesh. *AMBIO: A Journal of the Human Environment*. 32(2), 118–123.

Brito, J., Rizzo, L.V., Morgan, W.T., Coe, H., Johnson, B., Haywood, J., Longo, K., Freitas, S., Andreae, M.O. and Artaxo, P. 2014. Ground-based aerosol characterization during the South American Biomass Burning Analysis (SAMBBA) field experiment. *Atmospheric Chemistry and Physics*. 14(22), 12069–12083.

Bruun, T.B., De Neergaard, A., Lawrence, D. and Ziegler, A.D. 2009. Environmental consequences of the demise in Swidden cultivation in Southeast Asia: Carbon storage and soil quality. *Human Ecology*. 37(3), 375–388.

CMAQ, 2020. https://www.epa.gov/cmaq.

Dozier, J. 1981. A method for satellite identification of surface temperature fields of subpixel resolution. *Remote Sensing of the Environment*. 11, 221–229.

de Neergaard, A., Magid, J. and Mertz, O. 2008. Soil erosion from shifting cultivation and other smallholder land use in Sarawak, Malaysia. *Agriculture, Ecosystems & Environment*. 125(1–4), 182–190.

Dhandapani, S. and Evers, S. 2020. Oil palm 'slash-and-burn'practice increases post-fire greenhouse gas emissions and nutrient concentrations in burnt regions of an agricultural tropical peatland. *Science of the Total Environment*. 742, 140648.

Eck, T.F., Holben, B.N., Ward, D.E., Dubovik, O., Reid, J.S., Smirnov, A., Mukelabai, M.M., Hsu, N.C., O'Neill, N.T. and Slutsker, I. 2001. Characterization of the optical properties of biomass burning aerosols in Zambia during the 1997 ZIBBEE field campaign. *Journal of Geophysical Research: Atmospheres*. 106(D4), 3425–3448.

FAO, 2020. http://www.fao.org/faostat/en/.

Gabriel, A.G., De Vera, M. and Antonio, M.A.B. 2020. Roles of indigenous women in forest conservation: A comparative analysis of two indigenous communities in the Philippines. *Cogent Social Sciences*. 6(1), 1720564.

Giglio, L., Loboda, T., Roy, D.P., Quayle, B. and Justice, C.O. 2009. An active-fire based burned area mapping algorithm for the MODIS sensor. *Remote Sensing Environment*. 113, 408–420.

Giglio, L., Randerson, J.T. and Werf, G.R. 2013. Analysis of daily, monthly, and annual burned area using the fourth-generation global fire emissions database (GFED4). *Journal of Geophysical Research Biogeoscience*. 118, 317–328.

Giglio, L., Werf, G.R., Randerson, J.T., Collatz, G.J. and Kasibhatla, P. 2006. Global estimation of burned area using MODIS active fire observations. *Atmospheric Chemistry and Physics*. 6, 957–974.

Grandstaff, T. B. 1980. Shifting cultivation in northern Thailand. Possibilities for development. No.3. Cabdirect.org.

Gupta, P.K., Prasad, V.K., Kant, Y., Sharma, C., Ghosh, A.B., Sharma, M.C., Sarkar, A.K., Jain, S.L., Tripathi, O.P., Sharma, R.C. and Badarinath, K.V.S. 2001. Study of trace gases and aerosol emissions due to biomass burning at shifting cultivation sites in East Godavari District (Andhra Pradesh) during INDOEX IFP-99. *Current Science*. 80(10), 186–196.

Hoscilo, A., Page, S.E., Tansey, K.J. and Rieley, J.O. 2011. Effect of repeated fires on land-cover change on peatland in southern Central Kalimantan, Indonesia, from 1973 to 2005. *International Journal of Wildland Fire*. 20(4), 578–588.

Hoegh-Guldberg, O., Jacob, D., Bindi, M., Brown, S., Camilloni, I., Diedhiou, A., Djalante, R., Ebi, K., Engelbrecht, F., Guiot, J. and Hijioka, Y. 2018. Impacts of 1.5°C global warming on natural and human systems. *Global warming of 1.5°C. An IPCC Special Report*.

Inoue, Y., 2018. Ecosystem carbon *stock, atmosphere, and food* security in slash-and-burn land use: A geospatial study in mountainous region of Laos. In *Land-Atmospheric Research Applications in South and Southeast Asia*. Vadrevu, K.P., Ohara, T., and Justice, C. (Eds). Cham: Springer. pp. 641–665.

Immirzi, C.P. and Maltby, E. 1992. The global status of peatlands and their role in the carbon cycle. *Wetlands Ecosystems Research Group, Report 11*. Exeter, UK: University of Exeter.

IPCC. 2020. Climate Change and Land: an IPCC special report on climate change, desertification, land degradation, sustainable land management, food security, and greenhouse gas fluxes in terrestrial ecosystems. https://www.ipcc.ch/site/assets/uploads/sites/4/2020/02/SPM_Updated-Jan20.pdf

Jang, E., Kang, Y., Im, J., Lee, D.W., Yoon, J. and Kim, S.K. 2019. Detection and monitoring of forest fires using Himawari-8 geostationary satellite data in South Korea. *Remote Sensing*. 11(3), 271.

Jayachandran, S. 2009. Air quality and early-life mortality evidence from Indonesia's wildfires. *Journal of Human Resources*. 44(4), 916–954.

Johnston, F.H., Henderson, S.B., Chen, Y., Randerson, J.T., Marlier, M., DeFries, R.S., Kinney, P., Bowman, D.M. and Brauer, M. 2012. Estimated global mortality attributable to smoke from landscape fires. *Environmental Health Perspectives*. 120(5), 695–701.

Jones, D.S. 2006. ASEAN and transboundary haze pollution in Southeast Asia. *Asia Europe Journal*. 4(3), 431–446.

Justice, C.O., Giglio, L., Korontzi, S., Owens, J., Morisette, J.T., Roy, D., Descloitres, J., Alleaume, S., Petitcolin, F. and Kaufman, Y. 2002. The MODIS fire products. *Remote sensing of Environment.* 83(1–2), 244–262.

Justice, C.O., Kendall, J.D., Dowty, P.R. and Scholes, R.J. 1996. Satellite remote sensing of fires during the SAFARI campaign using NOAA advanced very high-resolution radiometer data. *Journal of Geophysical Research: Atmospheres.* 101(D19), 23851–23863.

Kaufman, Y.J., Hobbs, P.V., Kirchhoff, V.W.J.H., Artaxo, P., Remer, L.A., Holben, B.N., King, M.D., Ward, D.E., Prins, E.M., Longo, K.M. and Mattos, L.F. 1998. Smoke, clouds, and radiation-Brazil (SCAR-B) experiment. *Journal of Geophysical Research: Atmospheres.* 103(D24), 31783–31808.

Ketterings, Q.M., van Noordwijk, M. and Bigham, J.M. 2002. Soil phosphorus availability after slash-and-burn fires of different intensities in rubber agroforests in Sumatra, Indonesia. *Agriculture, Ecosystems and Environment.* 92(1), 37–48.

Ketterings, Q.M., Wibowo, T.T., van Noordwijk, M. and Penot, E. 1999. Farmers' perspectives on slash-and-burn as a land clearing method for small-scale rubber producers in Sepunggur, Jambi Province, Sumatra, Indonesia. *Forest Ecology and Management.* 120(1–3), 157–169.

Lin, N.H., Tsay, S.C., Maring, H.B., Yen, M.C., Sheu, G.R., Wang, S.H., Chi, K.H., Chuang, M.T., Ou-Yang, C.F., Fu, J.S. and Reid, J.S., 2013. An overview of regional experiments on biomass burning aerosols and related pollutants in Southeast Asia: From BASE-ASIA and the Dongsha Experiment to 7-SEAS. *Atmospheric Environment.* 78, 1–19.

Mukul, S.A. and Byg, A. 2020. What determines indigenous Chepang farmers' Swidden land-use decisions in the Central Hill Districts of Nepal? *Sustainability.* 12(13), 5326.

Naeher, L.P., Brauer, M., Lipsett, M., Zelikoff, J.T., Simpson, C.D., Koenig, J.Q. and Smith, K.R., 2007. Woodsmoke health effects: A review. *Inhalation Toxicology.* 19(1), 67–106.

Nguyen, T.L., Vien, T.D., Lam, N.T., Tuong, T.M. and Cadisch, G. 2008. Analysis of the sustainability with the composite Swidden agroecosystem: 1. Partial nutrient balance and recovery times of uplands Swiddens. *Agriculture, Ecosystems & Environment.* 128, 37–51.

Oanh, N.T.K., Permadi, D.A., Dong, N.P. and Nguyet, D.A. 2018. Emission of toxic air pollutants and greenhouse gases from crop residue open burning in Southeast Asia. In *Land-Atmospheric Research Applications in South and Southeast Asia.* Vadrevu, K.P., Ohara, T., and Justice, C. (Eds). Cham: Springer. pp. 47–66.

Page, S.E., Rieley, J.O., Shotyk, W. and Weiss, D. 1999. Interdependence of peat and vegetation in a tropical peat swamp forest. *Philosophical Transactions of the Royal Society of London. Series B, Biological Sciences.* 354, 1885–1897.

Pope III, C.A. and Dockery, D.W. 2006. Health effects of fine particulate air pollution: Lines that connect. *Journal of the Air & Waste Management Association.* 56(6), 709–742.

Prasad, V.K., Kant, Y., Gupta, P.K., Sharma, C., Mitra, A.A. and Badarinath, K.V.S. 2001. Biomass and combustion characteristics of secondary mixed deciduous forests in Eastern Ghats of India. *Atmospheric Environment.* 35(18), 3085–3095.

Ramachandran, S. 2018. Aerosols and climate change: Present understanding, challenges and future outlook. In: *Land-Atmospheric Research Applications in South/Southeast Asia.* Vadrevu, K.P., Ohara, T., and Justice, C. (Eds). Cham: Springer. pp. 341–378.

Randerson, J.T., Chen, Y., Werf, G.R., Rogers, B.M. and Morton, D.C. 2012. Global burned area and biomass burning emissions from small fires. *Journal of Geophysical Research Biogeoscience.* 117(G4), 1–23.

Reid, J.S., Hyer, E.J., Johnson, R.S., Holben, B.N., Yokelson, R.J., Zhang, J., Campbell, J.R., Christopher, S.A., Di Girolamo, L., Giglio, L. and Holz, R.E. 2013. Observing and understanding the Southeast Asian aerosol system by remote sensing: An initial review and analysis for the Seven Southeast Asian Studies (7SEAS) program. *Atmospheric Research.* 122, 403–468.

Roy, D.P., Jin, Y., Lewis, P.E. and Justice, C.O., 2005. Prototyping a global algorithm for systematic fire-affected area mapping using MODIS time series data. *Remote Sensing of Environment*. 97(2), 137–162.

Sastry, N. 2002. Forest fires, air pollution, and mortality in Southeast Asia. *Demography*. 39(1), 1–23.

Scheidel, A. and Work, C., 2016. Large-scale forest plantations for climate change mitigation? New frontiers of deforestation and land grabbing in Cambodia. *Global Governance/ Politics, Climate Justice and Agrarian/Social Justice: Linkages and Challenges*. 11.

Schroeder, W., Csiszar, I. and Morisette, J., 2008. Quantifying the impact of cloud obscuration on remote sensing of active fires in the Brazilian Amazon. *Remote Sensing of Environment*. 112(2), 456–470.

Schroeder, W., Oliva, P., Giglio, L. and Csiszar, I.A. 2014. The new VIIRS 375 m active fire detection data product: Algorithm description and initial assessment. *Remote Sensing Environment*. 143, 85–96.

Seiler, W. and Crutzen, P.J. 1980. Estimates of gross and net fluxes of carbon between the biosphere and the atmosphere from biomass burning. *Climate Changes*. 2, 207–247.

Stott, P.A., Goldammer, J.G. and Werner, W.L. 1990. The role of fire in the tropical lowland deciduous forests of Asia. In *Fire in the Tropical Biota*. Goldammer, J.G. (Ed.). Springer, Berlin, Heidelberg. pp. 32–44.

Suzuki, R., Takeda, S. and Thein, H.M. 2009. Effect of slash-and-burn on nutrient dynamics during the intercropping period of Taungya teak reforestation in the Bago Mountains, Myanmar. *Tropical Agriculture and Development*. 53(3), 82–89.

Swap, R.J., Annegarn, H.J., Suttles, J.T., Haywood, J., Helmlinger, M.C., Hely, C., Hobbs, P.V., Holben, B.N., Ji, J., King, M.D. and Landmann, T. 2002. The Southern African Regional Science Initiative (SAFARI 2000): Overview of the dry season field campaign. *South African Journal of Science*. 98(3–4), 125–130.

Toky, O.P. and Ramakrishnan, P.S. 1983. Secondary succession following slash and burn agriculture in North-Eastern India: I. Biomass, litterfall and productivity. *Journal of Ecology*. 71(3), 735–745.

Torres, O., Tanskanen, A., Veihelmann, B., Ahn, C., Braak, R., Bhartia, P.K., Veefkind, P. and Levelt, P. 2007. Aerosols and surface UV products from Ozone Monitoring Instrument observations: An overview. *Journal of Geophysical Research: Atmospheres*. 112(D24), 1–14.

Tsay, S.C., Hsu, N.C., Lau, W.K.M., Li, C., Gabriel, P.M., Ji, Q., Holben, B.N., Welton, E.J., Nguyen, A.X., Janjai, S. and Lin, N.H., 2013. From BASE-ASIA toward 7-SEAS: A satellite-surface perspective of boreal spring biomass-burning aerosols and clouds in Southeast Asia. *Atmospheric Environment*. 78, 20–34.

UCAR, 2020. https://www2.acom.ucar.edu/wrf-chem.

Uda, S.K., Hein, L. and Atmoko, D. 2019. Assessing the health impacts of peatland fires: A case study for Central Kalimantan, Indonesia. *Environmental Science and Pollution Research*. 26(30), 31315–31327.

Usup, A., Hashimoto, Y., Takahashi, H. and Hayasaka, H. 2004. Combustion and thermal characteristics of peat fire in tropical peatland in Central Kalimantan, Indonesia. *Tropics*. 14, 1–19.

Vadrevu, K. and Lasko, K., 2018. Intercomparison of MODIS AQUA and VIIRS I-Band fires and emissions in an agricultural landscape—Implications for air pollution research. *Remote Sensing*. 10(7), 978.

Vadrevu, K.P., Lasko, K., Giglio, L., Schroeder, W., Biswas, S. and Justice, C. 2019. Trends in vegetation fires in south and southeast Asian countries. *Scientific Reports*. 9(1), 1–13.

Vadrevu, K., Ohara, T. and Justice, C., 2017. Land cover, land use changes and air pollution in Asia: A synthesis. *Environmental Research Letters*. 12(12), 120201.

Van der Werf, G.R., Randerson, J.T., Giglio, L., Collatz, G.J., Mu, M., Kasibhatla, P.S., Morton, D.C., DeFries, R.S., Jin, Y.V. and van Leeuwen, T.T. 2010. Global fire emissions and the contribution of deforestation, savanna, forest, agricultural, and peat fires (1997–2009). *Atmospheric Chemistry and Physics*. 10(23), 11707–11735.

Varkkey, H. 2014. Regional cooperation, patronage and the ASEAN Agreement on transboundary haze pollution. *International Environmental Agreements: Politics, Law and Economics*. 14(1), 65–81.

Wettestad, J., 2018. *Clearing the Air: European Advances in Tackling Acid Rain and Atmospheric Pollution*. London: Routledge.

Wooster, M.J., Roberts, G., Perry, G.L.W. and Kaufman, Y.J. 2005. Retrieval of biomass combustion rates and totals from fire radiative power observations: FRP derivation and calibration relationships between biomass consumption and fire radiative energy release. *Journal of Geophysical Research Atmospheres*. 110(D24), 1–24.

Zaini, J., Susanto, A.D., Samoedro, E., Bionika, V.C. and Antariksa, B. 2020. Health consequences of thick forest fire smoke to healthy residents in Riau, Indonesia: A cross-sectional study. *Medical Journal of Indonesia*. 29(1), 58–63.

Zhang, T., Wooster, M.J., De Jong, M.C. and Xu, W. 2018. How well does the 'small fire boost' methodology used within the GFED4. 1s fire emissions database represent the timing, location and magnitude of agricultural burning? *Remote Sensing*. 10(6), 823.

Section I

Mapping and Monitoring of Fires, including Burned Areas

2 Identification of Smoldering Peatland Fires in Indonesia via Triple-Phase Temperature Analysis of VIIRS Nighttime Data

Christopher D. Elvidge
Colorado School of Mines, Colorado, USA

Mikhail Zhizhin
Colorado School of Mines, Colorado, USA
Russian Space Research Institute,
Russian Federation, Russia

Kimberly Baugh
University of Colorado, Colorado, USA

Feng-Chi Hsu
Colorado School of Mines, Colorado, USA

CONTENTS

INTRODUCTION

Smoldering peatland fires are a major source of transboundary smoke and greenhouse gas emissions from the islands of Sumatra and Borneo during drought years (Page et al., 2002; Siegert et al., 2001; Harrison et al., 2009; Tosca et al., 2011; Gaveau et al., 2014). It is well established that smoldering peat soil fires are lower in temperature than flaming-phase combustion and produce more smoke and partially oxidized gas (Akagi et al., 2011; Koppmann et al., 2005; Ohlemiller, 1995; Usup et al., 2004; Hungerford et al., 1995; Muraleedharan et al., 2000). Flaming biomass burning temperatures are in the range of 760–1400 K (Boonmee and Quintiere, 2002; Wotton et al., 2012). Smoldering, characterized by an orange glow, has temperatures in the 600–760 K range (Wotton et al., 2012; Boonmee and Quintiere, 2005; Broido and Nelson, 1975; Pastier et al., 2013). However, the smoldering producing the highest volume of smoke in peatlands is underground, warming the soil surface by conduction, with lower temperatures than the glowing ember smoldering combustion. Peatland surface with underground smoldering lacks the orange glow and has temperatures in the 320–500 K range (Usup et al., 2004; Hungerford et al., 1995). Having an ability to map smoldering peat soils in near real time would be quite useful for fire managers and in the definition of smoke source areas and deploying fire suppression teams.

Satellite remote sensing of smoldering peatland combustion is challenging for two reasons. First, the traditional satellite fire product, commonly referred to as hotspot data, uses a single midwave infrared (MWIR) channel, near 4 μm, for detection (Kaufman et al., 1998; Justice et al, 2002). While these algorithms use a long-wave infrared (LWIR) channel as a reference, the detection is essentially reliant on anomalies detected in the MWIR. No temperature can be calculated in this case because the Planck curve is being sampled at a single wavelength. With the wavelength of peak radiant emissions unknown, it is not possible to calculate the temperature with Wien's displacement law (Planck, 1901). Multispectral detections are required to model an infrared (IR) emitter's Planck curve and temperature. The second challenge to satellite remote sensing of smoldering is that the radiant emissions from flaming combustion are larger per unit area than those from smoldering. This is due to the T^4 term in the Stefan–Boltzmann law (Planck, 1901), which calculates radiant output based on temperature and source size. As a result, efforts to detect peatland source areas for transboundary smoke using satellite hotspots (Yulianti et al., 2012; Ismail et al., 2018) contend with the fact that flaming and smoldering peatland components cannot be separated and the flaming phase is disproportionally contributing to the radiant emissions.

There is a class of multispectral satellite thermal anomaly algorithms that do calculate temperatures based on Planck curve analyses (Dozier, 1981; Siegert et al., 2004; Elvidge et al., 2013 and 2019). The temperature of an IR emitter can be calculated based on the wavelength having peak radiant emissions based on Wien's displacement law. Studies conducted with bi-spectral data sources are limited to calculating temperatures and source sizes assuming there is a single IR emitter in the pixel (Dozier, 1981, Siegert et al., 2004). The original Visible Infrared Imaging Radiometer Suite (VIIRS) nightfire (VNF) algorithm, which uses up to six spectral bands, also assumes there is a single IR emitter (Elvidge et al., 2013). The single IR emitter assumption makes it possible to calculate a temperature and source size for subpixel IR emitters but is clearly not a realistic assumption given the pixel footprints in the range of 1 km². At that scale, there can be multiple IR emitters spanning a range of temperatures all contributing radiant energy to the satellite observed radiances.

Is it possible to resolve more than a single temperature phase inside individual pixel footprints? The answer to that question depends on the temperature separation of the phases, their source sizes, and spectral bands available. If the temperatures of the phases are separated far enough, their Planck curves are offset from each other, with certain spectral ranges dominated by a single phase. Planck curves for the hotter sources are shifted toward shorter wavelengths, and cooler sources are shifted to longer wavelengths. Thus, a sensor with wide separation of spectral bands may produce data suitable for the analysis of multiple temperature phases in individual pixels.

A study published several years ago demonstrated this concept, separating flaming and smoldering combustion phases in Indonesia with nighttime Landsat data (Elvidge et al., 2015). The method is based on the observation that the shortwave infrared (SWIR) radiances are dominated by the flaming phase and the smoldering results in thermal anomalies in the LWIR. The Landsat SWIR band radiances were used to model a Planck curve representing the flaming phase, and the LWIR radiances were used to model a lower temperature smoldering phase. In the Landsat study (Elvidge et al., 2015), a temperature threshold of 350 K was used to separate background from smoldering-dominated pixels. Both SWIR and LWIR thermal anomalies can be readily seen in the nighttime Landsat imagery, and the spatial distribution and intensity of these features vary between the two wavelength ranges. Field surveys conducted the day after a nighttime Landsat collection in Kalimantan confirmed that the LWIR thermal anomalies had smoldering peatland combustion. Comparing the distribution of flaming versus smoldering combustion, it was found that some pixels only had flaming, some only had smoldering, and others had a combination of flaming and smoldering.

In this chapter, we extend the nighttime Landsat flaming versus smoldering algorithm to nighttime VIIRS data within the VNF processing framework (Elvidge et al., 2013). At night, the VIIRS collects near-infrared (NIR), SWIR, and LWIR data that closely match the Landsat spectral bandpasses (Table 2.1). The Landsat algorithm assumes the detection pixels are 100% on fire. That is to say, the algorithm makes no attempt to model the Planck curve of the background. This is a reasonable simplifying assumption for Landsat, where pixel footprint sizes are small. However, the VIIRS M-band pixel footprints are about a thousand times larger than the Landsat spectral bands in the NIR and SWIR and about 100 times larger in the LWIR. Since the radiance detection limits are quite similar between Landsat and VIIRS, it is reasonable to expect that the larger source areas would be required to trigger detection in the VIIRS data. The VIIRS has a significant advantage over Landsat with the availability of two spectral bands in the MWIR, near 4 µm. The chapter outlines an augmentation to the VNF algorithm to un-mix flaming, smoldering, and background radiances in individual pixels and preliminary results obtained in Sumatra from a single VIIRS orbit collected over Sumatra in September of 2014. The new algorithm is referred to as VNF+.

TABLE 2.1
VIIRS Spectral Bands Collecting at Night

Band Designation	Range	Central Wavelength (μm)	VNF Detection Limit (Zhang and Kondragunta, 2008)	Landsat 8 Equivalent (μm)
M7	NIR	0.865	$0.034 \ W/m^2/sr/\mu m$ (VNF)	0.865
M8	NIR	1.24	0.088	1.37
M10	SWIR	1.61	0.036	1.61
M11	SWIR	2.25	0.023	2.2
M12	MWIR	3.7	0.041	
M13	MWIR	4.05	0.012	
M14	LWIR	8.55		
M15	LWIR	10.76		10.9
M16	LWIR	12.01		12.0

METHODS

The two SWIR and two LWIR channels available on Landsat were crucial for the success of the flaming versus smoldering analysis (Elvidge et al., 2015). Originally, the SNPP VIIRS only collected one of the SWIR bands at night, M10 – centered at 1.61 μm. Once the value of collecting the second SWIR band (M11 at 2.25 μm) was demonstrated, there was an approval process to begin operational SNPP VIIRS collections with both SWIR bands. This transition occurred early in 2018. The VIIRS on the NOAA-20 satellite has collected both SWIR bands at night from the start. There are a handful of nighttime VIIRS collections made prior to 2018 with the instrument in "day-mode" with all spectral bands collecting. For this study, we used an orbit from one of these "day-mode"-at-night orbits that passed over Sumatra on September 27, 2014, at 01:53 local time.

The analysis involves several modifications of the original VNF algorithm (Elvidge et al., 2013), which detects the presence of subpixel IR emitters in six spectral bands, two in the NIR, two in the SWIR, and two in the MWIR. The modified algorithm is referred to as VNF+. The first modification is the addition of an atmospheric correction using MODTRAN 5 (Berk et al., 2006) parameterized by atmospheric profiles for temperature, pressure, water vapor, CO_2, and O_3 derived from simultaneously acquired ATMS data (Boukabara et al., 2013). The second modification is to introduce a two-staged Planck curve fitting in the pixels where smoldering may be present.

VNF runs independent detection algorithms in four short-wavelength channels (NIR and SWIR) based on a characterization of the noise floor. The detection threshold is the mean plus four standard deviations. A second style of detector is required for the MWIR where the scene records radiances from the earth surface and clouds in addition to IR emitters. To find the hotspot pixels in the MWIR, we developed an algorithm that screens pixels against the diagonal M12 versus M13 data cloud that results from background temperature variations. Since background pixels far outnumber IR emitter pixels, the diagonal can be readily located and buffered to detect pixels containing IR emitters. In the standard VNF, detection in more than two spectral bands triggers a dual Planck curve fitting based on the observed radiances in the complete set of M-bands collecting at night (Table 2.1). In this study, the detection pixels are sorted into six types (Table 2.2) to define the algorithmic path and to explore whether multiple combustion types can be discerned. If the detection types yield the same temperature, that indicates that there is no possibility to discriminate combustion types.

The detection types and processing steps are listed below:

Type 0: Single-band detection, typically in M11. No further analysis is done since the Planck curve cannot be modeled. The M11 detection threshold is set low so that if there is detection in any other spectral bands, temperatures can be calculated. The result is a large number of M11-only false detections, where none of the other spectral bands had detection. The isolated M11 detections are filtered out with an adjacency test. That is to say, only clusters of two or more M11-only detections are preserved in the output.

Type 1: Detections in NIR and SWIR – with no MWIR detection, analyzed with Planck curve fitting based on the NIR and SWIR detection radiances, using 1000 K to initiate the interactive simplex Planck curve fitting. For Type 1, there is no difference from the standard VNF algorithm. Result is a temperature and source area for a single IR emitter.

Type 2: M11 SWIR and MWIR detections, analyzed with dual Planck curve fitting for a subpixel IR emitter and background. The Planck curve fitting is initiated with temperatures of 1000 K for the IR emitter and 3000 K for the background. For Type 2, there is no difference from the standard VNF. Results include temperature and source area estimates for a single IR emitter and background.

TABLE 2.2

Five Types of Planck Curve Fitting

Type	Spectral Bands	Fitting	Notes
0	M11	None	Filter out solitary pixels
1	M7, M8, M10, M11	Single Planck curve fitting	Flaming-phase combustion – primarily natural gas flares
2	M11, M12, and M13	Dual Planck curve fitting for IR emitter and background	Flameless glowing embers (smoldering)
3	M12 and M13	Dual Planck curve fitting for IR emitter and background	Rare occurrences
4	M7, M8, M10, M11, M12, and M13	Triple-phase fitting for two IR emitters and background	Solve for temperature and source area for a hot flaming and a cooler combustion phase plus background
5	M7, M8, M10, M11, M12, and M13	Dual Planck curve fitting for IR emitter and background	Assembled from Type 4 detections that yielded spurious results in triple-phase analysis

Type 3: MWIR (M12 and M13) detection only, analyzed with dual Planck curve fitting for a subpixel IR emitter and background. The Planck curve fitting is initiated with temperatures of 1000 K for the IR emitter and 300 K for the background. For Type 2, there is no difference from the standard VNF. Results are temperature and source area estimates for a single IR emitter and background.

Type 4: Detection in two SWIR and two MWIR bands, and maybe also in the NIR bands. In this case, the flaming-phase Planck curve is modeled with the SWIR and NIR radiances to calculate temperature, source area, and flaming-phase radiances at longer wavelengths for subtraction from the MWIR and LWIR channel radiances. The short-wavelength Planck curve fitting is initiated with a temperature of 1000 K. The residual MWIR and LWIR radiances are then analyzed with dual Planck curve fitting for a subpixel IR emitter and background. The second round of Planck curve fitting is initiated with temperatures of 400 K for the IR emitter and 300 K for the background. With two rounds of Planck curve fitting, the result includes two subpixel IR emitter temperatures and independently derived source area estimates, plus the background. The hotter temperature component is referred to as 4A, and the cooler IR emitter is 4B.

The Type 4 results are then filtered to reclassify pixels with spurious results in the 4B source areas. Most of the spurious results are either extremely small or extremely large 4B source areas. In both cases, this is an indication that the IR emitters present in the pixel are better described by a single temperature. The Type 4 pixels with spurious 4B source areas revert back to the standard VNF algorithmic path.

Type 5: The Type 4 pixels with spurious fit pixels are moved to the Type 5 set and analyzed with the standard VNF dual Planck curve analysis for a single IR emitter plus background.

Validation: Due to the retrospective nature of this prototyping study, we have no field data to present regarding validation. The only validation we are able to muster is comparisons of the low-temperature smoldering locations. However, we performed spatial comparisons of the smoldering pixel locations with two relevant spatial data products. First, the smoldering pixel locations were overlaid on peatland soil map of Sumatra (Miettinen et al., 2016) to check if they were located on known peatlands. Second, the smoldering temperature pixels were overlaid on Landsat 8 data of path 124 row 62 data collected the following morning of the same day to check if there is evidence that the locations are indeed sources of smoke with active combustion.

RESULTS

Numerical distribution of the types: The most abundant category is Type 0 prior to adjacency filtering with 749 pixels. Adjacency filtering thins this group to 84 pixels. The largest group is Type 1 with 159 pixels. The smallest group is Type 3 with 3 pixels. Type 2 has 125 pixels. Type 4 starts with 112 pixels and drops to 92 as 21 pixels shift to Type 5. Figure 2.1 shows a pie chart of the type tallies.

Temperature and source sizes of the types: Type 1, 2, 3, and 5 pixels are processed in the same way as the standard VNF algorithm. The resulting temperatures and source areas are shown in Figure 2.2. Type 1 pixels have the highest temperatures

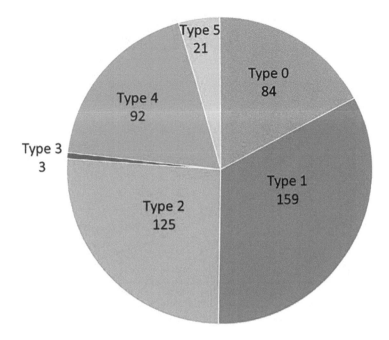

FIGURE 2.1   Pie chart showing the proportions of Types 1–5 from the study data set.

(1300–2300 K) and the smallest source sizes (10–0.2 m²). These are quite likely to be natural gas flares, which are known to burn hotter than biomass (Elvidge et al., 2013). Type 2 pixels are about half as hot as, and 10 times larger than, Type 1 pixels, ranging from 650 to 800 K and 100 to 2000 m² source areas. Type 3 pixels fall quite close to Type 2 pixels, with temperatures in the 800–900 K range. The largest cluster of Type 5 pixels is larger source area gas flares, 1600–2000 K and 5–100 m² source sizes. A second cluster of Type 5 pixels is spread from 650 to 1150 K.

The behavior of the Type 4 pixels is quite different from that of the other types in that temperatures and source areas are derived for two subpixel IR emitters (Figures 2.2 and 2.3). Type 4A is derived from the short-wavelength detection radiances from M7, M8, M10, and M11. This group has slightly higher temperatures and smaller source areas than the standard VNF analysis. Type 4B has a bimodal temperature distribution, with one in the range of 320–500 K and large source areas. A second set of Type 4B pixels is quite similar to Type 2 in terms of temperatures and source sizes. The third group of Type 4B pixels are in the range of 800–1400 K.

Validation: Overlay of the Type 4B pixel locations where temperatures are in the range of 320–500 K found that all are located on peatland soils. Examination of the Landsat 8 data collected later the same day shows a regional-scale smoke plume running diagonally across the scene (Figure 2.4). The smoldering pixels are clustered inside the source area for the smoke plume. In addition, the daytime Landsat SWIR and LWIR channels exhibit thermal anomalies in the smoldering found with VIIRS.

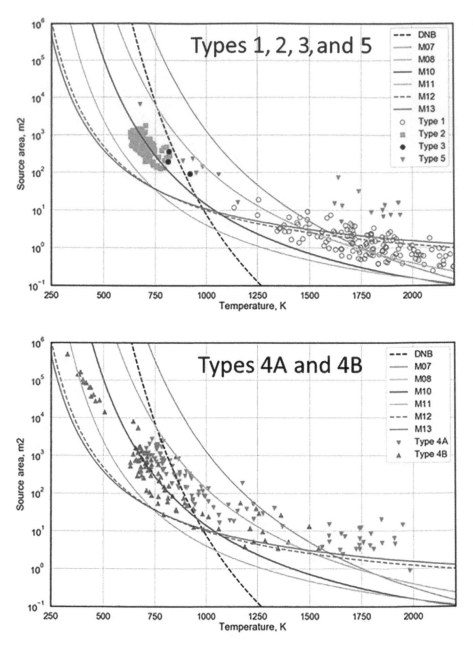

FIGURE 2.2 The upper chart shows a scattergram of temperature versus source area for Types 1, 2, 3, and 5. These types are processed with the standard VNF algorithm. The lower chart shows the Type 4A and 4B pixels, where the detection spans at least four spectral bands and the IR emitter is split between a hot phase and a cool phase. Derivation of the detection limit lines is described in Elvidge et al. (2013).

Combustion parameters:

Lat= -3.183881 Lon= 105.445885 deg.   UTC Time=2014/09/26 18:56:00
Radiant heat intensity=142.35 W/m²   Local time=2014/09/27 01:56:00
Flaming temperature=788 K   Radiant heat=26.38 MW
Smoldering temperature=432 K   Flaming footprint=831.46 m²
Background temperature=294 K   Smoldering footprint=65437.34 m²
Cloud state=clear   Atmosphere corrected=yes
File=SVM10_npp_d20140926_t1852060_e1857464_b15101_c20140927005747875402_noaa_ops.h5

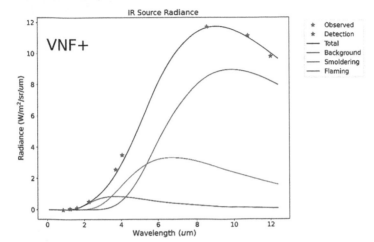

Combustion parameters:

Lat= -3.183881 Lon= 105.445885 deg.   UTC Time=2014/09/26 18:56:00
Radiant heat intensity=65.55 W/m²   Local time=2014/09/27 01:56:00
IR emitter temperature=588 K   Radiant heat=67.99 MW
Background temperature=303 K   IR emitter footprint=10034.13 m²
Cloud state=clear   Atmosphere corrected=yes
File=SVM10_npp_d20140926_t1852060_e1857464_b15101_c20140927005747875402_noaa_ops.h5

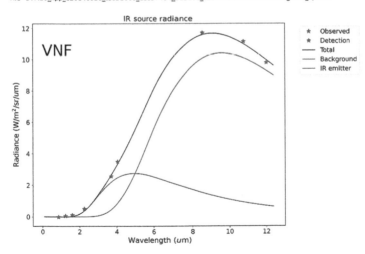

FIGURE 2.3   Results from VNF and VNF+ for a pixel with smoldering combustion. The IR emitter detection was in the SWIR and MWIR, bands M11–M13. The VNF Planck curve fitting produced a temperature of 588 K and source area of 10,034 m². VNF+ split the IR emitter into a flaming phase (788 K and 831 m²) and a smoldering phase (432 K and 65,437 m²).

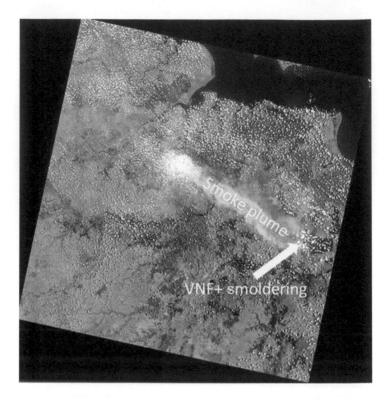

FIGURE 2.4   Landsat 8 path 124 row 62 image collected September 27, 2014, the morning after the nighttime VIIRS collection. This is a false color composite with bands 7, 5, and 3 as red, green, and blue. There is a regional-scale smoke plume running diagonally across the center of the image. The Type 4B VNF+ pixels with temperatures in the 320–500 K range are clustered in the smoke plume source area.

DISCUSSION

Previous studies have established that three combustion classes can be distinguished with satellite remote sensing data. The three can be distinguished based on temperature. Smoldering peat soils are the coolest of the three, with temperatures ranging from about 350 to 500 K (Elvidge et al., 2015, 2019; Siegert et al., 2001). Natural gas flaring is the hottest class, with temperatures in the 1400–2300 K range (Elvidge et al., 2013). Flaming biomass is found in a temperature range between these two, from 760 to 1400 K (Boonmee and Quintiere, 2002; Wotton et al., 2012).

To define the combustion classes resolved with the VNF+ algorithm, a temperature histogram was generated by combining the results from Type 1 to Type 5 (Figure 2.5). Our interpretation is that four types of combustion are present. At the low-temperature end, there is a smoldering peatland phase with temperatures in the range of 320–500 K and large source areas, arising from Type 4B. These temperatures are consistent with smoldering peat soils, where combustion is present in the subsurface soil column (Usup et al., 2004; Muraleedharan et al., 2000). A second smoldering phase is indicated by the histogram spike between 600 and 750 K, arising from Types 2, 4A,

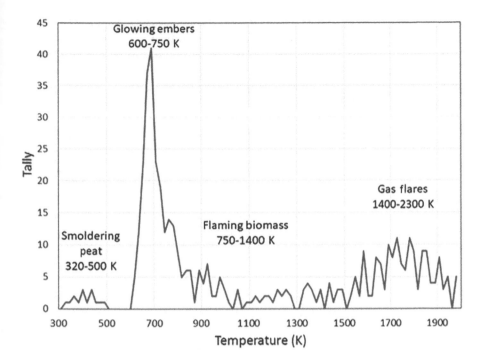

FIGURE 2.5   Temperature histogram of the VNF+ output from the Indonesia nighttime VIIRS test data from September 27, 2014.

and 4B. This matches the flameless, glowing wood ember combustion phase reported by several studies (Boonmee and Quintiere, 2002; Boonmee and Quintiere, 2005; Broido and Nelson, 1975; Pastier et al., 2013). There is a high-temperature gas flaring combustion phase from 1300 to 2300 K arising from Types 1, 4B, and 5. The fourth combustion phase is flaming biomass, from 750 to 1300 K, arising primarily from Types 4A and 4B, with minor contributions from Types 1, 2, 3, and 5.

When we started this study, we only expected to find three types of combustion: low-temperature smoldering peat soils, flaming biomass, and natural gas flaring. We were surprised to find the prominent temperature histogram spike from 600 to 750 K. This temperature range is consistent with the glowing wood embers (Harrison et al., 2009; Ismail et al., 2018; Kaufman et al., 1998), the primary style of smoldering producing volatile gases from soil wood, the process known as pyrolysis. Ohlemiller (1991) described this style of wood combustion this way: "During smoldering, the reacting surface of the wood char emitted an orange glow, which generally had a uniform brightness over the length of the oxidation zone."

CONCLUSION

A nighttime Landsat-based algorithm developed to detect and characterize flaming and smoldering combustion within single pixels has been adapted to nighttime VIIRS data. The development was conducted as an add-on to the existing VNF algorithm and is referred to as VNF+. The test data set is from an orbit of "day-mode" VIIRS data collected at night on September 27, 2014, and covering Sumatra and a portion of Borneo. The algorithm requires VNF detection in both SWIR bands (M10 and M11) and MWIR bands (M12 and M13). This condition was met in 23% of the VNF pixels in the test set. The analysis involves Planck curve modeling of a hot IR emitter using NIR and SWIR radiances, with subtraction of the modeled radiances from the MWIR and LWIR radiances. The residual MWIR and LWIR radiances are then analyzed for the temperatures and source areas for a subpixel low-temperature IR emitter and the background. Pixels yielding spurious results revert back to the standard VNF analysis.

The study data set yielded four distinct combustion classes: (1) smoldering peatland soils, where the smoldering is inside the soil column and the soil surface is being heated via conduction, and the detected temperatures are in the 320–500 K range and source areas are large; (2) a second low-temperature phase in the range of 600–750 K, which matches the temperature range reported for smoldering wood, which is flameless and characterized by orange glowing embers; (3) a flaming biomass phase from 750 to 1400 K; and (4) natural gas flaring from 1400 to 2000+ K.

Validation analyses confirm that the dense cluster of pixels from the two low-temperature classes are smoldering peatland fires. The first line of evidence in this regard is that the pixels fall on previously mapped peatlands. The second line of evidence is that the low-temperature pixels are tightly clustered in the source are for a regional smoke plume visible later the same day in Landsat imagery. A close examination of the Landsat reveals fresh burn scar and active fire in the zone where the VIIRS smoldering pixels are clustered.

The results to date indicate that nighttime VIIRS data have a sufficient number of spectral bands and detection limits set low enough to enable subpixel analysis of two combustion phases present inside individual pixels. This includes a hot phase associated with either flaming biomass burning or natural gas flaring. For biomass burning, the cooler phase may be either glowing embers or smoldering organic soil burning. For gas flares, the cooler phase may be the plume of heated black carbon present at certain flares?

The subpixel un-mixing of hot and cool IR emitters in individual pixels can be referred to as a triple-temperature phase analysis because the output includes three temperatures and three source areas: hot phase, cooler phase, and background. The procedure only works at night and requires data collection in at least two SWIR bands. The method's success with nighttime VIIRS data can be chalked up to the old adage that "the nighttime is the right time." With sunlight eliminated, it becomes possible to detect subpixel IR emitters in the NIR and SWIR and fully attribute the observed radiances to IR emitters. In addition, it is well known that flaming-phase combustion tends to subside at night when temperatures are cooler and humidity is higher (Zhang and Kondragunta, 2008). The diurnal reduction in flaming-phase combustion increases the proportion of radiant emissions coming from lower temperature combustion phases such as smoldering and flameless glowing embers.

Over the next year, it is our intention to deploy VNF+ for near-real-time product generation, displacing the original VNF algorithm. Clearly, further validation should be conducted to better understand the lower temperature combustion phases reported in this study. However, these preliminary results are encouraging and suggest that areas of smoldering and glowing ember combustion can be separated from flaming combustion for improved modeling of smoke and greenhouse gas emissions from biomass burning.

ACKNOWLEDGMENTS

This research was sponsored by the NOAA Joint Polar Satellite System (JPSS) Proving Ground program's Smoke and Fire Initiative.

REFERENCES

Akagi, S.K., Yokelson, R.J., Wiedinmyer, C., Alvarado, M.J., Reid, J.S., Karl, T., Crounse, J.D. and Wennberg, P.O., 2011. Emission factors for open and domestic biomass burning for use in atmospheric models. *Atmospheric Chemistry and Physics*. 11(9), 4039–4072.

Berk, A., Anderson, G.P., Acharya, P.K., Bernstein, L.S., Muratov, L., Lee, J., Fox, M., Adler-Golden, S.M., Chetwynd Jr, J.H., Hoke, M.L. and Lockwood, R.B. 2006. MODTRAN5: 2006 update. In *Algorithms and Technologies for Multispectral, Hyperspectral, and Ultraspectral Imagery XII* (Vol. 6233, p. 62331F). International Society for Optics and Photonics.

Boonmee, N. and Quintiere, J.G., 2002. Glowing and flaming autoignition of wood. *Proceedings of the Combustion Institute*. 29(1), 289–296.

Boonmee, N. and Quintiere, J.G., 2005. Glowing ignition of wood: the onset of surface combustion. *Proceedings of the Combustion Institute*. 30(2), 2303–2310.

Boukabara, S.A., Garrett, K., Grassotti, C., Iturbide-Sanchez, F., Chen, W., Jiang, Z., Clough, S.A., Zhan, X., Liang, P., Liu, Q. and Islam, T., 2013. A physical approach for a simultaneous retrieval of sounding, surface, hydrometeor, and cryospheric parameters from SNPP/ATMS. *Journal of Geophysical Research: Atmospheres*. 118(22), 12–600.

Broido, A. and Nelson, M.A., 1975. Char yield on pyrolysis of cellulose. *Combustion and Flame*. 24, 263–268.

Dozier, J. 1981. A method for satellite identification of surface temperature fields of sub-pixel resolution. *Remote Sensing Environment*. 11, 221–229.

Elvidge, C.D., Zhizhin, M., Hsu, F.C. and Baugh, K.E. 2013. VIIRS nightfire: satellite pyrometry at night. *Remote Sensing*. 5, 4423–4449.

Elvidge, C.D., Zhizhin, M., Hsu, F.C., Baugh, K., Khomarudin, M.R., Vetrita, Y., Sofan, P. and Hilman, D. 2015. Long-wave infrared identification of smoldering peat fires in Indonesia with nighttime Landsat data. *Environmental Research Letters*. 10(6), 065002.

Elvidge, C.D., Zhizhin, M., Baugh, K., Hsu, F.C. and Ghosh, T. 2019. Extending nighttime combustion source detection limits with short wavelength VIIRS data. *Remote Sensing*. 11(4), 395.

Gaveau, D.L., Salim, M.A., Hergoualc'h, K., Locatelli, B., Sloan, S., Wooster, M., Marlier, M.E., Molidena, E., Yaen, H., DeFries, R. and Verchot, L. 2014. Major atmospheric emissions from peat fires in Southeast Asia during non-drought years: evidence from the 2013 Sumatran fires. *Scientific Reports*. 4, 6112.

Harrison, M.E., Page, S.E. and Limin, S.H. 2009. The global impact of Indonesian forest fires. *Biologist*. 56(3), 156–163.

Hungerford, R.D., Frandsen, W.H. and Ryan, K.C., 1995. Ignition and burning characteristics of organic soils. In *Tall Timbers Fire Ecology Conference* (Vol. 19, pp. 78–91). Tallahassee, FL.

Ismail, P., Swastiko, W.A. and Fadlan, A. 2018, July. Forest fires hotspot detection in Indonesia using Himawari-8 satellite data. In *Proceeding of International Symposium for Sustainable Humanosphere* (pp. 121–127). Biomaterial LIPI.

Justice, C.O., Giglio, L., Korontzi, S., Owens, J., Morisette, J.T., Roy, D., Descloitres, J., Alleaume, S., Petitcolin, F. and Kaufman, Y. 2002. The MODIS fire products. *Remote sensing of Environment.* 83(1–2), 244–262.

Kaufman, Y.J., Justice, C.O., Flynn, L.P., Kendall, J.D., Prins, E.M., Giglio, L., Ward, D.E., Menzel, W.P. and Setzer, A.W. 1998. Potential global fire monitoring from EOS-MODIS. *Journal of Geophysical Research: Atmospheres.* 103(D24), 32215–32238.

Koppmann, R., von Czapiewski, K. and Reid, J.S. 2005. A review of biomass burning emissions, part I: gaseous emissions of carbon monoxide, methane, volatile organic compounds, and nitrogen containing compounds. *Atmospheric Chemistry and Physics Discussions.* 5, 10455–10516.

Miettinen, J., Shi, C. and Liew, S.C., 2016. Land cover distribution in the peatlands of Peninsular Malaysia, Sumatra and Borneo in 2015 with changes since 1990. *Global Ecology and Conservation.* 6, 67–78.

Muraleedharan, T.R., Radojevic, M., Waugh, A. and Caruana, A., 2000. Emissions from the combustion of peat: an experimental study. *Atmospheric Environment.* 34(18), 3033–3035.

Ohlemiller, T.J. 1991. Smoldering combustion propagation on solid wood. *Fire Safety Science.* 3, 565–574.

Ohlemiller, T.J. 1995. Smoldering combustion SFPE handbook of fire protection engineering Edited by M.A. Quincy, P.J. DiNenno, C.L. Beyler, R.L.P. Custer and W.D. Walton, 2nd edition (Quincy, MA: National Fire Protection Association) section 2, chapter 11, pp. 171–179.

Page, S.E., Siegert, F., Rieley, J.O., Boehm, H.D.V., Jaya, A. and Limin, S. 2002. The amount of carbon released from peat and forest fires in Indonesia during 1997. *Nature.* 420(6911), 61–65.

Pastier, M., Tureková, I., Turňová, Z. and Harangozó, J., 2013. Minimum ignition temperature of wood dust layers. Research Papers Faculty of Materials Science and Technology Slovak University of Technology. 21 (special issue), 127–131.

Planck, M., 1901. On the law of distribution of energy in the normal spectrum. *Annalen der physic.* 4(553), 1.

Siegert, F., Ruecker, G., Hinrichs, A. and Hoffmann, A.A. 2001. Increased damage from fires in logged forests during droughts caused by El Nino. *Nature,* 414(6862), 437–440.

Siegert, F., Zhukov, B., Oertel, D., Limin, S., Page, S. and Rieley, J.O. 2004. Peat fires detected by the BIRD satellite. *International Journal of Remote Sensing.* 25(16), 3221–3230.

Tosca, M.G., Randerson, J.T., Zender, C.S., Nelson, D.L., Diner, D.J. and Logan, J.A. 2011. Dynamics of fire plumes and smoke clouds associated with peat and deforestation fires in Indonesia. *Journal of Geophysical Research: Atmospheres.* 116(D08207). doi: 10.1029/2010JD015148

Usup, A., Hashimoto, Y., Takahashi, H. and Hayasaka, H.A. 2004. Combustion and thermal characteristics of peat fire in tropical peatland in Central Kalimantan, Indonesia. *Tropics.* 14(1), 1–19.

Wotton, B.M., Gould, J.S., McCaw, W.L., Cheney, N.P. and Taylor, S.W. 2012. Flame temperature and residence time of fires in dry eucalypt forest. *International Journal of Wildland Fire.* 21(3), 270–281.

Yulianti, N., Hayasaka, H. and Usup, A. 2012. Recent forest and peat fire trends in Indonesia the latest decade by MODIS hotspot data. *Global Environmental Research.* 16(1), 105–116.

Zhang, X. and Kondragunta, S., 2008. Temporal and spatial variability in biomass burned areas across the USA derived from the GOES fire product. *Remote Sensing of Environment.* 112(6), 2886–2897.

3 Evaluation of Sentinel-3 SLSTR Data for Mapping Fires in Forests, Peatlands, and Croplands – A Case Study over Australia, Indonesia, and India

Aditya Eaturu
University of Alabama Huntsville, USA

Krishna Prasad Vadrevu
NASA Marshall Space Flight Center, USA

CONTENTS

INTRODUCTION

Biomass burning is practiced in different regions of the world, including South/Southeast Asian countries (Streets et al., 2003). It is an important source of air pollution which results in the release of greenhouse gas (GHG) emissions and aerosols (Ito and Penner, 2004; Prasad et al., 2000; 2002; 2003; 2008; Gupta et al., 2001; Badarinath et al., 2008a, b; Kant et al., 2000; Vadrevu et al., 2006; 2008; Lasko and Vadrevu, 2018). The smoke particles released during the biomass burning can interact with the cloud droplets and alter Earth's radiation budget (Li et al., 2001; Ramachandran, 2018) by their light-scattering effects and by influencing cloud microphysical processes (Jones et al., 2007). The GHG emissions from bio-mass burning represent the largest source of interannual variability, in particular CO_2 fluxes (Szopa et al., 2007). It is estimated that biomass burning contributes to 7600 ± 359 Tg CO_2eq/year (FAOSTAT, 2020). Also, biomass burning is attributed to an increase in growth rates of CO, CO_2, and CH_4 during the El Niño–Southern Oscillation (ENSO) events (Langenfelds et al., 2002; Vadrevu, 2008). In addi-tion, biomass burning has been shown to influence a variety of land–atmospheric interactions at different scales, such as vegetation transpiration, soil erosion, and albedo (Crutzen and Andreae, 1990). Smoke-borne aerosols from fires disrupt nor-mal hydrological processes and reduce rainfall, potentially contributing to regional drought. In addition to these effects on Earth's radiation, atmosphere, climate, and ecosystems, the pollutants released from the biomass burning (Vadrevu et al., 2012; 2013; 2014a, b; Vadrevu and Lasko, 2015) can have adverse health effects such as asthma, acute respiratory illness, eye irritation, cardiovascular mortality, thrombo-sis, and in severe cases, mortality (Sigsgaard et al., 2015).

Most of the biomass burning events are driven by anthropogenic activities. In sev-eral regions of the world, fire is used as a management tool for land clearing (Justice et al., 2015). For example, fire is used to clear the forests for agriculture through slash-and-burn (Biswas et al., 2015a, b; Prasad et al., 2001,2002; Inoue, 2018) agri-cultural residues after crop harvest (Lasko et al., 2017; 2018a, b; Vadrevu and Lasko, 2018), to clear the land for the next crop, and to clear the forested lands for plantations (Israr et al., 2018), promoting the growth of grass in pasture lands for cattle (Stott et al., 1990), including reducing weeds prior to planting of crops (Simorangkir, 2007). Besides, the ignition source can also be intentional or accidental human activities. While most of these fires are anthropogenic, the drivers of fires can also be natu-ral, such as lightning and extreme and prolonged drought conditions. Regardless of the ignition source, in forested areas, the wildfires can spread rapidly and become uncontrollable due to the local meteorological and environmental conditions. As a result, fires can threaten human lives and cause severe economic damage. Thus, map-ping and monitoring of biomass burning events, fires, and impacts can help not only in understanding land–atmospheric interactions useful for climate change studies, but also in protecting human lives, ecosystems, and related functions (Vadrevu and Justice, 2011; Vadrevu et al., 2017; 2018, 2019). For mapping and monitoring of fires at large spatial scales and at repeated intervals, remote sensing technology can pro-vide valuable information at varied spatial scales. The literature review suggests that a variety of optical and Synthetic Aperture Radar (SAR) data can be effectively used to capture information on fire locations, fuel loads, burn frequency, burn intensity, and burned areas at various spatial and temporal scales. In particular, fire-related information at medium-to-high spatial resolution can be highly useful for quantify-ing environmental, social, and economic losses. Specific to fire mapping and moni-toring, three essential products are routinely developed from the satellites: (1) active fire products that contain information about the location and timing of fires burning at the time of the satellite overpass, with dates useful for fire control (Giglio et al., 2013; Schroeder et al., 2014); (2) burnt areas that are captured from the satellites after the event useful to assess the economic damage and emissions estimation (Justice et al., 2002); and (3) fire radiative power products that provide information on the radiant energy emitted from burning vegetation per unit time, i.e., the fire's strength that is linearly related to the biomass consumption rate (Wooster et al., 2005).

This study aims to assess the potential of Sea and Land Surface Temperature Radiometer (SLSTR) data for active fire mapping. To test these datasets' potential, we selected three different sites, focusing on forest fires in New South Wales, Australia; peatland fires in South Sumatra, Indonesia; and agriculture fires in Punjab, India. These sites were selected as they have varied fire behavior with respect to fuel ignition, flame development, and fire spread in peatland and forest ecosystems. The fire behavior characteristics influence the type of GHG emissions released during the biomass burning. For example, peatland burning is characterized by smoldering-type emissions (mostly CO and other hydrocarbons due to incomplete combustion). In contrast, intense forest fires with flaming conditions are dominated by the release of CO_2 due to complete oxidation. In the study, we first review the basic principles of active remote sensing from satellites and then use the SLSTR data to map fires in peatlands and forest fires. We compare the results in these ecosystems. The results highlight potential variations in fire characteristics in these ecosystems useful for fire mapping and monitoring.

ACTIVE FIRE DETECTION

The two important laws of physics that govern the remote sensing of active fire detection are Plank's law and Wien's displacement law. Forest fire flames are characterized by the high emissivity and thus are rich sources of infrared radiation (Robinson, 1991; Hua and Shao, 2017). The peak thermal radiant energy release of flaming fires is at much shorter wavelengths (Planck's law). For example, compared to the radiant energy of the ambient temperature background of nearly 300 K, the smoldering and flaming fire have relatively higher temperatures varying from 600 to 1000 K, respectively. Also, varying with the fire strength, the temperature curve peaks at different wavelengths with a shift in peak (Wien's law). In the visible region of the spectrum, mostly smoke obstructs the fire retrievals, whereas the strong infrared signal emitted by the flaming fires can be easily captured in the middle-infrared (MIR) region of the spectrum. In the MIR wavelength region of the spectrum, the spectral radiance emitted from flaming fires can be up to four orders of magnitude higher than the surrounding ambient background, with a significant increase in the signal; thus, mostly MIR channels are used for active fire detection (Robinson, 1991; Li et al., 2001; Matson and Dozier, 1981; Wooster et al., 2005). Also, for a given increase in temperature, the radiation of the MIR channel increases more rapidly than that of the thermal infrared (TIR) channel. For example, as noted by Hua and Shao (2017), an increase in temperature from 500 to 800 K will increase thermal radiation from 82 to 682 $W/m^2/\mu m/sr$ at the 4-μm channel. However, the radiation only increases from 58 to 135 $W/m^2/\mu m/sr$ at the 11-μm channel. Thus, in general, the temperature values of fire pixels in the MIR channel are significantly greater than those in the TIR channel and the MIR channel is more sensitive to fires than the TIR channel. However, in the MIR region, the fire signal, in some instances, can be confused with the sun glints as they can also generate high spectral radiances (Arino and Melinotte, 1998). Also, the bare ground might get heated due to intense sunlight in some places and the specular reflected sunlight can also increase the MIR channel signal, creating confusion and false signals as fires (Boles and Verbyla, 2000). One way to discriminate active fires from sun glints (due to clouds or water bodies) is through using visible and near-infrared (NIR) spectral bands as the subpixel fire signal is weak in this region of the spectrum, and clouds and water bodies show much robust signal in the visible region of the spectrum than MIR. Thus, the use of MIR channel as a threshold alone is not recommended for detecting active fires, and additional channels that can reduce the noise in a contextual way are always used in addition to the MIR (Justice et al., 1996; Flasse and Ceccato, 1996; Kaufman et al., 1998). Further, most of the active fire detection algorithms use brightness temperature values calculated from the spectral radiances using the inverse Planck function (Wooster et al., 2013) from the MIR or long-wave infrared (LWIR) channels. The brightness temperature values are used as thresholds to detect the fires. The whole process can include multiple steps to filter the noise from other background pixels; the initial step might be broad using a fixed threshold approach with lower values which may include non-fire pixels and fire pixels; and then as a subsequent step, testing each potential fire pixel against the background pixels to confirm the true nature of the fire pixel, based on multiple thresholds or steps contextually (Martín et al., 1999). Several of the active fire products are based on such contextual approach such as from Along-Track Scanning Radiometer (ATSR) (Arino et al., 1999), MODIS (Giglio et al., 2003), and the Geostationary Operational Environmental Satellite (GOES) Wildfire Automated Biomass Burning Algorithm (WF-ABBA) (Prins et al., 1998), the BIRD Hotspot Recognition Sensor (HSRS) fire products (Zhukov et al., 2006) and the Meteosat SEVIRI fire products (Roberts and Wooster, 2008), and Visible Infrared Imaging Radiometer Suite (VIIRS) fire product (Schroeder et al., 2014).

STUDY AREA

We focused on three different types of fires: intense forest fires in New South Wales, Australia; the peatland fires in the province of South Sumatra, Indonesia; and agricultural crop residue burning fires in Punjab, India (Figure 3.1). We selected three different cases as they differ on fire regime characteristics (fire frequency, severity, extent, seasonality, and relationship with climate) (Agee, 1993), thus allowing us to test the potential of S3 SLSTR datasets in fire detection.

DATASETS

S3 SLSTR is a dual-scan temperature radiometer in the low Earth orbit (800–830 km altitude) on board the S3 satellite. The mean global coverage revisit time for dual-view SLSTR observations is 1.9 days at the equator (ESA Sentinel online, 2020). SLSTR employs the along-track scanning dual-view (nadir and backward oblique) technique for 9 channels in the visible (VIS), thermal infrared (TIR), and shortwave infrared (SWIR) spectra (Table 3.1). It also provides two dedicated channels for fire and high-temperature event monitoring at 1 km resolution from dynamic range of the 3.74-μm channel and including dedicated detectors at 10.85 μm that are capable of detecting fires at ~650 K without saturation. We specifically use the S3 SLSTR Level 1B RBT products (https://scihub.copernicus.eu/dhus/#/home) for detecting active

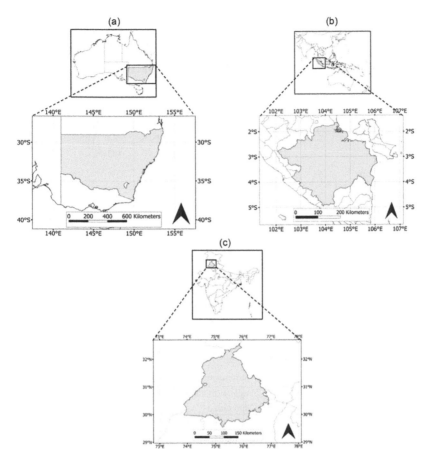

FIGURE 3.1 Study area location maps. (a) New South Wales, southern Australia; (b) South Sumatera, Indonesia; and (c) Punjab, India.

fires over Australia, Indonesia, and India. To compare our results from SLSTR, we used the Near real-time (NRT) Suomi National Polar-orbiting Partnership (Suomi NPP) VIIRS I Band 375 m and NRT MODIS Thermal Anomalies/Fire locations from MODIS 6.1 collection products processed by Fire Information for Resource Management System (https://firms.modaps.eosdis.nasa.gov/).

METHOD

The methodology followed for active fire detection using S3 SLSTR is summarized in the flowchart (Figure 3.2). The Level 1B SLSTR RBT product is used to detect fires which consist of the full-resolution geolocated radiometric measurements for each view (n – nadir; o – oblique) and for each channel as follows: a) on a 1-km grid for brightness temperatures (notation "_in" or "_io" – bands S6 to S9 and F1 and F2 "fire bands"); and b) on a 0.5-km grid for radiances (S1 to S5). In this case, three stripes are distinguished: A ("_an," "_ao"), B ("_bn," "_bo"), and Time Delay and

TABLE 3.1

Nine Different Bands in VNIR/SWIR/TIR and F1 and F2 Thermal Infrared Bands (Sentinel-3 SLSTR User Guide, European Space Agency)

Band	Central Wavelength (nm)	Bandwidth (nm)	Function	Comments	Resolution (m)
S1	554.27	19.26	Cloud screening, vegetation monitoring, aerosol	**VNIR** **Solar reflectance bands**	**500**
S2	659.47	19.25	NDVI, vegetation monitoring, aerosol		
S3	868.00	20.60	NDVI, cloud flagging, pixel co-registration		
S4	1374.80	20.80	Cirrus detection over land	**SWIR**	
S5	1613.40	60.68	Cloud clearing, ice, snow, vegetation monitoring		
S6	2255.70	50.15	Vegetation state and cloud clearing		
S7	3742.00	398.00	SST, LST, active fire	**Thermal IR ambient bands (200 K–320 K)**	**1000**
S8	10854.00	776.00	SST, LST, active fire		
S9	12022.50	905.00	SST, LST		
F1	3742.00	398.00	Active fire	**Thermal IR fire emission bands**	
F2	10854.00	776.00	Active fire		

Integration (TDI) ("_cn," "_co"), with TDI being a derived product from A and B stripes. The nadir bands with suffix "_in" (corresponding to 1-km brightness temperature bands S6–S9 and fire bands F1 and F2) or "_an" (corresponding to radiance bands S1–S5 in stripe A) in their name are used for fire detection.

A. Daytime product preprocessing

One of the critical steps of preprocessing is cloud masking. Clouds can obstruct the detection of fire pixels. Additionally, solar-reflected MIR radiation from certain clouds can create false alarms as fire signals if they are not masked out before fire detection (Wooster et al., 2012). The Level 1B product contains a cloud mask; however, some cloud mask algorithms also identify optically thick smoke as a cloud, even though fire detections can

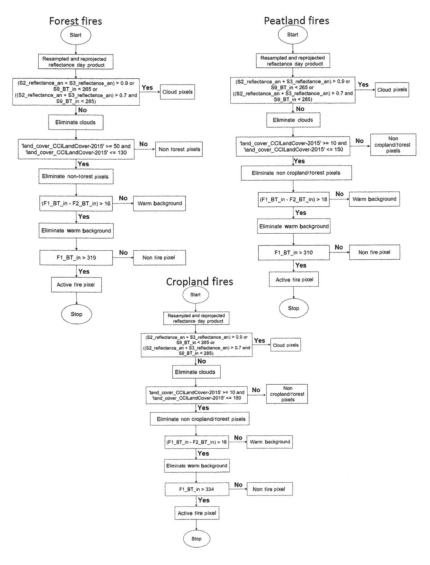

FIGURE 3.2 Algorithms used for active fire detection using Sentinel-3 SLSTR data for forest fires (Australia), peatland fires (Indonesia), and cropland fires (Punjab, India).

typically be made through the smoke. Smoke is generally relatively transparent at MIR wavelengths, unlike the meteorological cloud. Thus, the cloud mask available in the product is unsuitable for active fire detection, and a better cloud mask is necessary. We used a simple cloud test developed for daytime fire detection by Giglio et al. (2003) as follows:

$$\{(\rho^{0.65} + \rho^{0.86} > 0.9) \text{ OR } (T^{12} < 265\,\text{K})\} \text{ OR}$$
$$\{(\rho^{0.65} + \rho^{0.86} > 0.7) \text{ AND } (T^{12} < 285\,\text{K})\},$$

where ρ and T correspond to reflectance and thermal bands at certain wavelength (λ [μm]).

B. Nighttime product preprocessing

Except for the radiance to reflectance conversion, the preprocessing steps are the same for the night product. To derive the night cloud mask, only the brightness temperature bands are required; thus, only resampling, reprojection, and sunset steps were applied. Similar to the daytime product, nighttime pixels are flagged as cloud if the single condition is satisfied following Giglio et al. (2003).

C. Cloud mask for day product

(S2_reflectance_an + S3_reflectance_an) > 0.9   or   S9_BT_in < 265   or ((S2_reflectance_an + S3_reflectance_an) > 0.7 and S9_BT_in < 285).

D. Cloud mask for night product

S9_BT_in < 265.

E. Fire detection

F1_BT_in > 325 and (F1_BT_in − F2_BT_in) > 18 and cloud_mask == 0 and "land_cover_CCILandCover-2015" >= 50 and "land_cover_ CCILandCover-2015" <= 130.

F. Active fire detection

We first imported the S3 SL_1_RBT SLSTR product and determined whether the SLSTR product is day or night. If day, we first converted the radiance bands to derive reflectance. If night, no conversion was done. The reflectance (and radiance for the night) product is then resampled to acquire all bands to the same resolution. S3 Level 1 products are only geocoded but not projected. Thus, the resampled product is reprojected to the corresponding UTM zone projection. Cloud masks and land-cover bands are then applied, and we inspected the fire bands' pixels, and then, we recorded the threshold fire temperature. A new fire detection band was created by querying an expression to detect all the fire pixels over the threshold limit in the next step. Further, warm background, clouds, and unwanted land-cover pixels were eliminated. The final output raster was clipped to the study areas in Indonesia and Australia.

G. Comparison of Sentinel SLSTR with MODIS and VIIRS fire data

We first created a 10×10 km fire density grid for all the Sentinel, MODIS, and VIIRS datasets. In the process, we filtered the datasets to match the same date and boundary of the study area. For S3, we excluded all the non-fire pixels by setting an additional No Data value to 0. Since the S3 output was in raster format, we converted the same to vector points and matched with the MODIS and VIIRS point datasets. We then counted the number of fire points for all the three datasets in a 10x10 km polygon grid for relative accuracy.

RESULTS

The three different types of case studies representing forest fires (Australia), peatland fires (Indonesia), and agricultural crop residue burning fires (India) can be broadly categorized as crown fires, ground fires, and surface fires, respectively. For example, crown fires are characterized by the burning of trees up their entire length to the top; thus, they are highly intense in nature, and burning can last for days depending on the amount of area burned (NRCAN, 2020). The agricultural fires, in contrast, are mostly surface fires that consume mostly crop residues that are left on the ground after harvest and may last only for a few hours. In contrast to crown fires and surface fires, the ground fires occur in deep accumulations of humus, peat, and similar dead vegetation; the fires move very slowly but can become difficult to fully put out or suppress. They are also the longest burning fires lasting for months and can smolder all winter underground and then emerge at the surface again in spring (Usup et al., 2004; Page et al., 2011). Due to these varying fire characteristics, a single threshold could not be used for active fire detection.

The MODIS Terra and VIIRS-I Band daily active fires during forest burning in New South Wales, southern Australia, from 12/20/2019 to 01/10/2020 are shown in Figure 3.3. The active fire detections obtained from S3 SLSTR data suggest higher detections than those obtained from MODIS Terra for all the five dates. Further, the MIR for both S3 SLSTR and MODIS Terra was almost similar, whereas LWIR BT was slightly higher for S3 SLSTR than for MODIS (Table 3.2). Comparatively, VIIRS had a relatively higher T1–T2 BT difference than both the S3 SLSTR and MODIS Terra. Comparison of active fires within a 10×10 grid for different dates for

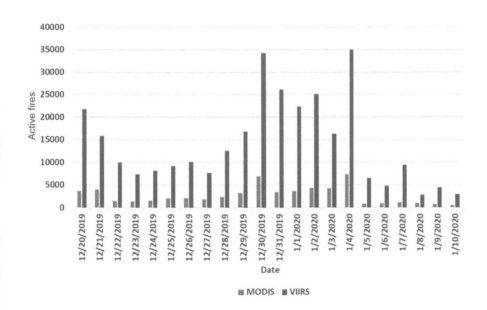

FIGURE 3.3 MODIS Terra and VIIRS-I Band daily active fires during forest burning in New South Wales, southern Australia.

TABLE 3.2
Forest Fire Characteristics Captured from Different Satellites, Southern Australia

Date	S3-T1 (MIR)-BT	S3-T2 (LWIR)-BT	TERRA-T1 (MIR20,21, 22,23)-BT	TERRA-T2 (LWIR31)-BT	VIIRS-14-T1 (MIR)-BT	VIIRS-15-T2 (LWIR)-BT	S3 (T1–T2)	TERRA (T1–T2)	VIIRS (T1–T2)	S3 Count	TERRA Count	VIIRS Count
7/28/2018	315	297	318.9	301.2	330.1	301.4	18	17.7	28.7	146	142	397
9/3/2018	322	307	No Fires	No Fires	330	304.3	15	No Fires	25.7	300	No Fires	591
9/4/2018	321	305	330.8	308.8	334.97	308.15	16	22	26.82	23	4	54
10/20/2018	317	299	307.4	291.8	325	291.2	18	15.6	33.8	267	98	977
8/23/2020	319	301	313	296	330	293.6	18	17	36.4	186	228	660
Average	318.8	301.8	317.525	299.45	330.014	299.73	17	18.075	30.284	184.4	118	535.8

S3-Sentinel-3 SLSTR middle-infrared (MIR) channel (denoted as T1); S3-Sentinel-3 SLSTR long-wave infrared (LWIR) (T2); MODIS Terra (20, 21, 22, 23 (MIR) channels (denoted as T1); MODIS Terra (31) LWIR channel (denoted as T2) ; VIIRS-14 MIR (denoted as T1); VIIRS-I5 TIR (denoted as T2). BT denotes brightness temperature (in kelvin). All fires are daytime. Differences in BT for MIR and LWIR channels for different satellites including active fire counts that were detected are also shown. On average, Sentinel-3 detected more forest fires than MODIS Terra, both of which have a similar equatorial satellite time of pass. Relatively, VIIRS detected more active fires than both Sentinel and MODIS.

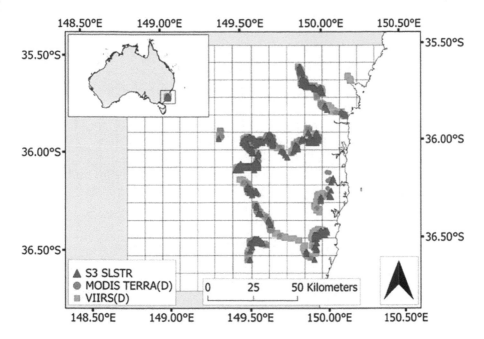

FIGURE 3.4 Comparison of Sentinel-3 SLSTR, MODIS Terra, and VIIRS (375 m) active fires over 10×10 km grid cells in New South Wales, southern Australia, on January 3, 2020. On average, Sentinel-3 SLSTR- and MODIS Terra-detected active fires were almost similar. In contrast, VIIRS active fire detections were much higher than the active fire detections of both Sentinel SLSTR and MODIS Terra and Aqua (combined data).

Sentinel SLSTR, MODIS Terra, and VIIRS respectively is as follows: 12/29/2019: 20, 13, 65; 12/30/2019: 27, no fires, 72; 1/1/2020: 6, 2, 15; 1/2/2020: 31, 12, 131; and 1/3/2020: 17, 20, 49 (Figure 3.4). The overall mean number of fires detected from Sentinel SLSTR, MODIS Terra, and VIIRS for all the above dates is 20, 12.0, and 66, respectively.

The MODIS Terra and VIIRS-I Band daily active fires during peatland burning in southern Sumatera, Indonesia, from 7/20/2018 to 10/27/2018 are shown in Figure 3.5. Results from S3 SLSTR data for peatland fires are shown in Table 3.3, which suggests higher detections than MODIS Terra for all the five dates. Further, the MIR BT was relatively higher for S3 SLSTR and MODIS Terra BT, whereas TIR BT was almost similar for both, including the T1–T2 BT difference. Comparatively, VIIRS had a relatively higher T1–T2 BT difference than both the S3 SLSTR and MODIS Terra. Comparison of active fires within a 10×10 km grid for different dates for Sentinel SLSTR, MODIS Terra, and VIIRS respectively is as follows: 7/8/2018: 4, 1, 6; 9/3/2018:4, 1, 3; 9/4/2018: 2, no fires, 9; 10/20/2018: 5, 4, 7; and 8/23/2020: 5, 2, 5 (Figure 3.6). The overall mean number of fires detected from Sentinel SLSTR, MODIS Terra, and VIIRS for all the above dates is 4.0, 2.0, and 6.0, respectively.

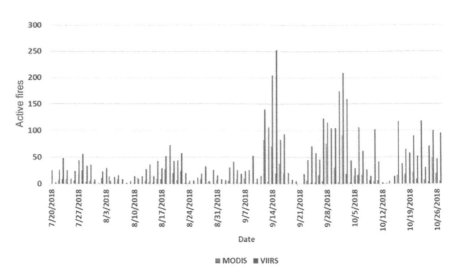

FIGURE 3.5   MODIS Terra and VIIRS-I Band daily active fires during peatland burning in southern Sumatera, Indonesia.

The MODIS Terra and VIIRS-I Band daily active fires during agricultural crop residue burning for a sample site in Moga, Punjab, India, from 7/20/2018 to 10/27/2018 are shown in Figure 3.7. The results obtained from S3 SLSTR data for agricultural fires in Punjab, India, suggested higher detections than those obtained from MODIS Terra data for all three dates (Table 3.4). Further, the MIR BT and LWIR BT were relatively higher for S3 SLSTR than for MODIS Terra, including the T1–T2 difference. Comparatively, VIIRS had a relatively higher T1–T2 BT difference than both the S3 SLSTR and MODIS Terra. Comparison of active fires within a 10×10 grid for Sentinel SLSTR, MODIS Terra and VIIRS respectively is as follows: 5/17/2020: 7, 4, 30; 5/18/2020: 5, 4, 44; and 5/21/2020: 2, 2, 10 (Figure 3.8). The overall mean number of fires detected from Sentinel SLSTR, MODIS Terra, and VIIRS for all the above dates is 8, 3.0, and 28, respectively.

A closer examination of the fire detection algorithm (Figure 3.3) for all the three different case studies suggests four important criteria in delineating fires: (1) masking of clouds; (2) fire detection over specific land-cover type; (3) brightness temperature (BT) thresholding in the MIR channel (3.74 μm) mainly to detect potential fires; and (4) using BT difference between the MIR (3.75 μm) versus LWIR (10.75 μm) channels to isolate fires from a warm background other than fires, such as bare soil, rock, and road (Qian et al., 2009). We used a variety of thresholds to account for these characteristics. The thresholding of the MIR channel is very important for fire detection because the active fires show maximum spectral emission in the MIR spectral range. In addition, the BT difference between T1 and T2 can be influenced by a variety of effects such as unequal atmospheric effects, unequal emissivities, solar reflection in MIR, and non-uniform fire scenes (Li et al., 2001). Thus, we carefully fine-tuned our algorithm to ensure that the difference of brightness temperature of MIR and LWIR channels should be larger enough. In contrast to the forest fires, for the agricultural fires, the temperatures can vary within a pixel,

TABLE 3.3

Peatland Fire Characteristics Captured from Different Satellites, Sumatera, Indonesia

Date	S3-T1 (MIR)-BT	S3-T2 (LWIR)-BT	TERRA-T1 (MIR20,21, 22,23)-BT	TERRA-T2 (LWIR31)-BT	VIIRS-14-T1 (MIR)-BT	VIIRS-15-T2 (LWIR)-BT	S3 (T1–T2)	TERRA (T1–T2)	VIIRS (T1–T2)	S3 Count	TERRA Count	VIIRS Count
7/28/2018	310	292	309.1	291.5	329.15	288.89	18	17.6	40.26	8	2	40
9/3/2018	310	290	308.4	288.9	330.08	295.63	20	19.5	34.45	19	1	24
9/4/2018	313	295	No Fires	No Fires	331.38	304.26	18	No Fires	27.12	18	No Fires	40
10/20/2018	318	300	311.8	291.2	330.14	298.29	18	20.6	31.85	57	10	74
8/23/2020	314	296	310.4	294.2	330.7	294.7	18	16.2	36	19	6	36
Average	**310**	**291**	**308.75**	**290.2**	**329.615**	**292.26**	**19**	**18.55**	**37.355**	**25.5**	**4.75**	**42.8**

For abbreviations, see Table 3.2. On average, Sentinel-3 detected more fires than MODIS Terra, both of which have almost a similar satellite time of pass. Relatively, VIIRS detected more active fires than both Sentinel and MODIS.

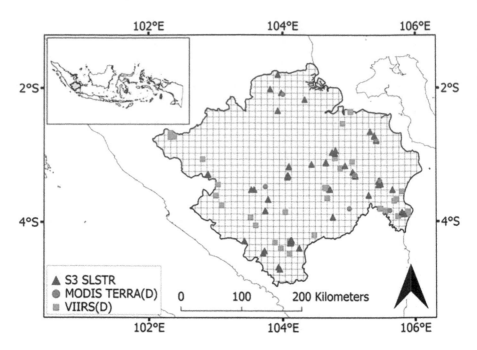

FIGURE 3.6 Comparison of Sentinel-3 SLSTR, MODIS Terra, and VIIRS (375 m) active fires over 10×10km grid cells in southern Sumatera, Indonesia, on October 20, 2018. On average, Sentinel-3 SLSTR detected more active fires than MODIS Terra. VIIRS active fire detections were much higher than the active fire detections of both Sentinel and MODIS.

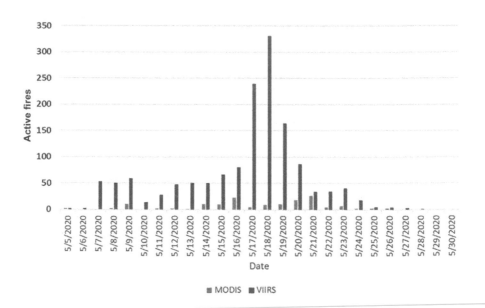

FIGURE 3.7 MODIS Terra and VIIRS-I Band daily active fires during agricultural crop residue burning in Moga District, Punjab, India.

TABLE 3.4

Agricultural Fire Characteristics Detected from Different Satellites, Punjab, India

Date	S3-T1 (MIR)-BT	S3-T2 (LWIR)-BT	TERRA-T1 (MIR20,21,22,23)-BT	TERRA-T2 (LWIR31)-BT	VIIRS-14-T1 (MIR)-BT	VIIRS-15-T2 (LWIR)-BT	S3 (T1–T2)	TERRA (T1–T2)	VIIRS (T1–T2)	S3 Count	TERRA Count	VIIRS Count
5/17/2020	331	313	325.9	301.1	330.02	302.77	18	24.8	27.25	125	41	1371
5/18/2020	331	313	327.2	311.4	330.05	304.76	18	15.8	25.29	86	84	2904
5/21/2020	340	322	324.9	312.2	340.06	314.64	18	12.7	25.42	30	10	253
Average	**334**	**316**	**326.00**	**308.23**	**333.38**	**307.39**	**18.00**	**17.77**	**25.99**	**80.33**	**45.00**	**1509.33**

For abbreviations, see Table 3.2. On average, Sentinel-3 detected more agricultural fires than MODIS Terra, both of which have a similar equatorial satellite time of pass. Relatively, VIIRS detected more active fires than both Sentinel and MODIS.

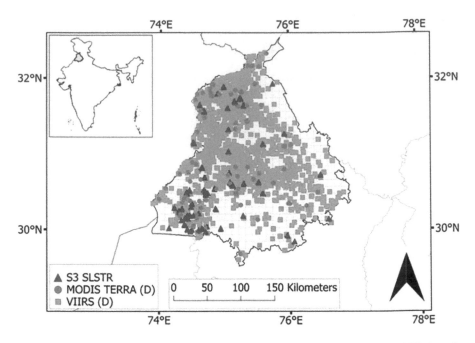

FIGURE 3.8 Comparison of Sentinel-3 SLSTR, MODIS Terra, and VIIRS (375 m) active fires over 10×10 km grid cells in Punjab, India, on May 18, 2020. On average, Sentinel-3 SLSTR detected more active fires than MODIS Terra. VIIRS active fire detections were much higher than the active fire detections of both Sentinel and MODIS.

resulting in unusually large differences in brightness temperature between the two channels. One of the challenges of detecting these fires during daytime is to account for soil reflectance, which can be high, in addition to solar reflection from the cloud and other bright surfaces. Unlike forest fires, which may be covered with a dense canopy, agricultural burning sites are open/bare areas with a minimum amount of vegetation, i.e., crop residues that are interspersed with the bare soil, which can create relatively higher MIR reflectance, thus creating confusion between the fires versus non-fires. For example, Giglio et al. (1999) noted that a soil surface with a temperature of 300 K and a reflectance of 0.28 could saturate the 3.75-μm channel, even when no fire is present (Giglio et al., 1999). Further, the solar reflection depends mainly on the amount of incident solar radiation and the albedo; thus, agricultural burning sites such as in Punjab can have relatively more albedo than the forested sites in Australia and the peatlands, the latter mostly being interspersed with mixed forests in Sumatera, Indonesia. Thus, the thresholds need to be carefully fine-tuned when quantifying the agricultural fires compared to the forest fires. In the case of peatlands, fires burn not only the surface vegetation but also the peat deposits up to 100 cm below the surface (Boehm et al., 2001). In terms of burning characteristics, peatland fires stand in between the forest and agricultural fires, as sites are interspersed with trees and peat. For example, between 30 and 122 tree species greater than 10 cm in diameter were reported in the 1-ha plots of peat swamp forest in Indonesia, a number lower than most dryland rainforests in the same region

(100–290 species) but similar to heath forests (29–129 species) (Kartawinata et al., 2008; Corlett, 2009). Thus, in the peatlands, the MIR BT was relatively less than agricultural fires due to less soil exposure and higher than forest fires due to more open canopies interspersed with peat. However, peatland fires can produce a large amount of smoke and haze, which can deter fire detection from optical satellites. In our case, we selected specific cases where there was relatively less cloud problem for testing the fire algorithm; thus, it was not a major issue. In addition, the satellite resolution, as well as the equatorial time of the pass, can also influence the fire detections. For example, the S3 orbit is a near-polar, sun-synchronous orbit with a descending node equatorial crossing at 10:00 hours mean local solar time, whereas the MODIS Terra has a 10:30 AM and AQUA a 1:30 PM equatorial crossing time, both viewing the entire Earth's surface every 1–2 days. Further, both the S3 and MODIS Terra and Aqua thermal bands have a 1 km resolution. Thus, in our study, we compared the S3 SLSTR against MODIS Terra active fires and found S3 SLSTR performing better than the MODIS Terra for all cases, i.e., forest, peatland, and agricultural fires. In contrast to both the above satellites, *VIIRS* satellite observes Earth's entire surface twice each day and has an equatorial crossing time of approximately 1:30 AM and 1:30 PM (local time). Further, the I-Band data have a relatively higher resolution of 375 m; due to these characteristics, the VIIRS-detected fires are relatively higher than the S3 and MODIS datasets. It may be noted that MODIS Terra has far exceeded its design life and has a strong chance of operating successfully into 2021; however, considering more than 20 years of its launch (December 1999), it might degenerate anytime. Thus, there is a need to explore other satellite datasets with similar characteristics as MODIS Terra useful for various earth science applications. Specific to fire studies, we infer that S3 SLSTR can fill such a gap in the coming years. We also infer the need for strong validation of active fire products through ground-truthing, which is not performed in the current study. In addition, additional testing might be required with temporal data to infer the robustness of the Sentinel SLSTR for fire studies. Nevertheless, the work presented in this study provides a quantitative evaluation of S3 SLSTR, MODIS Terra, and VIIRS-I Band datasets for active fire detection in forests, peatlands, and agricultural landscapes useful for mapping and monitoring of fires using remote sensing datasets.

ACKNOWLEDGMENTS

The authors thank the Copernicus, the European Space Agency, for Sentinel-3 SLSTR datasets. The authors are grateful to MODIS and VIIRS fire product developers for freely sharing the datasets.

REFERENCES

Agee, J.K. 1993. *Fire Ecology of Pacific Northwest Forests.* Washington, D.C., Island Press.
Arino, O., and Melinotte, J.M. 1998. The 1993 Africa fire map. *International Journal of Remote Sensing.* 19(11), 2019–2023.
Arino, O., Rosaz, J.M., and Goloub, P. 1999. The ATSR world fire atlas-a synergy with 'Polder' aerosol products. *Earth Observation Quarterly.* 64(1), 8.

Badarinath, K.V.S., Kharol, S.K., Krishna Prasad, V., Kaskaoutis, D.G., and Kambezidis, H.D. 2008a. Variation in aerosol properties over Hyderabad, India during intense cyclonic conditions. *International Journal of Remote Sensing.* 29(15), 4575–4597.

Badarinath, K.V.S., Kharol, S.K., Prasad, V.K., Sharma, A.R., Reddi, E.U.B., Kambezidis, H.D., and Kaskaoutis, D.G. 2008b. Influence of natural and anthropogenic activities on UV Index variations–a study over tropical urban region using ground based observations and satellite data. *Journal of Atmospheric Chemistry.* 59(3), 219–236.

Biswas, S., Lasko, K.D., and Vadrevu, K.P. 2015a. Fire disturbance in tropical forests of Myanmar—Analysis using MODIS satellite datasets. *IEEE Journal of Selected Topics in Applied Earth Observations and Remote Sensing.* 8(5), 2273–2281.

Biswas, S., Vadrevu, K.P., Lwin, Z.M., Lasko, K., and Justice, C.O. 2015b. Factors controlling vegetation fires in protected and non-protected areas of Myanmar. *PLoS One.* 10(4), e0124346.

Boehm, H.D.V., Siegert, F., Rieley, J.O., Page, S.E., Jauhiainen, J., Vasander, H., and Jaya, A. 2001. November. Fire impacts and carbon release on tropical peatlands in Central Kalimantan, Indonesia. In *Proceedings of the 22nd Asian Conference on Remote Sensing* (pp. 5–9), Singapore.

Boles, S.H., and Verbyla, D.L. 2000. Comparison of three AVHRR-based fire detection algorithms for interior Alaska. *Remote Sensing of Environment.* 72(1), 1–16.

Corlett, R.T. 2009. The Ecology of Tropical East Asia. Oxford University Press, Oxford.

Crutzen, P.J., and Andreae, M.O. 1990. Biomass burning in the tropics: Impact on atmospheric chemistry and biogeochemical cycles. *Science.* 250(4988), 1669–1678.

ESA, Sentinel online. 2020. https://sentinel.esa.int/web/sentinel/user-guides/sentinel-3-slstr /coverage.

FAOSTAT, 2020. http://www.fao.org/faostat/en/.

Flasse, S.P., and Ceccato, P. 1996. A contextual algorithm for AVHRR fire detection. *International Journal of Remote Sensing.* 17(2), 419–424.

Giglio, L., Kendall, J.D., and Justice, C.O. 1999. Evaluation of global fire detection algorithms using simulated AVHRR infrared data. *International Journal of Remote Sensing.* 20(10), 1947–1985.

Giglio, L., Descloitres, J., Justice, C.O., and Kaufman, Y.J. 2003. An enhanced contextual fire detection algorithm for MODIS. *Remote Sensing of Environment.* 87(2–3), 273–282.

Giglio, L., Randerson, J.T., and Van Der Werf, G.R. 2013. Analysis of daily, monthly, and annual burned area using the fourth-generation global fire emissions database (GFED4). *Journal of Geophysical Research: Biogeosciences.* 118(1), 317–328.

Gupta, P.K., Prasad, V.K., Sharma, C., Sarkar, A.K., Kant, Y., Badarinath, K.V.S., and Mitra, A.P. 2001. CH_4 emissions from biomass burning of shifting cultivation areas of tropical deciduous forests–experimental results from ground-based measurements. *Chemosphere-Global Change Science.* 3(2), 133–143.

Hua, L., and Shao, G. 2017. The progress of operational forest fire monitoring with infrared remote sensing. *Journal of Forestry Research.* 28(2), 215–229.

Inoue, Y. 2018. Ecosystem carbon stock, atmosphere and food security in slash-and-burn land use: A geospatial study in mountainous region of Laos. In: Vadrevu, K.P., Ohara, T., and Justice, C. (Eds). *Land-Atmospheric Research Applications in South/Southeast Asia.* Springer, Cham. pp. 641–666.

Israr, I., Jaya, S.N.I, Saharjo, H.S., Kuncahyo, B., and Vadrevu, K.P. 2018. Spatio-temporal analysis of land and forest fires in Indonesia using MODIS active fire dataset. In: Vadrevu, KP, Ohara, T., and Justice, C. (Eds). *Land-Atmospheric Research Applications in South/Southeast Asia.* Springer, Cham. pp. 105–128.

Ito, A., and Penner, J.E. 2004. Global estimates of biomass burning emissions based on satellite imagery for the year 2000. *Journal of Geophysical Research: Atmospheres.* 109(D14S05), 1–18. doi:10.1029/2003JD004423.

Jones, A., Haywood, J.M., and Boucher, O. 2007. Aerosol forcing, climate response and climate sensitivity in the Hadley Centre climate model. *Journal of Geophysical Research: Atmospheres.* 112(D20211), 1–14. doi:10.1029/2007JD008688.

Justice, C.O., Kendall, J.D., Dowty, P.R., and Scholes, R.J. 1996. Satellite remote sensing of fires during the SAFARI campaign using NOAA advanced very high-resolution radiometer data. *Journal of Geophysical Research: Atmospheres.* 101(D19), 23851–23863.

Justice, C.O., Giglio, L., Korontzi, S., Owens, J., Morisette, J.T., Roy, D., Descloitres, J., Alleaume, S., Petitcolin, F., and Kaufman, Y. 2002. The MODIS fire products. *Remote Sensing of Environment.* 83(1–2), 244–262.

Justice, C., Gutman, G., and Vadrevu, K.P. 2015. NASA land cover and land use change (LCLUC): An interdisciplinary research program. *Journal of Environmental Management.* 148(15), 4–9.

Kant, Y., Ghosh, A.B., Sharma, M.C., Gupta, P.K., Prasad, V.K., Badarinath, K.V.S., and Mitra, A.P. 2000. Studies on aerosol optical depth in biomass burning areas using satellite and ground-based observations. *Infrared Physics & Technology.* 41(1), 21–28.

Kartawinata, K., Purwaningsih, P.T., Yusuf, R., Abdulhadi, R., and Riswan, S. 2008. Floristics and structure of a lowland dipterocarp forest at Wanariset Samboja, East Kalimantan, Indonesia. *Reinwardtia.* 12(4), 301–323.

Kaufman, Y.J., Justice, C.O., Flynn, L.P., Kendall, J.D., Prins, E.M., Giglio, L., Ward, D.E., Menzel, W.P., and Setzer, A.W. 1998. Potential global fire monitoring from EOS-MODIS. *Journal of Geophysical Research: Atmospheres.* 103(D24), 32215–32238.

Langenfelds, R.L., Francey, R.J., Pak, B.C., Steele, L.P., Lloyd, J., Trudinger, C.M., and Allison, C.E. 2002. Interannual growth rate variations of atmospheric CO2 and its $\delta^{13}C$, H_2, CH_4, and CO between 1992 and 1999 linked to biomass burning. *Global Biogeochemical Cycles.* 16(3), 21–21.

Lasko, K., and Vadrevu, K.P. 2018. Improved rice residue burning emissions estimates: Accounting for practice-specific emission factors in air pollution assessments of Vietnam. *Environmental Pollution.* 236(5), 795–806.

Lasko, K., Vadrevu, K.P., Tran, V.T., Ellicott, E., Nguyen, T.T., Bui, H.Q., and Justice, C. 2017. Satellites may underestimate rice residue and associated burning emissions in Vietnam. *Environmental Research Letters.* 12(8), 085006.

Lasko, K., Vadrevu, K.P., and Nguyen, T.T.N. 2018a. Analysis of air pollution over Hanoi, Vietnam using multi-satellite and MERRA reanalysis datasets. *PLoS One.* 13(5), e0196629.

Lasko, K., Vadrevu, K.P., Tran, V.T., and Justice, C. 2018b. Mapping double and single crop paddy rice with Sentinel-1A at varying spatial scales and polarizations in Hanoi, Vietnam. *IEEE Journal of Selected Topics in Applied Earth observations and Remote Sensing.* 11(2), 498–512.

Li, Z.H., Kaufman, Y.J., Ichoku, C.H., Fraser, R.O., Trishchenko, A.L., Giglio, L.O., Jin, J., Yu, X.I. 2001. A review of AVHRR-based active fire detection algorithms: Principles, limitations, and recommendations. In: Ahern, F.J., Goldammer, J.G., and Justice, C.O. (Eds). *Global and Regional Vegetation Fire Monitoring from Space, Planning and Coordinated International Effort.* SPB Academic Publishing, The Hague, The Netherlands. pp. 199–225.

Martín, M.P., Ceccato, P., Flasse, S., and Downey, I. 1999. Fire detection and fire growth monitoring using satellite data. In: Chuvieco, E. (Ed). *Remote Sensing of Large Wildfires.* Springer, Berlin, Heidelberg. pp. 101–122.

Matson, M., and Dozier, J., 1981. Identification of subresolution high temperature sources using a thermal IR sensor. *Photogrammetric Engineering and Remote Sensing.* 47, 1311–1318.

NRCAN, 2020. https://www.nrcan.gc.ca/our-natural-resources/forests-forestry/wildland-fires-insects-disturban/forest-fires/fire-behaviour/13145.

Page, S.E., Rieley, J.O., and Banks, C.J. 2011. Global and regional importance of the tropical peatland carbon pool. *Global Change Biology.* 17(2), 798–818.

Prasad, V.K., Gupta, P.K., Sharma, C., Sarkar, A.K., Kant, Y., Badarinath, K.V.S., Rajagopal, T., and Mitra, A.P. 2000. NO$_x$ emissions from biomass burning of shifting cultivation areas from tropical deciduous forests of India–estimates from ground-based measurements. *Atmospheric Environment.* 34(20), 3271–3280.

Prasad, V.K., Kant, Y., and Badarinath, K.V.S. 2001. CENTURY ecosystem model application for quantifying vegetation dynamics in shifting cultivation areas: A case study from Rampa Forests, Eastern Ghats (India). *Ecological Research.* 16(3), 497–507.

Prasad, V.K., Kant, Y., Gupta, P.K., Elvidge, C., and Badarinath, K.V.S. 2002. Biomass burning and related trace gas emissions from tropical dry deciduous forests of India: A study using DMSP-OLS data and ground-based measurements. *International Journal of Remote Sensing.* 23(14), 2837–2851.

Prasad, V.K., Lata, M., and Badarinath, K.V.S. 2003. Trace gas emissions from biomass burning from northeast region in India—estimates from satellite remote sensing data and GIS. *Environmentalist.* 23(3), 229–236.

Prasad, V.K., Badarinath, K.V.S., and Eaturu, A. 2008. Biophysical and anthropogenic controls of forest fires in the Deccan Plateau, India. *Journal of Environmental Management.* 86(1), 1–13.

Prins, E.M., Feltz, J.M., Menzel, W.P., and Ward, D.E. 1998. An overview of GOES-8 diurnal fire and smoke results for SCAR-B and 1995 fire season in South America. *Journal of Geophysical Research: Atmospheres.* 103(D24), 31821–31835.

Qian, Y., Yan, G., Duan, S., and Kong, X. 2009. A contextual fire detection algorithm for simulated HJ-1B imagery. *Sensors.* 9(2), 961–979.

Ramachandran, S. 2018. Aerosols and climate change: Present understanding, challenges and future outlook. In: Vadrevu, K.P., Ohara, T., and Justice, C. (Eds). *Land-Atmospheric Research Applications in South/Southeast Asia.* Springer, Cham. pp. 341–378.

Roberts, G.J., and Wooster, M.J. 2008. Fire detection and fire characterization over Africa using Meteosat SEVIRI. *IEEE Transactions on Geoscience and Remote Sensing.* 46(4), 1200–1218.

Robinson, J.M. 1991. Fire from space: Global fire evaluation using infrared remote sensing. *International Journal of Remote Sensing.* 12(1), 3–24.

Schroeder, W., Oliva, P., Giglio, L., and Csiszar, I.A. 2014. The New VIIRS 375 m active fire detection data product: Algorithm description and initial assessment. *Remote Sensing of Environment.* 143, 85–96.

Sigsgaard, T., Forsberg, B., Annesi-Maesano, I., Blomberg, A., Bølling, A., Boman, C., Bønløkke, J., Brauer, M., Bruce, N., Héroux, M.E., and Hirvonen, M.R. 2015. Health impacts of anthropogenic biomass burning in the developed world. *European Respiratory Journal.* 46(6), 1577–1588.

Simorangkir, D. 2007. Fire use: Is it really the cheaper land preparation method for large-scale plantations? *Mitigation and Adaptation Strategies for Global Change.* 12(1), 147–164.

Stott, P.A., Goldammer, J.G., and Werner, W.L. 1990. The role of fire in the tropical lowland deciduous forests of Asia. In: Goldammer, J.G. (Ed). *Fire in the Tropical Biota.* Springer, Berlin, Heidelberg. pp. 32–44.

Streets, D.G., Yarber, K.F., Woo, J.H., and Carmichael, G.R. 2003. Biomass burning in Asia: Annual and seasonal estimates and atmospheric emissions. *Global Biogeochemical Cycles.* 17(4).

Szopa, S., Hauglustaine, D.A., and Ciais, P. 2007. Relative contributions of biomass burning emissions and atmospheric transport to carbon monoxide interannual variability. *Geophysical Research Letters.* 34(18).

Usup, A., Hashimoto, Y., Takahashi, H., and Hayasaka, H. 2004. Combustion and thermal characteristics of peat fire in tropical peatland in Central Kalimantan, Indonesia. *Tropics.* 14(1), 1–19.

Vadrevu, K.P. 2008. Analysis of fire events and controlling factors in eastern India using spatial scan and multivariate statistics. *Geografiska Annaler: Series A, Physical Geography.* 90(4), 315–328.

Vadrevu, K.P., and Justice, C.O. 2011. Vegetation fires in the Asian region: Satellite observational needs and priorities. *Global Environmental Research.* 15(1), 65–76.

Vadrevu, K.P., and Lasko, K.P., 2015. Fire regimes and potential bioenergy loss from agricultural lands in the Indo-Gangetic Plains. *Journal of Environmental Management.* 148, 10–20.

Vadrevu, K.P., and Lasko, K. 2018. Intercomparison of MODIS AQUA and VIIRS I-Band fires and emissions in an agricultural landscape—Implications for air pollution research. *Remote Sensing.* 10(7), 978. doi:10.3390/rs10070978.

Vadrevu, K.P., Eaturu, A., and Badarinath, K.V.S. 2006. Spatial distribution of forest fires and controlling factors in Andhra Pradesh, India using spot satellite datasets. *Environmental Monitoring and Assessment.* 123(1–3), 75–96.

Vadrevu, K.P., Badarinath, K.V.S., and Anuradha, E. 2008. Spatial patterns in vegetation fires in the Indian region. *Environmental Monitoring and Assessment.* 147(1–3), 1. doi:10.1007/s10661-007-0092-6.

Vadrevu, K.P., Csiszar, I., Ellicott, E., Giglio, L., Badarinath, K.V.S., Vermote, E., and Justice, C. 2012. Hotspot analysis of vegetation fires and intensity in the Indian region. *IEEE Journal of Selected Topics in Applied Earth Observations and Remote Sensing.* 6(1), 224–238.

Vadrevu, K.P., Giglio, L., and Justice, C. 2013. Satellite based analysis of fire–carbon monoxide relationships from forest and agricultural residue burning (2003–2011). *Atmospheric Environment.* 64, 179–191.

Vadrevu, K.P., Ohara, T., and Justice, C. 2014a. Air pollution in Asia. *Environmental Pollution.* 12, 233–235.

Vadrevu, K.P., Lasko, K., Giglio, L., and Justice, C. 2014b. Analysis of Southeast Asian pollution episode during June 2013 using satellite remote sensing datasets. *Environmental Pollution.* 12, 245–256.

Vadrevu, K.P., Ohara, T., and Justice, C. 2017. Land cover, land use changes and air pollution in Asia: A synthesis. *Environmental Research Letters.* 12(12), 120201.

Vadrevu, K.P., Ohara, T., and Justice, C. eds., 2018. *Land-Atmospheric Research Applications in South and Southeast Asia.* Springer, Cham.

Vadrevu, K.P., Lasko, K., Giglio, L., Schroeder, W., Biswas, S., and Justice, C. 2019. Trends in vegetation fires in south and southeast Asian countries. *Scientific Reports.* 9(1), 7422. doi:10.1038/s41598-019-43940-x.

Wooster, M.J., Roberts, G., Perry, G.L.W., and Kaufman, Y.J., 2005. Retrieval of biomass combustion rates and totals from fire radiative power observations: FRP derivation and calibration relationships between biomass consumption and fire radiative energy release. *Journal of Geophysical Research: Atmospheres.* 110(D24).

Wooster, M.J., Xu, W., and Nightingale, T., 2012. Sentinel-3 SLSTR active fire detection and FRP product: Pre-launch algorithm development and performance evaluation using MODIS and ASTER datasets. *Remote Sensing of Environment.* 120, 236–254.

Wooster, M.J., Roberts, G., Smith, A.M., Johnston, J., Freeborn, P., Amici, S., and Hudak, A.T. 2013. Thermal remote sensing of active vegetation fires and biomass burning events. In: Kuenzer, C. and Dech, S. (Eds). *Thermal Infrared Remote Sensing.* Springer, Dordrecht. pp. 347–390. doi:10.1007/978-94-007-6639-6_18.

Zhukov, B., Lorenz, E., Oertel, D., Wooster, M., and Roberts, G., 2006. Spaceborne detection and characterization of fires during the bi-spectral infrared detection (BIRD) experimental small satellite mission (2001–2004). *Remote Sensing of Environment.* 100(1), 29–51.

4 An Assessment of Burnt Area Signal Variations in Laos Using Sentinel-1A&B Datasets

Krishna Prasad Vadrevu
NASA Marshall Space Flight Center, USA

Aditya Eaturu
University of Alabama Huntsville, USA

Sumalika Biswas
Smithsonian Conservation Biology Institute, USA

Chittana Phompila
Faculty of Forest Sciences (FFS), National University of Laos (NUoL), Laos

CONTENTS

INTRODUCTION

Fire is most often used as a land-clearing tool in South/Southeast Asian countries. For example, forests are mostly cleared through slash-and-burn agriculture as in north India, Myanmar, northern Thailand, Cambodia, Laos, Philippines, etc. (Vadrevu et al., 2006; 2008; 2011; Prasad et al., 2008; Vadrevu and Justice, 2011; Hayasaka et al., 2014; Israr et al., 2018; Ramachandran, 2018; Oanh et al., 2018; Inoue, 2018; Tariq and Ul-Haq, 2018). In Indonesia, fires are mostly used to clear forests and peatlands for planting oil palm (Justice et al., 2015). In addition, the highly intensive agricultural production in South/Southeast Asia (S/SEA) is accompanied by the generation of significant amounts of crop residues (Lasko et al., 2017, 2018a, b; Vadrevu and Lasko, 2018). To clear the land for faster crop rotation, farmers prefer to burn the residues directly in the field after harvesting, most commonly in India, Pakistan, Myanmar, Vietnam, Thailand, Laos, and Cambodia. Some of the advantages of using fires as a management tool for these practices include saving labor costs and controlling insects, disease, and invasive weeds. However, most of these biomass burning activities have been reported to emit large amounts of greenhouse gases such as CO_2, CO, nonmethane hydrocarbons (NMHC), NO_x, SO_2, particulate matter, and other gases (Kant et al., 2000; Gupta et al., 2001; Prasad et al., 2000, 2001,2002,2003; Kharol et al., 2012) which may significantly deteriorate the ambient air quality and contribute to the regional transport of air pollution (Vadrevu et al., 2012; 2013; 2014a, b; 2015; 2017; 2018; 2019; Oanh et al., 2018). Also, the biomass burning is known to cause severe health effects such as chronic obstructive pulmonary disease, pneumoconiosis, bronchitis, cataract, corneal opacity, and blindness (Salvi and Barnes, 2010; Kim et al., 2011). Further, the biomass burning activities release a significant amount of smoke which can reduce visibility, thus causing road accidents (Singh and Sidhu, 2014).

Specifically, in the forested ecosystems, mapping burnt areas and impacts at regular intervals is important for planning future mitigation options related to fire hazards and biodiversity conservation (Biswas et al., 2015a, b). The ecosystem functions that are most often disturbed due to the fires include carbon loss, surface litter removal, and disturbance in soil nutrients, which may further impact the biodiversity of the forested ecosystems, including shift in plant species composition (Bradstock et al., 2012). The impact of fires on the ecosystems can vary based on the "fire severity" or "burn severity" (Key & Benson, 2006). Satellite remote sensing has been effectively used for mapping burnt areas and severity at different spatiotemporal scales and ecosystems, including agricultural areas. Several researchers attempted to systematically map burned areas including date of burn using low (30 m)-, moderate (~250–500 m)-, and coarse (~1 km)-resolution optical data (Pereira et al., 1999; Grégoire et al., 2003; Roy et al., 2008; Giglio, 2020). The burnt area estimates are then used to estimate greenhouse gases and aerosol emissions. Specific to burnt agricultural areas and emissions estimation using remote sensing data, large uncertainties exist in the literature (Badarinath et al., 2006; 2008a, b; Lasko et al., 2017; Roy et al., 2008). There is a strong need to evaluate different satellite data to map burnt areas. In addition to emissions, accurate and rapid mapping of burnt areas is fundamental to support fire management, design planning strategies for mitigation, and help and restore vegetation.

Compared to the above studies, which focused on the optical data, very few studies exist on the mapping of burnt areas using Synthetic Aperture Radar (SAR) data in Asia. The earlier studies on burnt area mapping using SAR data focusing on the Mediterranean region suggest backscatter signal differences before and after the fire due to changes in the dielectric constant of the scattering surfaces due to the moisture content (French et al., 1996). Gimeno and San-Miguel-Ayanz (2004), using RADARSAT data, demonstrated that topography influences the level of discrimination of burnt areas since areas affected by forest fires on face-slopes present a higher backscatter coefficient than those on back-slopes. Donezar et al. (2019) used the Sentinel-1 data to derive the burnt areas in the Mediterranean region by object-based image analysis (OBIA) and photo interpretation of SPOT-6 images. They highlighted the importance of the incidence angle, with lower angles resulting in higher accuracies. Also, they conclude that VH polarization performed slightly better than VV polarization; however, those differences are not significant. Engelbrecht et al. (2017) used a Normalised Difference α-Angle (NDαI) approach to burn-scar mapping using C-band data Polarimetric decompositions α-angles from pre-burn and post-burn scenes. The premise for using the NDαI as burnt area index is that α-angle provides an indication of scattering mechanism with lower than 40° associated with a higher contribution by surface scatterers (Czuchlewski and Weissel, 2005), and α-angles of between 40° and 50° associated with volume-scattering mechanisms. Thus, vegetation pre-fire will exhibit relatively higher angles than post-fire. Using this approach, Engelbrecht et al. (2017) report high overall accuracies of 97.4% (kappa = 0.72) and 94.8% (kappa = 0.57) for RADARSAT-2 and Sentinel-1A, respectively. Other studies that focused on burnt area mapping using SAR data include Gimeno and Ayanz (2004), Menges et al. (2004), Liew et al. (1999), Bernhard et al. (2012), Stroppiana et al. (2015), Imperatore et al. (2017), Zhang et al. (2019), and Ban et al. (2020).

In contrast to several studies that were conducted in the Mediterranean, African, or North American ecosystems, very few studies focused on burnt area mapping using Sentinel-1A&B data in Asia, especially Laos. The country is one of the major biodiversity hotspots in Asia and also has one of the highest forest cover. The deforestation rates in the country are high due to dependence of local people on forests. Most of the burning is attributed to slash-and-burn agriculture, including agricultural residue burning. In this study, we evaluated the potential of Sentinel-1A&B SAR data for burnt area mapping at a country scale. We used both the VV and VH polarization data to assess the burnt area and other land cover signal variations. We also compared the burnt area backscatter signal variations with incident angles. The results highlight the burnt area characteristics from Sentinel-1A&B SAR data in Laos useful for mapping and monitoring purposes.

STUDY AREA

We focused on Laos for the burnt area mapping. The country is landlocked with a total area of 236,800 km² and water area of 6000 km². Forests cover almost two-thirds of the country. The northern part of the region is dominated by hills with tropical rain forests with broad evergreen-to-mixed evergreen species, whereas the deciduous forest species occupy the southern region with mixed bamboo undergrowth. Mekong River is located in the west and forms a natural border with the neighboring countries. The climate is tropical with a pronounced rainy season from May through October, a cool dry season from November through February, and a hot dry season from March till April. In the latter hotter months, the biomass burning is prevalent with a peak during March. Most biomass burning is attributed to slash-and-burn agriculture and crop residue burning (Roder et al., 1997; Inoue, 2018).

DATASETS

We used the Sentinel-1 data from the European Space Agency, C-band SAR (5.4 GHz) data that have global coverage with a 6- to 12-day revisit time depending on the availability of Sentinel-1B imagery. The Sentinel-1 imagery is provided as dual-polarized Interferometric Wide (IW) swath data with vertical transmit, vertical receive (VV), and vertical transmit, horizontal receive (VH) polarizations. Each polarization is at a nominal spatial resolution of 5 m×20 m before preprocessing. Figure 4.1 depicts the 2018 raw VV data for the entire Laos. A monthly time-series stack of Sentinel-1A&B images with VV, VH, and angle information was acquired for 2 years from 2018 and 2019 from March to April. The Level 1 ground-range-detected, descending-mode, IW images were processed using the free and open-source Sentinel-1 toolbox. The ground-range-detected images were processed following guidelines, including applying restituted orbit files, multi-look azimuthal compressions to 20 m, terrain correction using SRTM 30 m version 4DEM, radiometric calibration adjustments to correct for viewing geometry effects, and refined lee speckle filter to reduce constructive and destructive interference, all resulting in sigma-nought backscatter data logarithmically scaled in decibels (Lasko and Vadrevu, 2018). The final data had VV and VH, including incidence angle data, all stacked as layers for different months (March, April, and May only) and years (2018 and 2019). Using these data, we assessed the backscatter signal variations for different land cover classes and burnt areas.

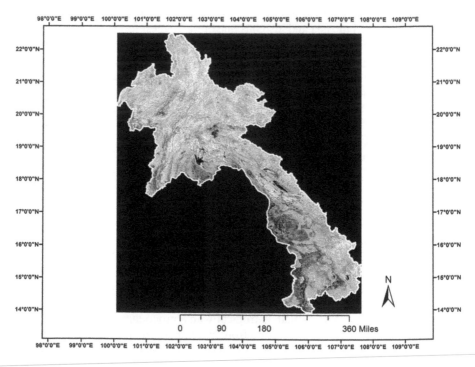

FIGURE 4.1   Laos Sentinel-1A VV polarization, April 2019.

METHODOLOGY

From the Sentinel-1A&B datasets, we extracted the VV, VH, and VH signal values for different land cover classes. We used the MODIS 1km land cover map as a mask for six different land cover classes, which included water, barren areas, croplands, forests, and urban areas. The MODIS land cover didn't include burnt area class; thus, we used a MODIS burnt area product (MCD64A1.v006) (Giglio, 2020) for extracting the VV and VH signals. The MODIS burned area product is a Terra and Aqua combined monthly data product available at 500 m resolution from 2003. Since the Sentinel data are at a nominal resolution of 5 m×20 m and the MODIS land cover and burned area products are at a 1 km and 500 m, respectively, to avoid confusion due to the inherent resolutions, we also used the PlanetScope data at 3 m resolution to infer the land cover types including the burnt areas. More specifically, we used the PlanetScope Analytic Ortho Scene surface reflectance (SR) Level 3B data, which have four spectral bands (blue, green, red, and near-infrared). All PlanetScope data were acquired for 2018 for March, April, and May biomass burning months with less than 50% cloud cover. Each PlanetScope image covers an area of 24.6×16.4 km², and we downloaded multiple images covering more than 30% of the country area. Through the visual interpretation, low-quality PlanetScope images affected by haze were removed, and only good-quality data were used to extract the land cover signal. A minimum of 30 points per class per month were created through visual interpretation of the PlanetScope data, and we exported the data as shapefiles. We then used the exported shapefiles to extract the VV, VH, and angle information from the Sentinel-1A&B monthly stacked-layer data. The extracted VV, VH, and corresponding angle data were analyzed using descriptive statistics, box plots, and scatter plots. Various backscatter thresholds for burnt areas were identified for different months (March, April, and May) and used for burnt versus no-burnt area classification. The threshold-based classification served our purpose as our focus is on a single class mapping in the study, i.e., Sentinel-1A&B-derived burnt areas. We then compared the burnt area maps using both MODIS burnt areas and PlanetScope data. While comparing, we used independent validation data; i.e., the data that were used for extracting the land cover signal were not used for validation. Although we used 2 years of data, the final burnt area (in km²) was averaged to infer overall values. Also, since the results are almost similar for both years, due to the limited space, we provide figures for either 2018 or 2019, but not both.

RESULTS AND DISCUSSION

Results on VV and VH characteristics for different land cover classes retrieved using Sentinel-1 data are shown as box plots (Figure 4.2). The box plots are useful as they provide a visual summary of the data to assess distribution, skewness, and other statistical measures. The plots represent variability or dispersion of the data with various components embedded in them, i.e., "minimum," first quartile (Q1), median, third quartile (Q3), and "maximum" with whiskers. The median marks the midpoint of the data denoted by a horizontal line in the box, and it divides the box into two parts; half the scores are greater than or equal to this value and half are

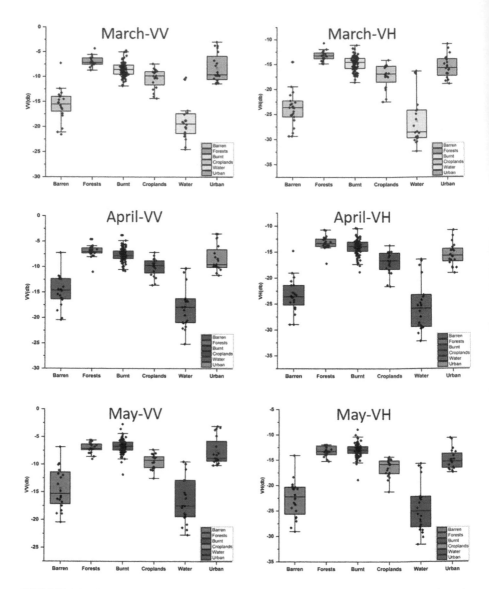

FIGURE 4.2 Box plots of burnt areas in Laos from Sentinel-1 VV and VH polarizations for March, April, and May 2019. Relatively, VV signal had higher backscatter than VH for all months. The signal for water and barren areas was unique in both VV and VH polarizations.

less. Results from both the VV and VH backscatter box plots for various land cover classes (Figure 4.2) suggested the following: (1) VV having a relatively higher back-scatter signal than VH for all land cover classes and months (March, April, May); (2) water and barren areas with a distinct VV and VH backscatter signal compared to the other land cover classes; (3) a consistent pattern in the VV and VH backscatter signal with the highest to lowest backscatter in the following order: forests>burnt areas>urban areas>croplands>barren>water; and (4) VV signal having much lesser differences (<−2 dB) in backscatter signal for urban, croplands, and burnt areas than the VH signal. The longer the box, the more dispersed the data. Thus, comparison of the interquartile ranges suggested the following order for VV: in March: Urban >Water>Barren>Croplands>Burnt>Forests; in April: Water>Barren>Urban>Crop lands>Burnt>Forests; and in May: Water>Barren>Urban>Croplands>Burnt>Forests. Although the order was the same for both April and May, VV-derived water and barren areas showed relatively higher dispersion during May than during April. For the VH-polarized data, the interquartile range was in the following order: for March: Water>Urban>Croplands>Burnt>Forests; for April: Water>Barren>Croplands> Urban>Burnt>Forests; and for May: Water>Barren>Croplands>Urban>Forests> Burnt. The larger whiskers indicate wider distribution, i.e., more scattered data, which can be seen for both water and barren areas in both VV and VH polarizations.

For March, the mean VV signal difference between the forest and burnt areas was 1.62 dB, and between burnt versus croplands, a 1.77 dB was noted. Relatively, less difference has been observed between the burnt areas versus urban classes (−0.28 dB). For the VH signal in March, forest and burnt areas had a mean difference of 1.55 dB and burnt versus croplands had a mean difference of 2.51 dB. Relatively, less difference has been observed between the burnt areas versus urban classes (0.71 dB). Thus, the VH signal difference was much distinct between burnt areas–croplands and burnt areas–urban areas in March.

For April, the mean VV signal difference between the forest and burnt areas was 0.68 dB, and between burnt versus croplands, a 2.33 dB was noted. Relatively, less difference has been observed between the burnt areas versus urban classes (0.69 dB). For the VH signal in March, forest and burnt areas had a mean difference of 0.80 dB and burnt versus croplands had a mean difference of 2.92 dB. Relatively, more difference has been observed between the burnt areas versus urban classes (1.41 dB), i.e., a much higher difference than the forest versus burnt area signal. Thus, VH signal differences were much distinct than VV during April.

For May, the mean VV signal difference between the forest and burnt areas was −0.31 dB; between the burnt areas versus croplands, a 2.79 dB was noted; and between the burnt areas versus urban classes, a 0.99 dB was noted. For the VH signal in May, forest and burnt areas had a mean difference of −0.27 dB, burnt versus croplands 3.59 dB, and burnt versus urban 1.88 dB. In this month, VH burnt area signal was distinct for both croplands and urban lands compared to VV.

We also analyzed the burnt area signal differences varying with the incident angles in addition to the above. The scatter plots for different months with varying VV, VH (vertical axis), and incident angles (horizontal axis) are shown in Figure 4.3. A clear distinction between the VV and VH signals can be seen with the varying incident angles; i.e., VV had a much higher backscatter than VH at all angles. Further, during March, the VV signal for burnt areas was in the range of −7 to −10 dB, whereas it was −13 dB to −16.5 dB for VH centered around 37.5°–39.5° incident angles. As the incident angle increased beyond 40°, the data were highly scattered for both VV and VH during March.

In April, the burnt area VV signal was mostly in the range of −6 to −8.5 dB, whereas it was −12 dB to −14.5 dB for VH centered around 37°–40° incident angles. Beyond 40° incident angles, both the VV and VH signals were weak and scattered.

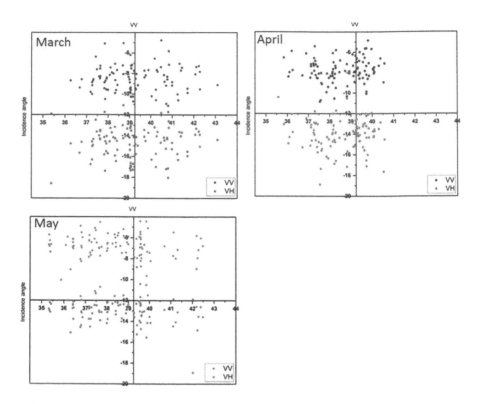

FIGURE 4.3 A scatter diagram of Sentinel-1 VV- and VH-derived burnt area signal variations in relation to incident angles for different months. The VV backscatter was relatively higher at all incident angles. Relatively less scatter of VV and VH signals can be seen during April, which suggests it as the best month for mapping burnt areas.

In May, the burnt area VV signal was mostly in the range of −6 to −8 dB, whereas it was −12 dB to −13.8 dB for VH. The data were centered around 37°–40° incident angles, beyond which both the VV and VH signals were weak and scattered. A relative comparison of these differences suggests a clear and distinct VH signal in April than in March and May, useful for delineating Laos' burnt areas.

We used the mean VV and VH burnt area values reported earlier as thresholds for mapping burnt versus non-burnt areas. The VV- and VH-derived burnt area maps for 2019 are shown in Figure 4.4. The VV and VH burnt area statistics (in km²), including MODIS burnt areas, are shown in Figure 4.5. Compared to the MODIS-derived burnt areas of 5422.82 km², VV-mapped areas were 40,547 and 9493 km², respectively. To infer these areas' accuracy, we randomly used the PlanetScope data to verify the burnt area statistics. An example is an overlay of VV-derived burnt area pixels on PlanetScope data shown in Figure 4.6. Compared to the VV signal, which had more commission burnt area pixels on PlanetScope, VH had very little commission and omission pixels. The VH signal accounted for more than 70% burnt areas in the PlanetScope data. Thus, we inferred that VH signal estimates were more accurate than the VV-derived burnt areas. We infer that more validation is required through ground-truthing.

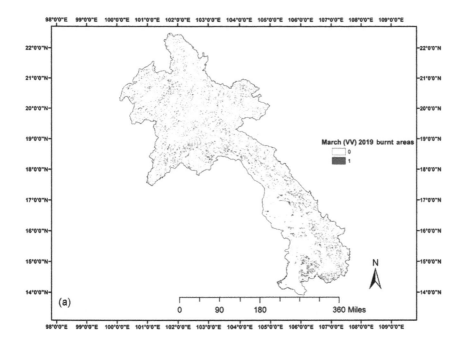

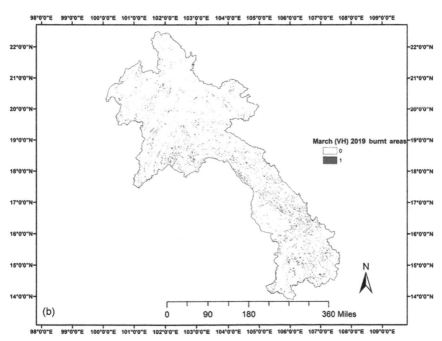

FIGURE 4.4 Burnt areas mapped using the Sentinel-1 VV (a) and VH (b) polarizations for March 2019, Laos; 1 indicates burnt areas and 0 indicates other classes or no data. Relatively, the VH (b)-derived burnt areas matched more closely with the PlanetScope very high-resolution datasets than the VV (a)-derived burnt areas.

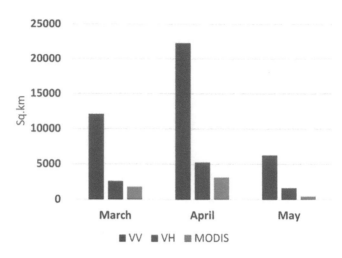

FIGURE 4.5 Burnt areas in Laos mapped using the Sentinel-1 VV and VH polarizations for March, April, and May 2019. Both VV and VH showed relatively higher burned areas than the MODIS (MCD64A1) burned area product. The VH-derived burnt areas compared more closely with the PlanetScope very high-resolution datasets than the VV-derived burnt areas.

The SAR-based burnt areas are mainly influenced by the resultant changes in radar backscattering before and after the fire. The total amount of energy scattered back to radar is influenced by a variety of parameters such as signal wavelength and polarization in addition to the properties such as vegetation type burnt, how much burnt, season, dielectric properties, soil moisture, and surface roughness (Imperatore et al., 2017). The terrain roughness, soil moisture, and vegetation structure alterations can strongly influence the SAR backscatter signal. A variety of SAR backscatter responses, either increase or decrease, have been reported in the literature with varying burnt conditions, sensor characteristics, and surface conditions (Menges et al., 2004; Engelbrecht et al., 2017). For example, as noted by Belenguer-Plomer et al. (2019), in boreal forests, higher backscatter values were reported for burned areas than for unburned areas at C-band VV polarization when soil moisture was high, whereas lower backscatter values were reported for sites with better drainage (Bourgeau-Chavez et al., 2002; Huang and Siegert, 2006). In temperate and Mediterranean regions, lower backscatter values were reported for burned areas than for unburned forests from C-band SAR data (Imperatore et al., 2017). A decrease in backscatter for both co- and cross-polarized C-band channels was reported in African open forests (Verhegghen et al., 2016). Also, a substantial backscatter decrease was reported for burned tropical forests at C-band VV polarization under dry weather conditions due to the decreased volume scattering and increased heat flux (Ruecker and Siegert, 2000; Lohberger et al., 2018). Also, discrimination from the unburned surrounding forests was difficult as the backscatter coefficient over burned areas increased after rainfall (Siegert and Ruecker, 2000). The lower backscatter values in both VV and VH polarizations for burned areas than for the forests observed in the current study are consistent with the results reported for tropical regions and African open forests. In particular, in Laos, the peak biomass burning occurs during March and April,

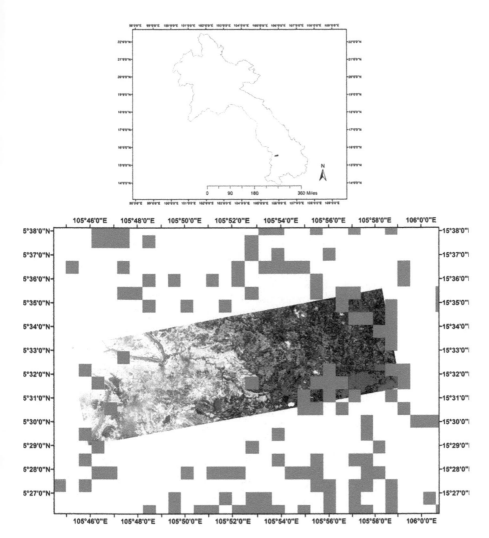

FIGURE 4.6 Sentinel-1 VV burnt area pixels (in blue) overlaid on PlanetScope (3.0 m) raw RGB data, March 3, 2018. The PlanetScope data swath is $24.6 \times 16.4 \, km^2$.

slowly decreasing in May. During these months, the temperatures are highly elevated with low moisture content in the soil; thus, backscatter is low in burnt areas. Also, our results on VH cross-polarization as a better indicator for burnt areas than single polarized VV match with the results of Zhang et al. (2019). Further, our results on VV and VH variations due to incident angles also match with the earlier reported studies of Gimeno and San-Miguel-Ayanz (2004) that the backscatter values at the lower incident angles perform better for discriminating undisturbed forests and fire-affected areas. In this study, we didn't study the impact of topography on the back-scatter signal in burnt areas, which seems to be another important factor impacting the signal (Soja et al., 2012) and thus needs further investigation. We also infer the need to use temporal datasets, including both amplitude and phase signal variations, through the InSAR techniques for effective mapping of burnt areas. Finally, ground validation of the results is highly recommended, which we could not perform due to the ongoing COVID-19 situation. Despite these limitations, our results highlight unique burnt area signal variations from VV, VH, and incident angles useful for burnt area mapping and monitoring in Laos and elsewhere.

ACKNOWLEDGMENTS

The authors gratefully acknowledge the NASA Land Cover/Land Use Change program for funding the South/Southeast Asia Research Initiative (SARI) under which the current study has been undertaken.

REFERENCES

Badarinath, K.V.S., Chand, T.K. and Prasad, V.K. 2006. Agriculture crop residue burning in the Indo-Gangetic Plains–a study using IRS-P6 AWiFS satellite data. *Current Science.* 91(8), 1085–1089.

Badarinath, K.V.S., Kharol, S.K., Krishna Prasad, V., Kaskaoutis, D.G. and Kambezidis, H.D. 2008a. Variation in aerosol properties over Hyderabad, India during intense cyclonic conditions. *International Journal of Remote Sensing.* 29(15), 4575–4597.

Badarinath, K.V.S., Kharol, S.K., Prasad, V.K., Sharma, A.R., Reddi, E.U.B., Kambezidis, H.D. and Kaskaoutis, D.G. 2008b. Influence of natural and anthropogenic activities on UV Index variations–a study over tropical urban region using ground based observations and satellite data. *Journal of Atmospheric Chemistry.* 59(3), 219–236.

Ban, Y., Zhang, P., Nascetti, A., Bevington, A.R. and Wulder, M.A. 2020. Near real-time wildfire progression monitoring with Sentinel-1 SAR time series and deep learning. *Scientific Reports.* 10(1), 1–15.

Belenguer-Plomer, M.A., Tanase, M.A., Fernandez-Carrillo, A. and Chuvieco, E. 2019. Burned area detection and mapping using Sentinel-1 backscatter coefficient and thermal anomalies. *Remote Sensing of Environment.* 233, 111345.

Bernhard, E.M., Twele, A. and Gähler, M., 2012, July. Burnt area mapping in the European-Mediterranean: SAR backscatter change analysis and synergistic use of optical and SAR data. In *2012 IEEE International Geoscience and Remote Sensing Symposium* (pp. 2141–2143), Seoul, South Korea.

Biswas, S., Lasko, K.D. and Vadrevu, K.P. 2015a. Fire disturbance in tropical forests of Myanmar—Analysis using MODIS satellite datasets. *IEEE Journal of Selected Topics in Applied Earth Observations and Remote Sensing.* 8(5), 2273–2281.

Biswas, S., Vadrevu, K.P., Lwin, Z.M., Lasko, K. and Justice, C.O. 2015b. Factors controlling vegetation fires in protected and non-protected areas of Myanmar. *PLoS One.* 10(4), e0124346.

Bourgeau-Chavez, L., Kasischke, E., Brunzell, S., Mudd, J. and Tukman, M. 2002. Mapping fire scars in global boreal forests using imaging radar data. *International Journal of Remote Sensing.* 23, 4211–4234.

Bradstock, R.A., Cary, G.J., Davies, I., Lindenmayer, D.B., Price, O.F. and Williams, R.J., 2012. Wildfires, fuel treatment and risk mitigation in Australian eucalypt forests: Insights from landscape-scale simulation. *Journal of Environmental Management.* 105, 66–75.

Czuchlewski, K.R. and Weissel, J.K., 2005, July. Synthetic Aperture Radar (SAR)-based mapping of wildfire burn severity and recovery. In *Proceedings. 2005 IEEE International Geoscience and Remote Sensing Symposium,* 2005. IGARSS'05. (Vol. 1, pp. 4).

Donezar, U., De Blas, T., Larrañaga, A., Ros, F., Albizua, L., Steel, A. and Broglia, M. 2019. Applicability of the multi temporal coherence approach to Sentinel-1 for the detection and delineation of burnt areas in the context of the Copernicus Emergency Management Service. *Remote Sensing*. 11(22), 2607.

Engelbrecht, J., Theron, A. and Vhengani, L. 2017, July. A normalised difference alpha-angle approach to burn scar extraction on multiple-polarisation Sar data. In *2017 IEEE International Geoscience and Remote Sensing Symposium (IGARSS)* (pp. 4274–4277), Fort Worth, TX.

French, N.H.F., Kasischke, E.S., Bourgeau-Chavez, L.L. and Harrell, P.A., 1996. Sensitivity of ERS-1 SAR to variations in soil water in fire-disturbed boreal forest ecosystems. *International Journal of Remote Sensing*, 17(15), pp. 3037–3053.

Giglio, L. 2020. https://lpdaac.usgs.gov/products/mcd64a1v006/.

Gimeno, M. and San-Miguel-Ayanz, J. 2004. Evaluation of RADARSAT-1 data for identification of burnt areas in Southern Europe. *Remote Sensing of Environment*. 92(3), 370–375.

Grégoire, J-M., Tansey, K. and Silva, J.M.N. 2003. The GBA2000 initiative: Developing a global burnt area database from SPOT-VEGETATION imagery. *International Journal of Remote Sensing*. 24(6), 1369–1376.

Gupta, P.K., Prasad, V.K., Sharma, C., Sarkar, A.K., Kant, Y., Badarinath, K.V.S. and Mitra, A.P. 2001. CH_4 emissions from biomass burning of shifting cultivation areas of tropical deciduous forests–experimental results from ground-based measurements. *Chemosphere-Global Change Science*. 3(2), 133–143.

Huang, S. and Siegert, F. 2006. Land cover classification optimized to detect areas at risk of desertification in North China based on SPOT VEGETATION imagery. *Journal of Arid Environments*. 67(2), 308–327.

Hayasaka, H., Noguchi, I., Putra, E.I., Yulianti, N. and Vadrevu, K. 2014. Peat-fire-related air pollution in Central Kalimantan, Indonesia. *Environmental Pollution*. 195, 257–266. doi: 10.1016/j.envpol.2014.06.031.

Imperatore, P., Azar, R., Calo, F., Stroppiana, D., Brivio, P.A., Lanari, R. and Pepe, A. 2017. Effect of the vegetation fire on backscattering: An investigation based on Sentinel-1 observations. *IEEE Journal of Selected Topics in Applied Earth Observations and Remote Sensing*. 10(10), 4478–4492.

Inoue, Y. 2018. Ecosystem carbon stock, atmosphere and food security in slash-and-burn land use: A geospatial study in mountainous region of Laos. In: Vadrevu, K.P., Ohara, T., and Justice, C. (Eds). *Land-Atmospheric Research applications in South/Southeast Asia*. Springer, Cham. pp. 641–666.

Israr, I., Jaya, S.N.I, Saharjo, H.S., Kuncahyo, B. and Vadrevu, K.P. 2018. Spatio-temporal analysis of land and forest fires in Indonesia using MODIS active fire dataset. In: Vadrevu, KP, Ohara, T., and Justice, C. (Eds). *Land-Atmospheric Research Applications in South/Southeast Asia*. Springer, Cham. pp. 105–128.

Justice, C., Gutman, G. and Vadrevu, K.P. 2015. NASA land cover and land use change (LCLUC): An interdisciplinary research program. *Journal of Environmental Management*. 148(15), 4–9.

Kant, Y., Ghosh, A.B., Sharma, M.C., Gupta, P.K., Prasad, V.K., Badarinath, K.V.S. and Mitra, A.P. 2000. Studies on aerosol optical depth in biomass burning areas using satellite and ground-based observations. *Infrared Physics & Technology*. 41(1), 21–28.

Key, C.H. and Benson, N.C. 2006. Landscape assessment (LA). In: Lutes, Duncan C.; Keane, Robert E.; Caratti, John F.; Key, Carl H.; Benson, Nathan C.; Sutherland, Steve; Gangi, Larry J. 2006. *FIREMON: Fire effects monitoring and inventory system. Gen. Tech. Rep. RMRS-GTR-164-CD*. Fort Collins, CO: US Department of Agriculture, Forest Service, Rocky Mountain Research Station. p. LA-1-55, 164.

Kharol, S.K., Badarinath, K.V.S., Sharma, A.R., Mahalakshmi, D.V., Singh, D. and Prasad, V.K. 2012. Black carbon aerosol variations over Patiala city, Punjab, India—a study during agriculture crop residue burning period using ground measurements and satellite data. *Journal of Atmospheric and Solar-Terrestrial Physics*. 84, 45–51.

Kim, K.H., Jahan, S.A. and Kabir, E. 2011. A review of diseases associated with household air pollution due to the use of biomass fuels. *Journal of Hazardous Materials*. 192(2), 425–431.

Lasko, K. and Vadrevu, K.P. 2018. Improved rice residue burning emissions estimates: Accounting for practice-specific emission factors in air pollution assessments of Vietnam. *Environmental Pollution*. 236(5), 795–806.

Lasko, K., Vadrevu, K.P., Tran, V.T., Ellicott, E., Nguyen, T.T., Bui, H.Q. and Justice, C. 2017. Satellites may underestimate rice residue and associated burning emissions in Vietnam. *Environmental Research Letters*. 12(8), 085006.

Lasko, K., Vadrevu, K.P. and Nguyen, T.T.N. 2018a. Analysis of air pollution over Hanoi, Vietnam using multi-satellite and MERRA reanalysis datasets. *PLoS One*. 13(5), e0196629.

Lasko, K., Vadrevu, K.P., Tran, V.T. and Justice, C. 2018b. Mapping double and single crop paddy rice with Sentinel-1A at varying spatial scales and polarizations in Hanoi, Vietnam. *IEEE Journal of Selected Topics in Applied Earth Observations and Remote Sensing*. 11(2), 498–512.

Lasko, K., Vadrevu, K.P., Tran, V.T., Ellicott, E., Nguyen, T.T., Bui, H.Q. and Justice, C. 2017. Satellites may underestimate rice residue and associated burning emissions in Vietnam. *Environmental Research Letters*. 12(8), 085006.

Liew, S.C., Kwoh, L.K., Padmanabhan, K., Lim, O.K. and Lim, H. 1999. Delineating land/forest fire burnt scars with ERS interferometric synthetic aperture radar. *Geophysical Research Letters*. 26(16), 2409–2412.

Lohberger, S., Stängel, M., Atwood, E.C. and Siegert, F. 2018. Spatial evaluation of Indonesia's 2015 fire-affected area and estimated carbon emissions using Sentinel-1. *Global Change Biology*. 24(2), 644–654.

Menges, C.H., Bartolo, R.E., Bell, D. and Hill, G.E. 2004. The effect of savanna fires on SAR backscatter in northern Australia. *International Journal of Remote Sensing*. 25(22), 4857–4871.

Oanh, N.T.K., Permadi, D.A., Dong, N.P. and Nguyet, D.A. 2018. Emission of toxic air pollutants and greenhouse gases from crop residue open burning in Southeast Asia. In: Vadrevu, K.P., Ohara, T. and Justice, C. (Eds). *Land-Atmospheric Research applications in South/Southeast Asia*. Springer, Cham. pp. 47–68.

Pereira, J. M., Sá, A. C., Sousa, A. M., Silva, J. M., Santos, T. N. and Carreiras, J. M. 1999. Spectral characterisation and discrimination of burnt areas. In: Chuvieco, E. (Ed). *Remote Sensing of Large wildfires*. Springer, Berlin, Heidelberg. pp. 123–138.

Prasad, V.K., Gupta, P.K., Sharma, C., Sarkar, A.K., Kant, Y., Badarinath, K.V.S., Rajagopal, T. and Mitra, A.P. 2000. NO_x emissions from biomass burning of shifting cultivation areas from tropical deciduous forests of India–estimates from ground-based measurements. *Atmospheric Environment*. 34(20), 3271–3280.

Prasad, V.K., Kant, Y. and Badarinath, K.V.S. 2001. CENTURY ecosystem model application for quantifying vegetation dynamics in shifting cultivation areas: A case study from Rampa Forests, Eastern Ghats (India). *Ecological Research*. 16(3), 497–507.

Prasad, V.K., Kant, Y., Gupta, P.K., Elvidge, C. and Badarinath, K.V.S. 2002. Biomass burning and related trace gas emissions from tropical dry deciduous forests of India: A study using DMSP-OLS data and ground-based measurements. *International Journal of Remote Sensing*. 23(14), 2837–2851.

Prasad, V.K., Lata, M. and Badarinath, K.V.S. 2003. Trace gas emissions from biomass burning from northeast region in India—estimates from satellite remote sensing data and GIS. *Environmentalist*. 23(3), 229–236.

Prasad, V.K., Badarinath, K.V.S. and Eaturu, A. 2008. Biophysical and anthropogenic controls of forest fires in the Deccan Plateau, India. *Journal of Environmental Management.* 86(1), 1–13.

Ramachandran, S. 2018. Aerosols and climate change: Present understanding, challenges and future outlook. In: Vadrevu, K.P., Ohara, T. and Justice, C. (Eds). *Land-Atmospheric Research Applications in South/Southeast Asia.* Springer, Cham. pp. 341–378.

Roder, W., Phengchanh, S. and Maniphone, S. 1997. Dynamics of soil and vegetation during crop and fallow period in slash-and-burn fields of northern Laos. *Geoderma.* 76(1–2), 131–144.

Roy, D.P., Boschetti, L., Justice, C.O. and Ju, J. 2008. The collection 5 MODIS burned area product—Global evaluation by comparison with the MODIS active fire product. *Remote Sensing of Environment.* 112(9), 3690–3707.

Ruecker, G. and Siegert, F. 2000. Burn scar mapping and fire damage assessment using ERS-2 SAR images in East Kalimantan, Indonesia. *The International Archives of the Photogrammetry, Remote Sensing.* 33, 1286–1293.

Salvi, S. and Barnes, P.J. 2010. Is exposure to biomass smoke the biggest risk factor for COPD globally? *Chest.* 138(1), 3–6.

Siegert, F. and Ruecker, G., 2000. Use of multitemporal ERS-2 SAR images for identification of burned scars in south-east Asian tropical rainforest. *International Journal of Remote Sensing.* 21(4), 831–837.

Singh, Y. and Sidhu, H.S. 2014. Management of cereal crop residues for sustainable rice-wheat production system in the Indo-Gangetic plains of India. *Proceedings of the Indian National Science Academy.* 80(1), 95–114.

Soja, M.J., SandBerg, G. and Ulander, L.M. 2012. Regression-based retrieval of boreal forest biomass in sloping terrain using P-band SAR backscatter intensity data. *IEEE Transactions on Geoscience and Remote Sensing.* 51(5), 2646–2665.

Stroppiana, D., Azar, R., Calò, F., Pepe, A., Imperatore, P., Boschetti, M., Silva, J., Brivio, P.A. and Lanari, R. 2015. Integration of optical and SAR data for burned area mapping in Mediterranean Regions. *Remote Sensing.* 7(2), 1320–1345.

Tariq, S. and Ul-Haq, Z. 2018. Satellite remote sensing of aerosols and gaseous pollution over Pakistan. In: Vadrevu, K.P., Ohara, T., and Justice, C. (Eds). *Land-Atmospheric Research applications in South/Southeast Asia.* Springer, Cham. pp. 523–552.

Vadrevu, K.P. and Justice, C.O. 2011. Vegetation fires in the Asian region: Satellite observational needs and priorities. *Global Environmental Research.* 15(1), 65–76.

Vadrevu, K.P. and Lasko, K. 2018. Intercomparison of MODIS AQUA and VIIRS I-Band fires and emissions in an agricultural landscape—Implications for air pollution research. *Remote Sensing.* 10(7), 978. doi:10.3390/rs10070978.

Vadrevu, K.P., Eaturu, A. and Badarinath, K.V.S. 2006. Spatial distribution of forest fires and controlling factors in Andhra Pradesh, India using spot satellite datasets. *Environmental Monitoring and Assessment.* 123(1–3), 75–96.

Vadrevu, K.P., Badarinath, K.V.S. and Anuradha, E. 2008. Spatial patterns in vegetation fires in the Indian region. *Environmental Monitoring and Assessment.* 147(1–3), 1.

Vadrevu, K.P., Ellicott, E., Badarinath, K.V.S. and Vermote, E. 2011. MODIS derived fire characteristics and aerosol optical depth variations during the agricultural residue burning season, north India. *Environmental Pollution.* 159(6), 1560–1569.

Vadrevu, K.P., Csiszar, I., Ellicott, E., Giglio, L., Badarinath, K.V.S., Vermote, E. and Justice, C. 2012. Hotspot analysis of vegetation fires and intensity in the Indian region. *IEEE Journal of Selected Topics in Applied Earth Observations and Remote Sensing.* 6(1), 224–238.

Vadrevu, K.P., Giglio, L. and Justice, C. 2013. Satellite based analysis of fire–carbon monoxide relationships from forest and agricultural residue burning (2003–2011). *Atmospheric Environment.* 64, 179–191.

Vadrevu, K.P., Ohara, T. and Justice, C. 2014a. Air pollution in Asia. *Environmental Pollution*. 12, 233–235.

Vadrevu, K.P., Lasko, K., Giglio, L. and Justice, C. 2014b. Analysis of Southeast Asian pollution episode during June 2013 using satellite remote sensing datasets. *Environmental Pollution*. 12, 245–256.

Vadrevu, K.P., Lasko, K., Giglio, L. and Justice, C. 2015. Vegetation fires, absorbing aerosols and smoke plume characteristics in diverse biomass burning regions of Asia. *Environmental Research Letters*. 10(10), 105003.

Vadrevu, K.P., Ohara, T. and Justice, C. 2017. Land cover, land use changes and air pollution in Asia: A synthesis. *Environmental Research Letters*. 12(12), 120201.

Vadrevu, K.P., Ohara, T. and Justice, C. Eds. 2018. *Land-Atmospheric Research Applications in South and Southeast Asia*. Springer, Cham.

Vadrevu, K.P., Lasko, K., Giglio, L., Schroeder, W., Biswas, S. and Justice, C. 2019. Trends in vegetation fires in south and southeast Asian countries. *Scientific Reports*. 9(1), 7422.

Verhegghen, A., Eva, H., Ceccherini, G., Achard, F., Gond, V., Gourlet-Fleury, S. and Cerutti, P.O. 2016. The potential of Sentinel satellites for burnt area mapping and monitoring in the Congo Basin forests. *Remote Sensing*. 8(12), 986.

Zhang, P., Nascetti, A., Ban, Y. and Gong, M. 2019. An implicit radar convolutional burn index for burnt area mapping with Sentinel-1 C-band SAR data. *ISPRS Journal of Photogrammetry and Remote Sensing*. 158, 50–62.

5 Peatland Surface Loss due to Fires in Central Kalimantan, Indonesia – A Case Study Using Differential Interferometry SAR (DInSAR)

Yessy Arvelyna
Remote Sensing Technology Center of Japan, Japan

Hidenori Takahashi
Hokkaido Institute of Hydro-climate,
Hokkaido University, Japan

Lies Indrayanti
Forestry Department, Faculty of Agriculture,
University of Palangka Raya, Indonesia

Hiroshi Hayasaka
Arctic Research Center, Hokkaido University, Japan

Krishna Prasad Vadrevu
NASA Marshall Space Flight Center, USA

Retno Maryani
Ministry of Forestry, Indonesia

Mitsuru Osaki
Research Faculty of Agriculture, Hokkaido University, Japan

Hirose Kazuyo
Japan Space Systems, Japan

CONTENTS

INTRODUCTION

Biomass burning is one of the important sources of greenhouse gas emissions (GHGs) and aerosols in South/Southeast Asia (S/SEA) (Gupta et al., 2001; Prasad et al., 2000; 2002; 2003; Kant et al., 2000; Badarinath et al., 2008a, b; Biswas et al., 2015a, b). Mostly, the sources of biomass burning are anthropogenic in S/SEA, which is attributed to slash-and-burn agriculture or clearing of forest lands for growing commercial plantations, urban development, etc., with significant forest cover loss including biomass and bioenergy (Justice et al., 2015; Vadrevu and Lasko, 2015; Vadrevu et al., 2014a, b; 2019).

Of the different natural factors, drought-induced fires due to the El Niño–Southern Oscillation (ENSO) in Southeast Asia and Indonesia are most common (Nurbaya and Agung, 2018; Vadrevu et al., 2013; 2014a, b; Israr et al., 2018). Fires in Indonesia are a recurrent problem causing haze and smog across the region, with air pollutants reaching hazardous levels in Indonesia, neighboring countries of Singapore, Malaysia, and southern Thailand. Specifically, in Indonesia, peatland burning is most common with smoldering fires, which can last from days to months, releasing incompletely oxidized products such as CO, CH_4, and OC (Vadrevu et al., 2014b; Hayasaka et al., 2014; 2015). In Indonesia, smallholder agriculture farms account for ~ 15 million hectares (Mha), which covers 12% of forest land located at Sumatra, Kalimantan, Sulawesi, and Papua. In Kalimantan, peatlands cover 8.39 Mha. As the largest tropical peatlands globally, peatlands in Indonesia play an essential role in the global carbon cycle as they store a substantial amount of carbon. Various issues, such as forest destruction, land fires, and peat drainage, contribute to peatland degradation. Of the different factors, fires in peatlands cause the most destruction as they can burn for a long time, depending on the peat and dry conditions' depth. The Government of Indonesia has been carrying various peatland management and conservation measures to prevent degradation through new policies and law enforcement (Hayasaka et al., 2015; Nurbaya and Agung, 2018).

In 2015, forest and peatland fires in Indonesia became a severe problem causing immense pollution through greenhouse gas (GHG) emissions. The GHGs emitted caused severe health issues to not only residents but also in neighboring countries. The burned area was estimated at about 2.61 Mha, and the economic losses were evaluated to be ~ IDR 221 trillion (Endrawati, 2016). The disaster happened during the El Niño event characterized by a long dry season (about 150 days) from late May to late October and low precipitation (e.g., 1 mm/day) lower than average rainfall in the dry season (e.g., 3.9 mm). The dry season resulted in the lowering of groundwater level, thus providing a conducive environment for forest and peatland fires in Kalimantan. The forest and peatland fires increased by the mid of August 2015, which triggered severe air pollution reaching hazardous levels ($PM_{10} > 420 \times 10^{-6}$ g/m^3) with a peak on October 21, 2015, with the highest PM_{10} concentrations (3760×10^{-6} g/m^3). Local ground truth data indicated that most of the emissions are from peatlands that occurred after a long dry season due to El Niño (Hayasaka et al., 2015).

In Kalimantan, most biomass is burnt to clear the lands and reduce pests and weeds and increase soil fertility for subsequent crops (Hayasaka et al., 2014; 2015). During the peatland fires, the vegetation cover and peat deposits are the main fuels. The average depth of burnt areas at Kalampangan District after the peatland fire event in 2009 is estimated at 22.03 cm, and a maximum depth is ~42.3 cm. The ignition of peat deposits creates smoldering wildfires that can release several GHGs (Davies et al., 2013). For one of the experimental peatland fires in Central Kalimantan, the peat ignition reached 50 cm depth with a maximum speed of fire spread about 2.5 cm/h (Hayasaka et al., 2015). The burnt area's depth far away from the canal decreased with the increase in fire frequency, for example, 4–18 cm for 1–4 fires or more (Page, 2018).

The release of CO_2 and other GHGs from forest fires and peatland became one of the biggest concerns impacting the global climate. In 2015, the level of emissions from the peatland fires in Indonesia was estimated at 712,602 Gg CO_2e during peak El Niño (Figure 5.1). After implementing the national government's mitigation measures, the emission level was successfully decreased to 12,513 Gg CO_2e (Nurbaya and Agung, 2018). To support such effective mitigation measures, it is necessary to

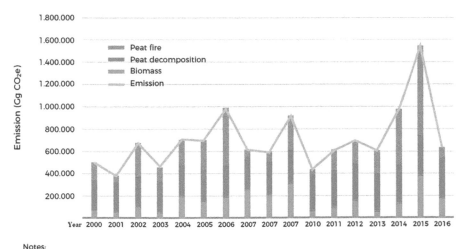

Notes:
Gg = gigagram
1 Gg CO₂e = 1.000 ton CO₂e

FIGURE 5.1   Emissions from peatlands in Indonesia (KLHK, 2018).

monitor the peatland fires without the limitation by weather condition and smoke cover as they can hinder remote detection of fires and thus underestimation of burnt areas, GHGs, and aerosols (Lasko et al., 2017; Lasko and Vadrevu, 2018; Lasko et al., 2018a; Vadrevu and Lasko, 2015; 2018).

Monitoring of peatland fires can be done using optical satellites such as MODIS and VIIRS (Vadrevu and Justice, 2011; Vadrevu and Lasko, 2015; 2018; Vadrevu et al., 2015; 2017; 2018). However, persistent cloud cover in the tropics and the smoke released during the burning can inhibit fire detection from optical sensors. Thus, the use of Synthetic Aperture Radar (SAR) data is recommended as the data can penetrate clouds and smoke (Arvelyna et al., 2015; 2018a, b; Lasko et al., 2018b). In this study, we use the ALOS-2 PALSAR-2 image data to characterize the peatland surface-level changes (subsidence) in the west of Kalimantan using the differential interferometry SAR (DInSAR) method from 2014 to 2018. DInSAR has been successfully used to measure landslide activity in some regions (Catani et al., 2005; Bulmer et al., 2006); however, very few studies utilized the DInSAR for peatland fire monitoring. DInSAR datasets provide spatially extensive estimates of surface deformation relative to the satellite look direction; thus, changes before and after a fire in the peatland cover loss and depth can be assessed. This study's objective is to assess the fluctuation in peatland surfaces before and after fires, more specifically peatland surface height difference, and to measure peatland burnt area depth. We implemented our study in Central Kalimantan, Indonesia.

STUDY AREA

We focused on the southeast region of the Palangka Raya city, Central Kalimantan, where peatland fires occurred extensively during 2014–2016. More specifically, we selected three different sites located at Pulang Pisau District representing swamp shrub, barren land, and secondary swamp, respectively. Also, we selected site 4 located on swamp shrub at Mantangai District, and site 5 and site 6 on dryland agriculture at Kalampangan District (KLHK 2018). The groundwater observation stations are located at dryland agriculture sites with grassland (Taka-1) and regenerated trees (Base-1 and Kalteng-1) and a peat depth of about 3 m.

The fire hotspots are observed on 2014/10/9 (Site 6), 2015/10/23 (Site 2), 2015/10/25 (Site 1, Site 2, Site 4), 2016/3/10 (Site 5), and 2016/7/30 (Site 3); coinciding with these dates, PALSAR-2 data were also obtained. At the Kalampangan site, the distance of fire hotspots to groundwater measurement stations is about 2.5 km; the peat depth at this site is taken from the field observations (Indrayani et al., 2011).

DATASETS AND METHODS

ALOS-2 PALSAR-2 dataset over Central Kalimantan (Table 5.1) is acquired through the 1st Earth Observation Research Announcement Collaborative Research (JFY2017–2018) from JAXA. The study area's fire hotspot data during 2015–2016 are retrieved from the LAPAN website for MODIS Aqua and Terra and SNPP VIIRS data

TABLE 5.1
SAR and Optical Data Used in the Study

ID	Data/Date	Parameter	Info
PALSAR-2 Data			
Resolution: 6 m			
1	140828 & 150409	Ascending, SM2	Before/after peatland fire at study area
		Resolution: 6 m	
2	151008 & 161006	Ascending	Before/after peatland fire at study area
3	161006 & 171005	SM3 mode	After peatland fire incident at study area
4	171005 & 180906	Resolution: 10 m	
5	150325 & 160504		Before/after peatland fire at study area
6	160504 & 170503	Descending	After peatland fire incident at study area
7	170503 & 170920	SM3 mode	
8	170920 & 180919	Resolution: 10 m	
Optical Data			
9	2015/9/20	Landsat-8	Before peatland fire incident at study area
		Resolution: 30 m	
10	2015/9/14		Before peatland fire incident at study area
11	2015/9/24	Sentinel-2	
12	2015/10/24	Resolution: 10 m	During peatland fire incident at study area
13	2016/8/29		After peatland fire incident at study area

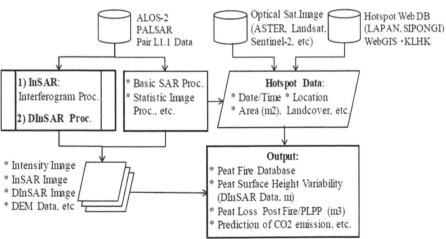

FIGURE 5.2   Methodology flowchart.

(Roswintiarti, 2016). Sentinel-2 optical data (10 m resolution) are acquired separately from the United States Geological Survey Earth Explorer for fire hotspot validation.

The DInSAR method is applied on each ALOS-2 L1.1 pair data before and after peat fire incidents using ENVI SARscape software (Figure 5.2) to derive peatland surface height differences that represent "peat loss post-fire (PLPF)" data along the line of sight of the satellite. A multi-temporal intensity image of the PALSAR-2 pair was then processed to delineate the coverage of burnt peatland areas. The vertical fluctuation of peatland surface before and after peat fire from DInSAR analysis was then compared with the groundwater and surface-level variability derived from the SESAME system using the submersible pressure transducer and the laser range finder installed nearby peatland fire study area. We also obtained the rainfall data for Tjilik Riwut Airport (BMKG), located approximately 7–40 km to NW of fire hotspots; the data were used to monitor precipitation events during the SAR data acquisition.

RESULTS AND DISCUSSION

FLUCTUATION OF PEATLAND SURFACE LEVEL

Measurements from the SESAME observation system during 2013–2018 suggested peatland ground surface level (GSL) variations at station Taka-1 as 17.6–24.2 cm and station Base-1 as 15.7–30.6 cm. During the El Niño period and the peatland fire peak in 2015, the groundwater level dropped from July to early November with a maximum difference of −96 cm on 2015/10/27, followed by decrease in GSL of 3.7 cm (Figure 5.3). Compared to the previous year, the change of GSL data followed the change of Ground Water Level (GWL) during the dry season (June to November 2013). GSL data are comparatively stable when GWL is above the ground surface during the rainy season from December to April (Takahashi et al., 2018).

DInSAR

DInSAR processing is applied on PALSAR-2 data pairs, before and after fire incidents for selected fire hotspots with 80% level coincidence (Figure 5.4). The difference of DInSAR data on the vertical direction with GSL data for stations Taka-1 and Base-1 is shown in Table 5.2. The difference data retrieved for Taka-1 are about ± 2.87 cm for Ascending − SM2 mode (6 m resolution); ± 3.45 ~ 8.6 cm for Ascending − SM3 mode (resolution 10 m); and ± 0.14 ~ 4.1 cm for Descending − SM3 mode (Figure 5.5).

At Site 1 into Site 4 (T1–T9), the maximum height difference of peatland surface in the downward direction is about −2.9 cm before fire incidents (data pair Id.1) and −23.5 cm after fire incidents (data pair Id.2–3), suggesting the possibility of peat loss after fire (Figure 5.6). Peat loss is higher at Hbd points around hotspot Aq3 at Site 2 of the barren land. Considering that the rainfall data from BMKG (2018) are higher on 2016/2/25 (74 mm) than on 2016/10/6 (2.5 mm), the peatland surface is expected to be higher due to more water absorption, shown by DInSAR data for data pair Id.2 at T3 and T5 (swamp shrub) and T6 (secondary swamp forest).

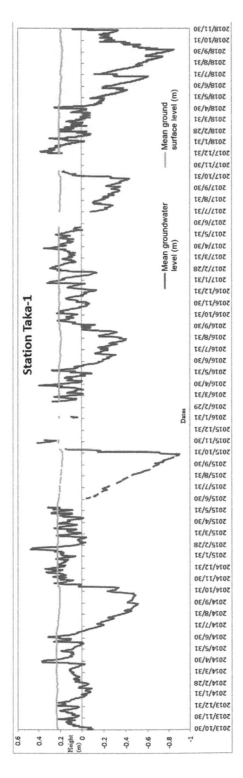

FIGURE 5.3   Fluctuation of groundwater at station Taka-1, Central Kalimantan, during 2013–2018.

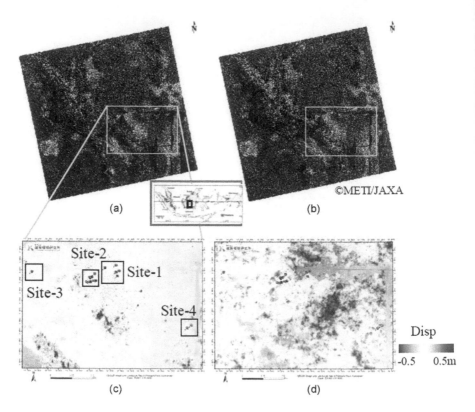

FIGURE 5.4 PALSAR-2 pair data: (a) intensity image of master data (15/10/8) and (b) slave data (16/10/6), and (c) DInSAR image of study area at Pulang Pisau and Mantangai districts before the fire incident and (d) after the fire incident.

TABLE 5.2

The Difference between DInSAR and GSL Data

Parameter	DInSAR (Vd) – GSL	
	Taka-1	Base-1
Ascending, SM2	−0.028726	0.04918
Resolution: 6 m		
Ascending	0.034529	0.069635
SM3, resolution: 10 m	0.051218	−0.036832
	−0.086004	-
	0.012226	0.018368
Descending	0.0014	0
SM3, resolution: 10 m	−0.03146	0
	−0.041032	0

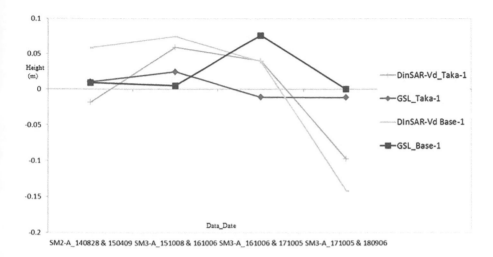

FIGURE 5.5 DInSAR versus GSL data.

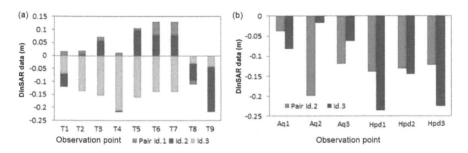

FIGURE 5.6 Peatland surface height difference before/after fires from DInSAR data for (a) T1–T9 and (b) Site 2 at Pulang Pisau District.

Peat depths at 12 observation points in Kalampangan District varied about 3.72–6 m, categorized as deep peat with the maturity level of hemic and sapric. Shallower peat depth has more maturity level. Deep peat has a higher moisture content, bearing capacity, and total copper (Cu) than shallow peat. The moisture content in the 100 cm depth of peatland with 6 m thickness is higher than that in the one with 4 m thickness (Indrayani et al., 2011). The hotspots from the peatland fire in 2014 (e.g., PLF-K1 and PLF-P5) are located in the northern part of the field survey site.

The fluctuations in the peatland surface at observation points are shown in Figure 5.7. The DInSAR analysis suggested lower vertical fluctuation in peatland surface during the El Niño season during 2014–2015 than during the following year. Generally, the vertical fluctuation movement is larger for deep peat with a maximum difference of about 2.2 cm for peat depth of 6.2 m, compared with 3.72 m depth, possibly due to larger water-bearing capacity and more extensive moisture content. The tree cover exists at points 5 and 8, which might have contributed to increased vertical fluctuation variability. The vertical downward movement at hotspot points showed the lowest measurement of

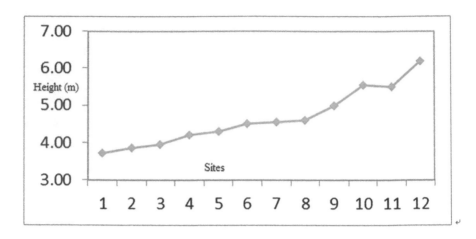

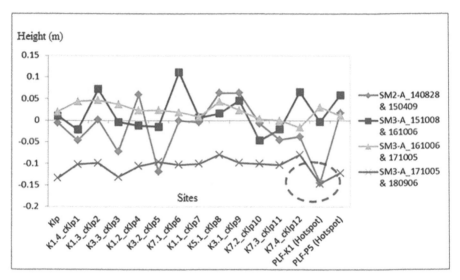

FIGURE 5.7   Peat depth of observation points in the Kalampangan site and its vertical surface fluctuation from DInSAR analysis.

about −14.4 cm (PLF-K1) and 18.4 cm (PLF-K2) for data pair before/after peatland fire
during 2014–2015. These data suggest peat loss after the fire. The general fluctuation
for peatland surface during the 2017–2018 period was lowest compared to the previous
year, which possibly showed the effect of peatland fire on the peatland surface's subsidence in the long term. About 15 cm downward movement is observed at hotspot points
during 2017–2018 for the data pair. The yearly subsidence rate at the Kalampangan site
is measured by about 4.1 cm during 2015–2016, 1.24 during 2016–2017, and 7.4 cm during 2017–2018. Besides, the proposed method of a false-color composite of SAR image
using the combination of standard deviation, gradient, and mean data of SAR intensity
image of ALOS-2 pair data before/after peatland fire incidents showed the burnt peatland area as whitish area distributed around the fire hotspots.

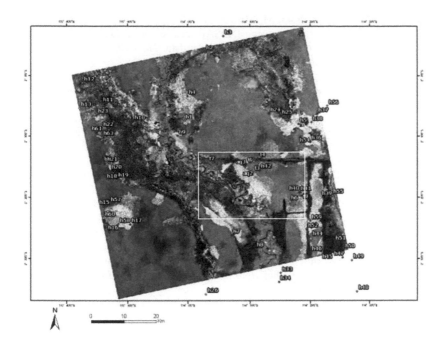

FIGURE 5.8 Peatland burnt areas depicted as whitish spots on PALSAR-2 data with fire hotspots.

CONCLUSION

DInSAR analysis results on ALOS-2 data before/after fire incidents showed that peat loss after fire incidents could be derived using peatland surface height difference analysis. The proposed methods are useful for peatland fire monitoring and can increase the accuracy of existing monitoring methods using optical satellite images, limiting when clouds and smoke cover exist.

ACKNOWLEDGMENTS

We would like to thank JAXA for providing the ALOS-2 PALSAR image data through the 1st Earth Observation Research Announcement from JFY2017–2018 (PI: PRA1A001). We would also like to thank Dr. R. Rovani, Forestry Attaché of Indonesia, Embassy in Tokyo, for important suggestions.

REFERENCES

Arvelyna, Y., Hirose, K., Anshari, G., Kunihiro, T., Koichi, H. and Yanto, R. 2015. The application of interferometry SAR data analysis for peatland and land use monitoring in west Kalimantan. *The 59th Japan Remote Sensing Conference Proceeding* (pp. 207–208), Nagasaki, Japan.

Arvelyna, Y., Sato, T. and Hosomi, K. 2018a. The Application of DInSAR analysis for peatland fire monitoring using ALOS-2 PALSAR data. *The 64th Japan Remote Sensing Conference Proceeding* (pp. 37–38), Tokyo, Japan.

Arvelyna, Y., Takahashi, H., Osaki, M., Hirose, H. and Sato, T. 2018b. Monitoring of peat-land surface height variability using DInSAR analysis on ALOS-2 PALSAR data and field observation of ground water/surface level data. *The 64th Japan Remote Sensing Conference Proceeding* (pp. 111–112), Tokyo, Japan.

Badarinath, K.V.S., Kharol, S.K., Krishna Prasad, V., Kaskaoutis, D.G. and Kambezidis, H.D. 2008a. Variation in aerosol properties over Hyderabad, India during intense cyclonic conditions. *International Journal of Remote Sensing*. 29(15), 4575–4597.

Badarinath, K.V.S., Kharol, S.K., Prasad, V.K., Sharma, A.R., Reddi, E.U.B., Kambezidis, H.D. and Kaskaoutis, D.G. 2008b. Influence of natural and anthropogenic activities on UV Index variations–a study over tropical urban region using ground based observations and satellite data. *Journal of Atmospheric Chemistry*. 59(3), 219–236.

Biswas, S., Lasko, K.D. and Vadrevu, K.P. 2015a. Fire disturbance in tropical forests of Myanmar—Analysis using MODIS satellite datasets. *IEEE Journal of Selected Topics in Applied Earth Observations and Remote Sensing*. 8(5), 2273–2281.

Biswas, S., Vadrevu, K.P., Lwin, Z.M., Lasko, K. and Justice, C.O. 2015b. Factors controlling vegetation fires in protected and non-protected areas of Myanmar. *PLoS One*. 10(4), e0124346.

Bulmer, M.H., Petley, D.N., Murphy, W. and Mantovani, F. 2006. Detecting slope deformation using two-pass differential interferometry: Implications for landslide studies on Earth and other planetary bodies. *Journal of Geophysical Research: Planets*. 111(E6). doi:10.1029/2005JE002593.

Catani, F., Casagli, N., Ermini, L., Righini, G. and Menduni, G. 2005. Landslide hazard and risk mapping at catchment scale in the Arno River basin. *Landslides*. 2(4), 329–342.

Data Online Pusat Database – BMKG. 2018. http://dataonline.bmkg.go.id/.

Endrawati, S.H. 2016. *Analisa data titik panas (hotspot) & areal kebakaran hutan dan lahan*. Ministry of Environment and Forestry, Indonesia. pp. 1, 9–13.

Davies, G.M., Gray, A., Rein, G. and Legg, C.J. 2013. Peat consumption and carbon loss due to smouldering wildfire in a temperate peatland. *Forest Ecology and Management*. 308, 169–177.

Gupta, P.K., Prasad, V.K., Sharma, C., Sarkar, A.K., Kant, Y., Badarinath, K.V.S. and Mitra, A.P. 2001. CH_4 emissions from biomass burning of shifting cultivation areas of tropical deciduous forests–experimental results from ground-based measurements. *Chemosphere-Global Change Science*. 3(2), 133–143.

Hayasaka, H., Noguchi, I., Putra, E.I., Yulianti, N. and Vadrevu, K., 2014. Peat-fire-related air pollution in Central Kalimantan, Indonesia. *Environmental Pollution*, 195, 257–266.

Hayasaka, H., Takahashi, H., Limin, S. H., Yulianti, N. and Usup, A. 2015. Peat fire occurrence in tropical peatland ecosystems. In: *Tropical Peatland Ecosystems*. Osaki, M. and Tsuji, N. (Eds.) Springer, Japan. doi:10.1007/978-4-431-55681-7.

Hotspot Information as Forest/Land Fire's Alert. 2018. LAPAN. http://modis-catalog.lapan.go.id/monitoring/hotspot/index.

Indrayani, L., Marsoem, S.N., Prayitno, T.A., Supriyo, H. and Radjagukguk, B. 2011. Distribusi ketebalan gambut dan sifat-sifat tanah di hutan rawa gambut Kalampangan, Kalimantan Tengah. Wanatropika. University of Gadjah Mada. pp. 56–72.

Israr, I., Jaya, S.N.I, Saharjo, H.S., Kuncahyo, B. and Vadrevu, K.P. 2018. Spatio-temporal analysis of land and forest fires in Indonesia using MODIS active fire dataset. In: *Land-Atmospheric Research Applications in South/Southeast Asia*. Vadrevu, K.P., Ohara, T., and Justice, C. (Eds). Springer, Cham. pp. 105–128.

Justice, C., Gutman, G. and Vadrevu, K.P. 2015. NASA land cover and land use change (LCLUC): An interdisciplinary research program. *Journal of Environmental Management*. 148(15), 4–9.

Kant, Y., Ghosh, A.B., Sharma, M.C., Gupta, P.K., Prasad, V.K., Badarinath, K.V.S. and Mitra, A.P. 2000. Studies on aerosol optical depth in biomass burning areas using satellite and ground-based observations. *Infrared Physics & Technology*. 41(1), 21–28.

KLHK, 2018. Kementrian Lingkungan Hidup dan Kehutanan (KLHK). Interactive Maps 2018. http://webgis.dephut.go.id:8080/kemenhut/index.php/en/map/interactive-map-2.

Lasko, K., Vadrevu, K.P., Tran, V.T., Ellicott, E., Nguyen, T.T., Bui, H.Q. and Justice, C. 2017. Satellites may underestimate rice residue and associated burning emissions in Vietnam. *Environmental Research Letters*. 12(8), 085006.

Lasko, K. and Vadrevu, K.P. 2018. Improved rice residue burning emissions estimates: Accounting for practice-specific emission factors in air pollution assessments of Vietnam. *Environmental Pollution*. 236(5), 795–806.

Lasko, K., Vadrevu, K.P. and Nguyen, T.T.N. 2018a. Analysis of air pollution over Hanoi, Vietnam using multi-satellite and MERRA reanalysis datasets. *PLoS One*. 13(5), e0196629.

Lasko, K., Vadrevu, K.P., Tran, V.T. and Justice, C. 2018b. Mapping double and single crop paddy rice with Sentinel-1A at varying spatial scales and polarizations in Hanoi, Vietnam. *IEEE Journal of Selected Topics in Applied Earth Observations and Remote Sensing*. 11(2), 498–512.

Nurbaya, S. and Agung, R. 2018. *The State of Indonesia's Forests, 2018*. Ministry of Environment and Forestry, Republic of Indonesia, Jakarta. 157 pages.

Page, S. 2018. The Tropical Peatland Fire Dynamic. http://www.borneonaturefoundation.org/wp-content/uploads/2018/01/3-Page-Fire-in-tropical-peat-landscapes.pdf.

Prasad, V.K., Gupta, P.K., Sharma, C., Sarkar, A.K., Kant, Y., Badarinath, K.V.S., Rajagopal, T. and Mitra, A.P. 2000. NO_x emissions from biomass burning of shifting cultivation areas from tropical deciduous forests of India–estimates from ground-based measurements. *Atmospheric Environment*. 34(20), 3271–3280.

Prasad, V.K., Kant, Y., Gupta, P.K., Elvidge, C. and Badarinath, K.V.S. 2002. Biomass burning and related trace gas emissions from tropical dry deciduous forests of India: A study using DMSP-OLS data and ground-based measurements. *International Journal of Remote Sensing*. 23(14), 2837–2851.

Prasad, V.K., Lata, M. and Badarinath, K.V.S. 2003. Trace gas emissions from biomass burning from northeast region in India—estimates from satellite remote sensing data and GIS. *Environmentalist*. 23(3), 229–236.

Roswintiarti, O. 2016. Informasi titik panas (Hotspot) kebakaran hutan/lahan, Guide V.01. LAPAN. http://pusfatja.lapan.go.id/files_uploads_ebook/publikasi/Panduan_hotspot_2016%20 versi%20draft%201_LAPAN.pdf.

Takahashi, H., Shigenaga, Y., Hamada, Y., Osaki, M. and Setiadi, B. 2018. Near-real time remote monitoring of ground-water/surface levels and warning to carbon loss from tropical peatlands in Indonesia. In: *The Proceedings of the International Workshop on Forest Ecological Resources Security for Next Generation: Development and Routine Utilization of Forest Ecological Resources and their Domestication* (eds. Shiodera, S., Fujita, M., & Kobayashi, S.) pp. 80–90. Center for Southeast Asian Studies. Research Center, Kyoto University, Kyoto.

Vadrevu, K.P. and Justice, C.O. 2011. Vegetation fires in the Asian region: Satellite observational needs and priorities. *Global Environmental Research*. 15(1), 65–76.

Vadrevu, K.P., Giglio, L. and Justice, C. 2013. Satellite based analysis of fire–carbon monoxide relationships from forest and agricultural residue burning (2003–2011). *Atmospheric Environment*. 64, 179–191.

Vadrevu, K.P., Lasko, K., Giglio, L. and Justice, C. 2014a. Analysis of Southeast Asian pollution episode during June 2013 using satellite remote sensing datasets. *Environmental Pollution*. 12, 245–256.

Vadrevu, K.P., Ohara, T. and Justice, C. 2014b. Air pollution in Asia. *Environmental Pollution*. 12, 233–235.

Vadrevu, K.P. and Lasko, K.P., 2015. Fire regimes and potential bioenergy loss from agricultural lands in the Indo-Gangetic Plains. *Journal of Environmental Management*. 148, 10–20.

Vadrevu, K.P., Lasko, K., Giglio, L. and Justice, C. 2015. Vegetation fires, absorbing aerosols and smoke plume characteristics in diverse biomass burning regions of Asia. *Environmental Research Letters*. 10(10), 105003.

Vadrevu, K.P., Ohara, T. and Justice, C. 2017. Land cover, land use changes and air pollution in Asia: A synthesis. *Environmental Research Letters*. 12(12), 120201.

Vadrevu, K.P. and Lasko, K. 2018. Intercomparison of MODIS AQUA and VIIRS I-Band fires and emissions in an agricultural landscape—Implications for air pollution research. *Remote Sensing*. 10(7), 978. doi:10.3390/rs10070978.

Vadrevu, K.P., Ohara, T. and Justice, C. 2018. *Land-Atmospheric Research Applications in South and Southeast Asia*. Springer, Cham. 725 pages.

Vadrevu, K.P., Lasko, K., Giglio, L., Schroeder, W., Biswas, S. and Justice, C. 2019. Trends in vegetation fires in south and Southeast Asian countries. *Scientific Reports*, 9(1), 7422. doi:10.1038/s41598-019-43940-x.

6 Burnt Area Mapping in Nainital, Uttarakhand, India, Using Very High-Resolution PlanetScope Imagery

Krishna Prasad Vadrevu
NASA Marshall Space Flight Center, USA

Aditya Eaturu
University of Alabama Huntsville, USA

Sumalika Biswas
Smithsonian Conservation Biology Institute, USA

CONTENTS

INTRODUCTION

Biomass burning is most prevalent in tropics as fire is often used to clear the forests for cultivation (Biswas et al., 2015a, b; Prasad et al., 2001; 2008; Justice et al., 2015; Inoue, 2018). However, such burning can result in significant economic losses including adverse effects on the ecosystem functions such as loss of biodiversity, biomass, and nutrients, including soil disturbance (Hayasaka et al., 2014; Israr et al., 2018; Oanh et al., 2018; Vadrevu et al., 2017; 2018). Biomass burning can also significantly affect human health as it results in release of several pollutants (Sigsgaard et al., 2015). The greenhouse gas emissions released from biomass burning such as CO, CO_2, SO_2, hydrocarbons, and other aerosols can impact local as well as regional climate (Prasad et al., 2000; 2002; 2003; Prasad and Badarinath 2004; Kant et al., 2000; Gupta et al., 2001; Badarinath et al., 2008a, b; Vadrevu, 2008; Vadrevu et al.,

2015; 2018; 2019; Lasko and Vadrevu, 2018). Quantifying burnt areas in different regions of the world gains significance as the results can aid in fire management and mitigation efforts (Vadrevu et al., 2013; 2014a, b; Justice et al., 2015; Lasko et al., 2018a, b). Remote sensing has significant potential for mapping and monitoring of fires including burnt area assessment. Due to their synoptic coverage and multitemporal and multispectral and repetitive coverage capabilities, remote sensing data can be effectively used in fire and biomass burning emissions research (Vadrevu and Justice, 2011). Specifically, the burnt areas can be detected from the satellites as the spectral response of the land surface is changed after fires due to the loss of vegetation, greenness, and water content or changes in the moisture content, soil color, etc., (Lentile et al., 2006; Vadrevu et al., 2006; 2008; Chuvieco et al., 2018). Mostly, such changes are detected as a decrease in spectral reflectance in the visible–near-infrared and an increase in the mid-infrared wavelengths of the spectrum (Epting et al., 2005; Van Wagtendonk et al., 2004). Thus, one of the most widely used indices for burnt area mapping is Normalized Burn Ratio (NBR) (Key and Benson, 2005), which combines the reflectances in the near-infrared (ρNIR) and mid-infrared bands (ρMIR). However, several fire regime characteristics such as fire frequency, intensity, and temporal distribution can impact the fire signal (Vadrevu et al., 2012) including the NBR. Different fire regimes can result in varying burn-scar patterns that might create signal differences from the remote sensing satellites in a given region.

Specifically, forested landscapes in several South/Southeast Asian countries unlike in the North American continent or Brazilian Amazon may not involve canopy fires; rather, they can be small surface fires and might last for a couple of days. The burnt areas can be smaller patches and may not include large areas specifically in the agricultural lands where the field size is small in several South/Southeast Asian countries (Lasko et al., 2017; Vadrevu and Lasko, 2015; 2018). In general, the accuracy of burnt area products is influenced by differences between burned and unburned area reflectances, temporal persistence of the burned area signal, spectral changes not caused by fire, and the size and spatial distribution of the burned areas (Chuvieco et al., 2004; Roy et al., 2005; Schroeder et al., 2014; Giglio et al., 2016). For example, the charcoal residue and the vegetation scar produced after fire can be confused with the shadows from clouds. In agricultural landscapes, the burnt patches can be relatively short-lived and the resulting ash left on the ground can be scattered or washed out by rain in a short amount of time. Irrespective of the landscapes, if using only optical data, the clouds can hinder the remote sensing of burnt areas. These uncertainties can impact the burnt area mapping and product accuracies. Thus, there is an immense need to evaluate multiple satellite remote sensing datasets for mapping and monitoring the burnt areas in different regions of the world.

Some of the latest burnt area products include MCD45 (500 m) (Roy et al., 2005), MCD64 (500 m) (Giglio et al., 2018), and Fire Climate Change Initiative (cci) (250 m) (Chuvieco et al., 2012). Compared to these products, burnt areas can be mapped using relatively higher resolution satellites such as Landsat (30 m) or Sentinel-2A&B (10-m visible and near-infrared bands and 20 m). In addition, very high-resolution satellite data less than 3 m resolution from commercial vendors such as Digital Globe and PlanetScope are readily available. The very high resolution combined with the daily temporal coverage offered by the commercial vendors relative to the existing freely available Sentinel and Landsat data is quite enticing. However, not many studies focused on these very high-resolution datasets in the Indian subcontinent. Thus, in this study, we explored the usefulness of PlanetScope data for burnt area mapping in Nainital, Uttarakhand, India, in 2019.

Uttarakhand is located in northern India and is mostly hilly with thirteen different districts (Figure 6.1) occupying an area of 53,483 km² (20,650 mi²), of which 86% is mountainous and 65% is covered by forest. The following vegetation type dominates at different altitudes: alpine shrubs (3000–5000 m); subalpine conifer forests (3000–2600 m); broadleaf forests (2600–1500 m); subtropical pine forests (1500 m); and moist-deciduous forests (below 1500 m). Uttarakhand witnessed a massive wildfire during May 2019 during which more than 900 ha of forest was burned. Incidents of fire have been reported from all 13 districts of Uttarakhand, the worst affected districts being Nainital and Almora which are mostly comprised of oak and pine

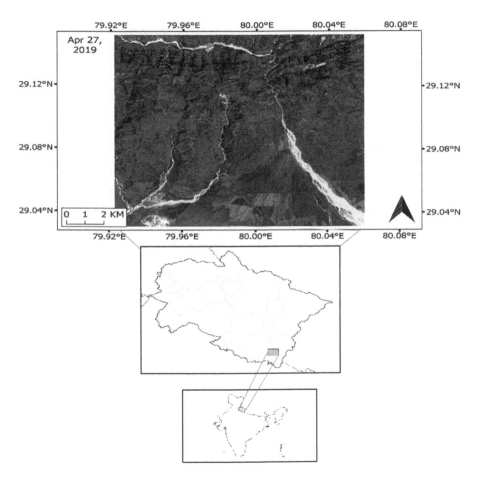

FIGURE 6.1   Study area location map with the PlanetScope data for April 27, 2019, Nainital, Uttarakhand, India.

trees. The primary driver of fires was anthropogenic and due to negligence. The main objective of this study is to assess the potential of very high-resolution 3-m PlanetScope data for burnt area delineation. To capture the fire events, we also used the active fire data (VIIRS 375 m and MODIS collection 6.1) during the end of April till May 23, 2019, to identify the peak burning dates. The burnt areas derived from the PlanetScope were compared with the Landsat (30 m) data for relative comparison. The results identify important differences between the Landsat (30 m) and PlanetScope (3 m) data useful for burnt area mapping studies.

DATASETS

A. VIIRS active fires

The first VIIRS was launched in October 2011 aboard the Suomi National Polar-orbiting Partnership (S-NPP) satellite. The VIIRS instrument carries two separate sets of multispectral channels providing full global coverage at both 375 and 750 m nominal resolutions every 12 hours or less depending on the latitude. The VIIRS satellite incorporates fire-sensitive channels, including a dual-gain, high-saturation temperature 4-μm channel, enabling active fire detection and characterization. The active fire product, based on the 375-m-resolution I-bands and 750-m-moderate-resolution "M"-bands of VIIRS, is currently generated (Schroeder et al., 2014). In this study, we specifically used the VIIRS 375 m active fire product (VNP14IMG). The algorithm for this product builds on the well-established MODIS Fire and Thermal Anomalies product using a contextual approach to detect thermal anomalies. Due to its higher spatial resolution, the VNP14IMG active fire product captures more fire pixels than the MODIS MCDML product (Schroder et al., 2014). Specific to the fire radiative power (FRP), the VNP14IMG FRP is calculated through a combination of both VIIRS 375 and 750 m data. The former is used to identify fire-affected, cloud (solid blue), water (dashed blue), and valid background pixels. Then, colocated M13 channel radiance data (750 m) coinciding with fire pixels and valid background pixels are used in the FRP calculation. More details about the product can be found in Schroder et al. (2014).

B. MODIS collection 6 burnt area product (Objectives 1–3)

We used the MODIS collection 6.1 burnt area products to first assess the amount of burned area in the study area. As highlighted in Giglio (2020), the MCD64A1 collection 5 product was first released in late 2009 in conjunction with the version 3 Global Fire Emissions Database (GFED). Minor adjustments were made over the next several years, directed primarily at reducing (albeit only slightly) an excessive loss of small agricultural burns in croplands, culminating in the release of the global C5.1 MCD64A1 product in late 2012. The algorithm used for mapping is a hybrid algorithm that supplements daily surface reflectance imagery with daily active fire data. Compared to the C5 algorithm, the C6 algorithm-based MCD64A1 product is an improved one and supersedes the heritage Collection 5.1 (C5.1) MCD64A1 and MCD45A1 products. The new algorithm is designed to be extremely tolerant of cloud and aerosol contamination and offers better detection of small burns, burn-date temporal uncertainty reduction, and a large reduction in the extent of unmapped areas.

C. PlanetScope imagery

We explored the potential of PlanetScope (RGB, NIR) at 3 m resolution for mapping the burnt areas. Visual interpretation together with threshold-based classification and spatial analysis has been used to derive the burnt areas. The data have been compared with the MODIS burnt areas.

RESULTS AND DISCUSSION

A sample image of Landsat and PlanetScope for May 7, 2019, is shown in Figure 6.2a and b. Due to their higher resolution, PlanetScope data could capture relatively more variations in land cover features and burnt areas than the Landsat data. The results on MODIS and VIIRS active fires covering the PlanetScope scene for the Nainital study area from April 20 till May 23, 2020, are shown in Figure 6.3a. Relatively higher fire counts were captured by the VIIRS (375 m) data than by the MODIS (1 km) data. The peak fire was on May 2, 4, 5, and 6.

The Landsat-8 was available on April 21 and May 7, which were 16 days apart (Figure 6.3b and c). Most of the burning occurred during April 29 and May 6, 2020; thus, Landsat could not capture the burnt area dynamics. In contrast, with the daily PlanetScope coverage, nine different images were available from April 27

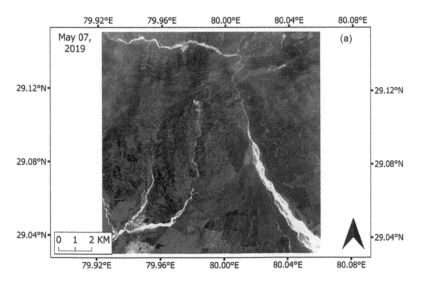

FIGURE 6.2   LANDSAT data for May 7 (a) on the top and corresponding PlanetScope data.

(*Continued*)

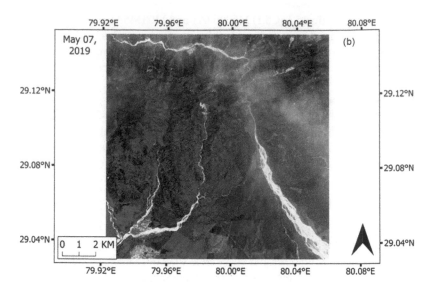

FIGURE 6.2 (CONTINUED) (b) at the bottom for the same date. Due to its higher resolution, the PlanetScope data could capture relatively more burnt areas and smoke variations than the Landsat data.

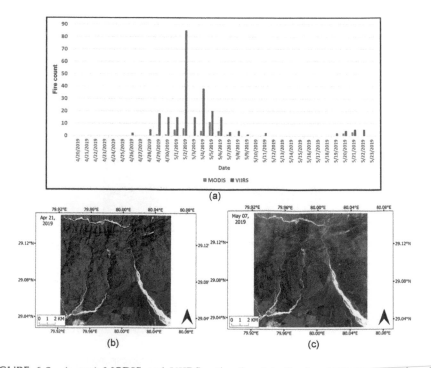

FIGURE 6.3 (a, top) MODIS and VIIRS active fire data for the study area covering the PlanetScope scene for different dates. LANDSAT scenes were available 16 days apart on April 21, 2019 (b, bottom left), and May 7, 2019 (c, bottom right). However, most of the burning occurred during April 29 and May 6, 2020; thus, Landsat could not capture the burnt area dynamics.

TABLE 6.1

Burnt Areas Derived from the PlanetScope Imagery Starting April 27, 2019, till May 19, 2019

Date	Burnt Area (km²)
April 27	2.39
April 30	8.18
May 1	4.15
May 4	26.19
May 5	17.93
May 7	35.69
May 8	44.46
May 13	30.30
May 19	19.16

Note that the burnt areas (km²) are for the sampled PlanetScope scene and not for entire Nainital, Uttarakhand, India.

till May 19. These data could capture burnt area progression with the maximum area of 35.69 km² on May 7 and 44.46 km² on May 8 (Table 6.1 and Figure 6.4). The PlanetScope data were not available for the other days of May 2, 3, and 6. Nevertheless, the PlanetScope data with a relatively higher number of scenes could capture more burnt areas. Although the increase in burnt areas was almost linear, on May 5, there was a decrease (17.93 km²), which was not expected compared to May 4 (26.19 km²). The decrease in the burnt areas on May 5 was attributed to the smoke that hindered the burnt area delineation by reducing the contrast with other land cover classes in the image. The burnt areas were higher on May 7 (35.69 km²), reaching a peak on May 8 (44.46 km²), which later decreased on May 13 (30.30 km²) and May 19 (19.16 km²). These results suggest the potential of PlanetScope data in capturing burnt areas more rigorously than the Landsat datasets.

The typical spectral reflectance of healthy vegetation is shown in Figure 6.5, suggesting that various factors govern the reflectance at different wavelengths. For example, in the visible region of the spectrum (400–700 nm; 1 micrometer (μm) = 1000 nanometers), chlorophyll (a and b) selectively absorbs blue (400–500 nm) and red (600–700 nm) light for photosynthesis and absorbs less over the green wavelengths (500–600 nm) and thus the green appearance of healthy vegetation. The yellow-to-orange-red pigment, carotene, has strong absorption in the blue wavelengths (400–500 nm). The red-and-blue pigment, xanthophyll, also absorbs strongly in the 400–500 nm range and is responsible for various deciduous leaf colors. In contrast to the visible domain (400–700 nm), where mostly pigments dominate the reflectance, in the near-infrared (NIR) region (700–1400 nm), the vegetation reflectance is most influenced by the spongy mesophyll and cellular structure of the leaves. In the shortwave infrared (SWIR) region, the water absorption dominates in the 1450 nm, 1950 nm, and 1250 nm (Figure 6.4). The overall reflectance in SWIR is governed by internal vegetation structure and water absorption. Typical healthy vegetation shows

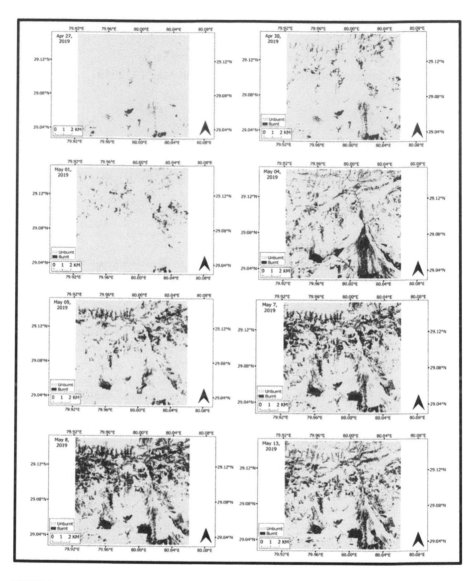

FIGURE 6.4 Burnt areas captured by the daily PlanetScope imagery starting April 27, 2019, till May 13, 2019, in Nainital, Uttarakhand, India. Due to the short repeat cycle, the PlanetScope data could capture relatively more burnt areas and dynamics than the Landsat data with a 16-day repeat cycle.

very high reflectance in the NIR and low reflectance in the SWIR region. Thus, when the vegetation is burnt, a reverse phenomenon can be observed, i.e., a decrease in NIR and an increase in SWIR; such differences can be used to distinguish burnt areas from the healthy vegetation. The popular NBR computed as NIR-SWIR/ NIR+SWIR is based on such a principle (Key and Benson, 1999; 2005). A high

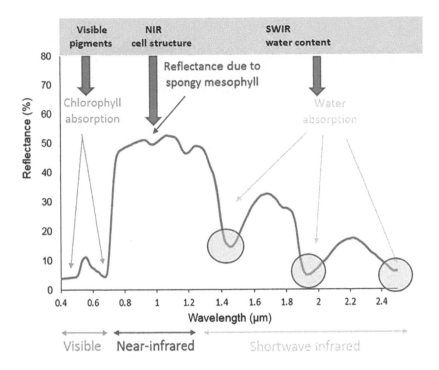

FIGURE 6.5 Spectral reflectance of healthy vegetation in the visible, near-infrared, and shortwave infrared regions of the electromagnetic spectrum.

NBR value indicates healthy vegetation, while a low value suggests bare ground and recently burnt areas. Non-burnt areas are normally attributed to values close to zero.

In contrast to the Landsat data, which have a SWIR band, the PlanetScope doesn't have a SWIR band; thus, for the delineation of the burnt areas, we had to rely on the red and NIR, which both show a decrease in reflectance in the burnt areas. NIR data from the PlanetScope imagery depict a decrease in NIR signal and slow vegetation recovery after fire for different dates in the study area (Figure 6.6).

The higher spatial and temporal resolution of the PlanetScope imagery was quite useful in delineating the burnt areas and capturing the progression more robustly than the Landsat data in our case. The results suggest that PlanetScope imagery can be effectively used to delineate burnt areas; however, it needs caution. For example, each PlanetScope tile typically covers $17 \times 21\,km^2$; thus, when analyzing the data for large areas, preprocessing steps might consume a significant amount of time. Despite such limitations, the PlanetScope data with the unique combination of daily coverage and high resolution can characterize changing ground conditions more rapidly (daily) than other datasets. In particular, in the areas affected by fires, robust information on burnt areas progression and ground conditions captured by the PlanetScope imagery can help fire prevention, suppression, and restoration; thus, the use of these datasets is highly recommended.

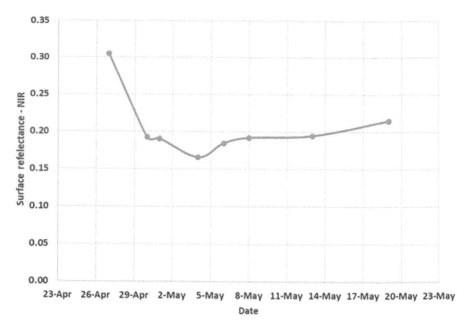

FIGURE 6.6 Burnt area signal retrieved from the PlanetScope data depicting a decrease in NIR and slow vegetation recovery after fire for different dates in Nainital, Uttarakhand, India.

ACKNOWLEDGMENTS

The authors thank the NASA LCLUC program for funding the South/Southeast Asia Research Initiative under which this study has been performed.

REFERENCES

Badarinath, K.V.S., Kharol, S.K., Krishna Prasad, V., Kaskaoutis, D.G., and Kambezidis, H.D. 2008a. Variation in aerosol properties over Hyderabad, India during intense cyclonic conditions. *International Journal of Remote Sensing.* 29(15), 4575–4597.

Badarinath, K.V.S., Kharol, S.K., Prasad, V.K., Sharma, A.R., Reddi, E.U.B., Kambezidis, H.D., and Kaskaoutis, D.G. 2008b. Influence of natural and anthropogenic activities on UV Index variations–a study over tropical urban region using ground based observations and satellite data. *Journal of Atmospheric Chemistry.* 59(3), 219–236.

Biswas, S., Lasko, K.D., and Vadrevu, K.P. 2015a. Fire disturbance in tropical forests of Myanmar—Analysis using MODIS satellite datasets. *IEEE Journal of Selected Topics in Applied Earth Observations and Remote Sensing.* 8(5), 2273–2281.

Biswas, S., Vadrevu, K.P., Lwin, Z.M., Lasko, K., and Justice, C.O. 2015b. Factors controlling vegetation fires in protected and non-protected areas of Myanmar. *PLoS One.* 10(4), e0124346.

Chuvieco, E., Cocero, D., Riano, D., Martin, P., Martınez-Vega, J., de la Riva, J., and Pérez, F. 2004. Combining NDVI and surface temperature for the estimation of live fuel moisture content in forest fire danger rating. *Remote Sensing of Environment.* 92(3), 322–331.

Chuvieco, E., Sandow, C., Günther, K.P., González-Alonso, F., Pereira, J.M., Pérez, O., Bradley, A.V., Schultz, M., Mouillot, F., and Ciais, P., 2012. Global burned area mapping from European satellites: The ESA fire_CCI project. *International Archives of the Photogrammetry, Remote Sensing and Spatial Information Sciences.* 39(B8), 13–16.

Chuvieco, E., Lizundia-Loiola, J., Pettinari, M.L., Ramo, R., Padilla, M., Tansey, K., Mouillot, F., Laurent, P., Storm, T., Heil, A., and Plummer, S., 2018. Generation and analysis of a new global burned area product based on MODIS 250 m reflectance bands and thermal anomalies. *Earth System Science Data.* 10(4), 2015–2031.

Epting, J., Verbyla, D., and Sorbel, B., 2005. Evaluation of remotely sensed indices for assessing burn severity in interior Alaska using Landsat TM and ETM+. *Remote Sensing of Environment.* 96(3–4), 328–339.

Giglio, L., Schroeder, W., and Justice, C.O. 2016. The collection 6 MODIS active fire detection algorithm and fire products. *Remote Sensing of Environment.* 178, 31–41.

Giglio, L., Boschetti, L., Roy, D.P., Humber, M.L., and Justice, C.O., 2018. The Collection 6 MODIS burned area mapping algorithm and product. *Remote Sensing of Environment,* 217, 72–85.

Giglio, L., Boschetti, L., Roy, D, Hoffmann, A., and Humber, M., 2020. MCD64 Burned Area Collection 6 User's Guide v1.3 (December 2020). https://modis-fire.umd.edu/guides.html

Gupta, P.K., Prasad, V.K., Sharma, C., Sarkar, A.K., Kant, Y., Badarinath, K.V.S., and Mitra, A.P. 2001. CH$_4$ emissions from biomass burning of shifting cultivation areas of tropical deciduous forests–experimental results from ground-based measurements. *Chemosphere-Global Change Science.* 3(2), 133–143.

Hayasaka, H., Noguchi, I., Putra, E.I., Yulianti, N., and Vadrevu, K. 2014. Peat-fire-related air pollution in Central Kalimantan, Indonesia. *Environmental Pollution.* 195, 257–266. DOI: 10.1016/j.envpol.2014.06.031.

Inoue, Y. 2018. Ecosystem carbon stock, atmosphere and food security in slash-and-burn land use: A geospatial study in mountainous region of Laos. In: Vadrevu, K.P., Ohara, T., and Justice, C. (Eds). *Land-Atmospheric Research applications in South/Southeast Asia.* Springer, Cham. pp. 641–666.

Israr, I., Jaya, S.N.I., Saharjo, H.S., Kuncahyo, B., and Vadrevu, K.P. 2018. Spatiotemporal analysis of land and forest fires in Indonesia using MODIS active fire dataset. In: Vadrevu, K.P., Ohara, T., and Justice, C. (Eds). *Land-Atmospheric Research Applications in South/Southeast Asia.* Springer, Cham. pp. 105–128.

Justice, C., Gutman, G., and Vadrevu, K.P. 2015. NASA land cover and land use change (LCLUC): An interdisciplinary research program. *Journal of Environmental Management.* 148(15), 4–9.

Kant, Y., Ghosh, A.B., Sharma, M.C., Gupta, P.K., Prasad, V.K., Badarinath, K.V.S., and Mitra, A.P. 2000. Studies on aerosol optical depth in biomass burning areas using satellite and ground-based observations. *Infrared Physics & Technology.* 41(1), 21–28.

Key, C.H., and Benson, N.C. 1999. The Normalized Burn Ratio, a Landsat TM radiometric index for burn severity. Available online at: http://nrmsc.usgs.gov/research/nbr.htm.

Key, C.H., and Benson, N.C. 2005. *Landscape Assessment: Remote Sensing of Severity, the Normalized Burn Ratio and Ground Measure of Severity, the Composite Burn Index.* FIREMON: Fire Effects Monitoring and Inventory System. USDA Forest Service, Rocky Mountain Res. Station, Ogden, Utah.

Lasko, K., and Vadrevu, K.P. 2018. Improved rice residue burning emissions estimates: Accounting for practice-specific emission factors in air pollution assessments of Vietnam. *Environmental Pollution.* 236(5), 795–806.

Lasko, K., Vadrevu, K.P., Tran, V.T., Ellicott, E., Nguyen, T.T., Bui, H.Q., and Justice, C. 2017. Satellites may underestimate rice residue and associated burning emissions in Vietnam. *Environmental Research Letters.* 12(8), 085006.

Lasko, K., Vadrevu, K.P., and Nguyen, T.T.N. 2018a. Analysis of air pollution over Hanoi, Vietnam using multi-satellite and MERRA reanalysis datasets. *PLoS One.* 13(5), e0196629.

Lasko, K., Vadrevu, K.P., Tran, V.T., and Justice, C. 2018b. Mapping double and single crop paddy rice with Sentinel-1A at varying spatial scales and polarizations in Hanoi, Vietnam. *IEEE Journal of Selected Topics in Applied Earth Observations and Remote Sensing.* 11(2), 498–512.

Lentile, L.B., Holden, Z.A., Smith, A.M., Falkowski, M.J., Hudak, A.T., Morgan, P., Lewis, S.A., Gessler, P.E., and Benson, N.C. 2006. Remote sensing techniques to assess active fire characteristics and post-fire effects. *International Journal of Wildland Fire.* 15(3), 319–345.

Oanh, N.T.K., Permadi, D.A., Dong, N.P., and Nguyet, D.A. 2018. Emission of toxic air pollutants and greenhouse gases from crop residue open burning in Southeast Asia. In: Vadrevu, K.P., Ohara, T., and Justice, C. (Eds). *Land-Atmospheric Research Applications in South/Southeast Asia.* Springer, Cham. pp. 47–68.

Prasad, V.K., and Badarinth, K.V.S. 2004. Land use changes and trends in human appropriation of above ground net primary production (HANPP) in India (1961–98). *Geographical Journal.* 170(1), 51–63.

Prasad, V.K., Gupta, P.K., Sharma, C., Sarkar, A.K., Kant, Y., Badarinath, K.V.S., Rajagopal, T., and Mitra, A.P. 2000. NO_x emissions from biomass burning of shifting cultivation areas from tropical deciduous forests of India–estimates from ground-based measurements. *Atmospheric Environment.* 34(20), 3271–3280.

Prasad, V.K., Kant, Y., and Badarinath, K.V.S. 2001. CENTURY ecosystem model application for quantifying vegetation dynamics in shifting cultivation areas: A case study from Rampa Forests, Eastern Ghats (India). *Ecological Research.* 16(3), 497–507.

Prasad, V.K., Kant, Y., Gupta, P.K., Elvidge, C., and Badarinath, K.V.S. 2002. Biomass burning and related trace gas emissions from tropical dry deciduous forests of India: A study using DMSP-OLS data and ground-based measurements. *International Journal of Remote Sensing.* 23(14), 2837–2851.

Prasad, V.K., Lata, M., and Badarinath, K.V.S. 2003. Trace gas emissions from biomass burning from northeast region in India—estimates from satellite remote sensing data and GIS. *Environmentalist.* 23(3), 229–236.

Prasad, V.K., Badarinath, K.V.S., and Eaturu, A. 2008. Biophysical and anthropogenic controls of forest fires in the Deccan Plateau, India. *Journal of Environmental Management.* 86(1), 1–13.

Roy, D., Jin, Y., Lewis, P., and Justice, C. 2005. Prototyping a global algorithm for systematic fire-affected area mapping using MODIS time-series data. *Remote Sensing Environment.* 97, 137–162.

Schroeder, W., Oliva, P., Giglio, L., and Csiszar, I.A. 2014. The New VIIRS 375 m active fire detection data product: Algorithm description and initial assessment. *Remote Sensing of Environment.* 143, 85–96.

Sigsgaard, T., Forsberg, B., Annesi-Maesano, I., Blomberg, A., Bølling, A., Boman, C., Bønløkke, J., Brauer, M., Bruce, N., Héroux, M.E., and Hirvonen, M.R. 2015. Health impacts of anthropogenic biomass burning in the developed world. *European Respiratory Journal.* 46(6), 1577–1588.

Vadrevu, K.P. 2008. Analysis of fire events and controlling factors in eastern India using spatial scan and multivariate statistics. *Geografiska Annaler: Series A, Physical Geography.* 90(4), 315–328.

Vadrevu, K.P., and Justice, C.O. 2011. Vegetation fires in the Asian region: Satellite observational needs and priorities. *Global Environmental Research.* 15(1), 65–76.

Vadrevu, K.P., and Lasko, K.P., 2015. Fire regimes and potential bioenergy loss from agricultural lands in the Indo-Gangetic Plains. *Journal of Environmental Management.* 148, 10–20.

Vadrevu, K.P., and Lasko, K. 2018. Intercomparison of MODIS AQUA and VIIRS I-Band fires and emissions in an agricultural landscape—Implications for air pollution research. *Remote Sensing.* 10(7), 978. doi:10.3390/rs10070978.

Vadrevu, K.P., Eaturu, A., and Badarinath, K.V.S. 2006. Spatial distribution of forest fires and controlling factors in Andhra Pradesh, India using spot satellite datasets. *Environmental Monitoring and Assessment.* 123(1–3), 75–96.

Vadrevu, K.P., Badarinath, K.V.S., and Anuradha, E. 2008. Spatial patterns in vegetation fires in the Indian region. *Environmental Monitoring and Assessment.* 147(1–3), 1. doi:10.1007/s10661-007-0092-6.

Vadrevu, K.P., Csiszar, I., Ellicott, E., Giglio, L., Badarinath, K.V.S., Vermote, E., and Justice, C. 2012. Hotspot analysis of vegetation fires and intensity in the Indian region. *IEEE Journal of Selected Topics in Applied Earth Observations and Remote Sensing.* 6(1), 224–238.

Vadrevu, K.P., Giglio, L., and Justice, C. 2013. Satellite based analysis of fire–carbon monoxide relationships from forest and agricultural residue burning (2003–2011). *Atmospheric Environment.* 64, 179–191.

Vadrevu K.P., Ohara T., and Justice, C. 2014a. Air pollution in Asia. *Environmental Pollution.* 12, 233–235.

Vadrevu, K.P., Lasko, K., Giglio, L., and Justice, C. 2014b. Analysis of Southeast Asian pollution episode during June 2013 using satellite remote sensing datasets. *Environmental Pollution.* 12, 245–256.

Vadrevu, K.P., Lasko, K., Giglio, L., and Justice, C. 2015. Vegetation fires, absorbing aerosols and smoke plume characteristics in diverse biomass burning regions of Asia. *Environmental Research Letters.* 10(10), 105003.

Vadrevu, K.P., Ohara, T., and Justice, C. 2017. Land cover, land use changes and air pollution in Asia: A synthesis. *Environmental Research Letters.* 12(12), 120201.

Vadrevu, K.P., Ohara, T., and Justice, C. 2018. *Land-Atmospheric Research Applications in South and Southeast Asia.* Springer, Cham. 725 pages.

Vadrevu, K.P., Lasko, K., Giglio, L., Schroeder, W., Biswas, S., and Justice, C. 2019. Trends in vegetation fires in south and Southeast Asian countries. *Scientific Reports,* 9(1), 7422. doi:10.1038/s41598-019-43940-x.

Van Wagtendonk, J.W., Root, R.R., and Key, C.H. 2004. Comparison of AVIRIS and Landsat ETM+ detection capabilities for burn severity. *Remote Sensing of Environment.* 92(3), 397–408.

7 Investigations on Land and Forest Fires in the North Indian Region over a Decade

Narendra Singh and Ashish Kumar
Aryabhatta Research Institute of Observational
Sciences (ARIES), India

CONTENTS

INTRODUCTION

Right through the beginning of human civilization, fires have played a distinct role in changing the landscapes and forest cover (Pausas and Keeley, 2009). Natural or man-made burning of dead or living vegetation that includes crop residues, grassland, forest, and burning of biomass for fuel, significantly contributes to the emission of a variety of gaseous and particulate pollutants in the atmosphere. Among the agriculture or forest fires around the world, savanna fires are one of the biggest sources of biomass burning and representing about 43% of aggregate worldwide emissions, while burning of agricultural wastes and forest fire contribute 23% and 18% of the aggregate emission, respectively (Yadav and Devi, 2018). Amazon rainforest fires of 2019 are another big example over decades which might have severely affected the regional air quality, global atmospheric chemistry, biodiversity, earth's radiative budget, and biogeochemical cycles (Crutzen and Andreae, 1990; Escobar, 2019) together with the persistent seasonal fires from different regions causing complex physical and chemical processes in the atmosphere. The seasonal fires from a number of places are being observed with the available remote sensing techniques mainly through satellites such as Moderate Resolution Imaging Spectroradiometer (MODIS), Visible and Infrared Scanner (VIRS), Along-Track Scanning Radiometer (ATSR), and Spinning Enhanced Visible and Infrared Imager (SEVIRI) (Kasischke et al., 2003; Calle et al., 2006; Giglio et al., 2000, 2016) and have been presented in various perspectives such as impacts on health, agriculture, and environment. Some of the hotspots of seasonal fires over the globe are South America, Africa's savanna (Andreae et. al., 1998), parts of the Australian continent, and the Asian region. South and Southeast Asian countries are also locations of fires and biomass burning that emit pollutants with huge impact through transport of transboundary air pollution. A recent study reported that in comparison with other South Asian countries, India has experienced the highest number of annual agricultural and recurrent forest fires (Vadrevu et al., 2019).

During the last few decades, the increasing population, the increasing demand of habitats and agricultural land, and the requirement of industries have been contributing to deforestation and thereby reducing the carbon sink as well. Among other countries across the globe, India, particularly the densely populated northern region, is also suffering from rapid shrinkage and denudation of forest covers due to high rate of deforestation, anthropogenic drainage, and repeated fire incidents. In the last few decades, the repeated fires in the north Indian states, primarily triggered due to uncontrolled burning of the crop residues, bushes, or vegetated lands or forests, including land encroachments using fire, have significantly contributed to the release of tons of hazardous trace gases in the air and altered the air quality and affected human health over the region. These pollutants potentially get transported to the Himalayan region through upslope winds developing along the mountain systems particularly in the morning hours, as well as with the vertical mixing in boundary layer, which subsequently move with mean flow to the other locations. As a consequence, the Himalayan ecosystem has spiflicated and threatened a large number of fragile and sensitive species and has affected floral wealth by reducing the pollination and flowering processes, and their seeding, hence resulting in natural imbalances and nonproductive biological regimes (e.g., Pant et al., 2018; Negi and Rawal, 2019). It is also frequently noticed that the implications of seasonal fires are severe on the air quality of Delhi and National Capital Region (NCR), particularly during the winter fires of Punjab and Haryana, when lower atmospheric dynamics plays a vital role in sustaining the smog for longer periods around this region. Therefore, it is essential to understand the distribution and intensity of the active fire/hotspots over this region as well, in order to plan the control and remedial actions to address and mitigate the future fires.

DATA AND METHODS

Study Region

Despite the worldwide growing body of information on biomass burning and pollutants, only a modest amount of data from the Himalayan location, which is also a forest fire-prone region, is available due to logistic difficulties and complex terrain for deployment of ground-based monitoring stations. An attempt is made here to study and present the decadal variability of land and forest fires over the north Indian region using satellite-derived products. Considering the importance of the Himalayan ecosystem to the global atmospheric regimes and sustainability of hydrological cycles of Indian subcontinent, this study also investigates the transport process, as how the pollutants resulting from fire events are taken to the higher altitudes in the Himalayas.

The study area encompasses the region between 24–38°N and 68–85°E that includes Punjab, Haryana, southern slopes of the Himalayan region, and adjacent foothill plain regions which are the part of Indo-Gangetic plains (IGP) as represented by Figure 7.1.

Based on the Tropical Rainfall Measuring Mission (TRMM) estimate (2010–2019), the selected region receives an average rainfall of ~50 mm/month with highest rainfall during July and August (~150 mm/month average). According to Modern-Era Retrospective Analysis for Research and Applications-2 (MERRA-2) reanalysis model, the surface wind speed varies little throughout the year with monthly mean of 4–5 m/s, while the monthly mean surface temperature varies between 1.2°C (January) and 26.5°C (June). The westerly and northwesterly winds, in general, are quite common throughout the year over this region.

DATA SOURCE

For the broader understanding on spatiotemporal distribution of hotspots over the region considered for this study, the satellite-derived data product MCD14ML from

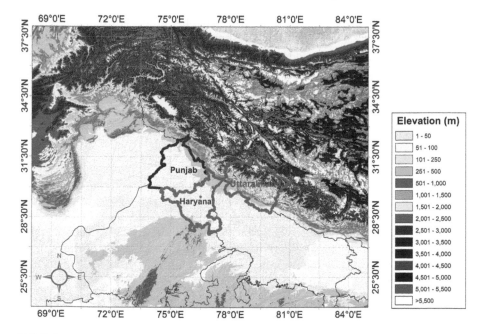

FIGURE 7.1   Topographic map of the study region (24–38°N and 68–85°E).

MODIS is used for the period from January 2010 to December 2019. This MODIS-detected Terra and Aqua sensor product provides the month-wise detection of active fire/hotspots at 1-km^2 footprint, and the detection is based on emission of the targets at about 4- and 11-μm wavelength bands (Giglio et al., 2009, 2016). The data used are extracted from the National Aeronautics and Space Administration (NASA) Fire Information for Resource Management System (FIRMS) database in the form of monthly files containing the geographic locations, date, brightness temperature, and fire radiative power (FRP), which describes the energy radiated by the fire per unit time in megawatts (MW).

The MCD14ML product has the potential to estimate the burned areas and can be used as a proxy in biomass burning-related investigation over larger regions (Devisscher et al., 2016). Additionally, it is also used as a critical input in different regional and global models (Giglio et al., 2006; Devisscher et al., 2016). Despite the wide application of the MCD14ML product, it possesses certain limitations and uncertainties. One possible limitation is that it does not provide the information on the source of fire that could be due to land surface conversion, maintenance, and wildfire (Devisscher et al., 2016). Further, it only provides the information of geographical coordinates where the fire had occurred and not on the burnt areas over which fire took place. The sources of error in active fire detection from MODIS sensor may be attributed to the prevailing atmospheric or environmental conditions (e.g., cloud cover, warm days, heavy smoke) at the time of detection, or sometimes the error is introduced due to detection from the exposed hot surfaces or obscuring surfaces (e.g., rocks, sandy soils, tree canopy). The errors associated with these sources when using the MCD14ML product could be minimized by considering only the active fire pixels at high confidence in the analysis (Giglio et al., 2009, 2016; Devisscher et al., 2016). Therefore, in this study only the occurrences of fire events with high confidence (80%–100%) levels from the metadata provided are taken into account for analysis, whereas the low (< 30%)- and nominal (30%–79%)-confidence cases have been discarded to reduce the uncertainty.

RESULTS AND DISCUSSION

SEASONAL ACTIVE FIRE/HOTSPOTS

A preliminary analysis on active fire pixels in the defined geographical region for the period 2010–2019 revealed that at a high confidence (80%–100%) level, about 85% of fire detections among all available datasets have collectively occurred only during April–May (summer) and October–November (post-monsoon), with considerable interannual variability. The active fire/hotspots detected in these contrasting months of two different seasons for the entire period of study (except October 2019 for which the fire data are not available) are shown in Figure 7.2a–d.

It is evident from Figure 7.2a–d that during summer, the fire counts are observed to be highest in major portion or subregions of Punjab, Haryana, and the plains of Uttarakhand over the decade 2010–2019. In summer, except over a portion of southwest Haryana, the fire pixels are widespread in entire Punjab and Haryana, which

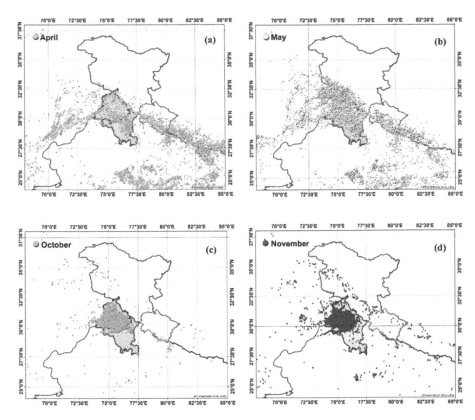

FIGURE 7.2 MODIS MCD14ML (combined Aqua and Terra) high-confidence (≥80%) active fire detected over the study region during 2010–2019 for (a) April (beryl green dots), (b) May (yellow dots), (c) October (green dots), and (d) November (red dots). The active fire/ hotspot data are not available for October 2019.

mainly consist of more than 60% of agriculture land. The intense fire episodes in these states are mainly attributed to crop residue burning, as the farmers in these states, before sowing the seeds, generally, clear their fields by burning the residues/leftover straws of last cropping, e.g., wheat crop in summer (Kumar et al., 2018). However, summer burning in Uttarakhand results from the agricultural fires and uncontrolled forest fires in the southern slopes of Uttarakhand Himalaya. In Uttarakhand, more than 48% of the lands are covered with medium-to-dense and open forests consisting of mainly oak, rhododendron, and chir-pine trees whose foliage has the tendency to catch fire easily during summer when favorable atmospheric conditions, primarily high-temperature, low-moisture, and high-wind conditions, prevail. Few independent studies on the forest fire in Uttarakhand alone have confirmed its recurrent occurrence during summer (e.g., Singh et al., 2016). This repeated forest fire in April and May as seen in Figure 7.2a and b ravages the rich variety of flora and fauna every year, hence causing long-term alterations to the Himalayan ecosystem.

Similarly, post-monsoon (October and November), the fire appears to be widespread in entire Punjab and over a small portion of northwestern Haryana, which may again be considered to be originating from the rice paddy burning in this region. Nevertheless, there could be some other fire sources such as industries, and the fire events in these states cannot be truly linked to agriculture-based burning. However, past studies have reported that more than 65% of the annual fire events occurring in these areas come from agricultural fires (Singh and Panigrahy, 2011; Vijayakumar et al., 2016). Post-monsoon, almost negligible active fires were observed in the state of Uttarakhand.

A quantitative picture on the interannual variability of the total number of active fire/hotspot counts that were detected in the three mentioned states in months of April–May and October–November, over the decade 2010–2019, is given in Figure 7.3a–d. It is noticeable that Uttarakhand witnessed the highest number of fires during April 2016 as compared to Punjab and Haryana, whereas a consistently high number of fires are observed in Punjab over the entire period of study. Since 2014, the month of May is showing a rising trend in fire counts for Punjab, while a decreasing trend is seen in the month of October, and again high fire counts are seen in November.

The statistics on FRP associated with the active fire detected in three states (Punjab, Haryana, and Uttarakhand) during summer and in Punjab and Haryana post-monsoon over the period 2010–2019 is presented in Table 7.1. From the table, it appears that in a decade (2010–2019), the maximum fire events over the north Indian region have occurred in 2016, during which a total of 3969 hotspots ($N_{\text{Aqua+Terra}}$) were detected during summer and post-monsoon. The average FRP associated with these hotspots is about 39.1 ± 28.3 MW, while the highest FRP (518.7 MW) is observed in Haryana during summer.

Implication of Agricultural Burning and Forest Fire to the Himalayan Region

As discussed in the earlier section, Punjab and Haryana are the hotspots of agricultural fire in summer and post-monsoon; however, Uttarakhand is the hotspot

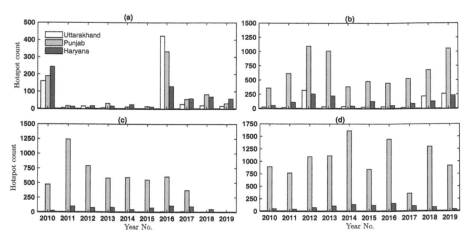

FIGURE 7.3 Histogram showing the total number of MCD14ML (combined Aqua and Terra) high-confidence (≥80%) hotspots during (a) April, (b) May, (c) October, and (d) November over the decade 2010–2019 for Uttarakhand, Punjab, and Haryana states. The active fire/hotspot data are not available for October 2019.

TABLE 7.1

Seasonal Variation, over the Period 2010–2019, in the MCD14ML (Combined Aqua and Terra) High-Confidence (≥ 80%) Hotspots (NAqua + Terra) and Associated Mean ± Standard Deviation (SD) and Maxima of Fire Radiative Power (FRP) in Megawatts (MW) for Punjab, Haryana, and Uttarakhand

Seasons		Summer			Post-monsoon	
		Punjab	*Haryana*	*Uttarakhand*	*Punjab*	*Haryana*
2010	$N_{Aqua+Terra}$	548	297	184	1364	71
	Mean ± SD	38.4 ± 32.8	43.3 ± 35	35.9 ± 29.7	33.6 ± 15.8	32.1 ± 12.8
	Maximum	341.2	256.6	189.9	158.6	84.9
2011	$N_{Aqua+Terra}$	632	122	25	2007	140
	Mean ± SD	38.1 ± 26.5	40.7 ± 33.9	28 ± 13.6	35.7 ± 18.4	31.2 ± 11.4
	Maximum	196.7	287.7	68.8	197.5	76.4
2012	$N_{Aqua+Terra}$	1100	273	336	1884	147
	Mean ± SD	46.1 ± 35.3	44.5 ± 43.3	33 ± 27.3	34.5 ± 16.7	33.8 ± 18.4
	Maximum	359	312.6	244	141.8	131.2
2013	$N_{Aqua+Terra}$	1041	237	37	1694	187
	Mean ± SD	44.7 ± 41.4	40.1 ± 26.2	24.4 ± 15.2	33.4 ± 17.1	34.6 ± 19.7
	Maximum	560.8	191.9	69	231.4	183.4
2014	$N_{Aqua+Terra}$	395	70	39	2205	180
	Mean ± SD	34.9 ± 19.5	34.9 ± 27.1	34.4 ± 27.5	38.4 ± 21.2	35.5 ± 17
	Maximum	162.7	170.9	121.4	287.3	128.8
2015	$N_{Aqua+Terra}$	492	136	27	1393	191
	Mean ± SD	33.4 ± 19.9	42.3 ± 27.7	30.3 ± 39.9	35.5 ± 19.7	34.8 ± 22.5
	Maximum	124	177	181.8	173.7	184.2
2016	$N_{Aqua+Terra}$	780	184	457	2283	265
	Mean ± SD	43.3 ± 34.8	46.1 ± 45.2	38.3 ± 33.4	36.9 ± 19.5	41.9 ± 41.2
	Maximum	387.4	518.7	300.9	221.3	498
2017	$N_{Aqua+Terra}$	581	148	48	949	198
	Mean ± SD	38.2 ± 30.6	43.4 ± 35.6	34.9 ± 29	33.7 ± 18.8	33.6 ± 14.2
	Maximum	383.8	300.6	145.4	192	90.4
2018	$N_{Aqua+Terra}$	761	196	236	1658	130
	Mean ± SD	43.1 ± 41.2	46.1 ± 40.8	38.9 ± 38.3	34 ± 17.2	35.5 ± 23.9
	Maximum	569.5	308.7	303	230.2	197.5
2019	$N_{Aqua+Terra}$	1084	289	277	914	38
	Mean ± SD	47.3 ± 46.6	57.7 ± 55.8	29.4 ± 20.6	36.1 ± 18.9	37.5 ± 28.4
	Maximum	538.8	461.4	163.5	162.8	141.1

The relevant data are missing for October 2019.

for summer forest fire. The fire occurring during these seasons not only affects the regional air quality but also has adverse impacts at the distant locations through transport processes in the atmosphere. The smoke plumes, resulting from agricultural burning and forest fire in the north Indian states, contain the fine-mode aerosols that are lifted up into the boundary layer and with the prevailing winds subsequently get transported to the surrounding regions including the IGP, foothill Himalaya, and toward high-altitude locations in Himalayan Uttarakhand (e.g., Kumar et al., 2018). Very often the smoke plumes, together with the urban and industrial pollutions, reach the southern slopes of Himalaya and contribute to the formation of haze, and hence reducing the visibility, in the Himalayan region (e.g., Ramanathan and Ramana, 2005). A typical case study presented in Figure 7.4a and b), from Aryabhatta Research Institute of Observational Sciences (ARIES) (29.36° N, 79.46°E), Nainital, Uttarakhand, a high-altitude station (~1958 m above mean sea level) in the central Himalayan region with ground-based light detection and ranging (LiDAR) observations (Kumar et al., 2018) on November 2, 2011 (20:00–23:30 hours local time). The figure clearly displays the high aerosol loading in the atmosphere up to about 2 km above ground level (AGL), with columnar mean aerosol extinction and aerosol optical depth (AOD) of $0.16 \pm 0.03\,km^{-1}$ and 0.34 ± 0.02, respectively. As a result of heavy aerosol loading in the vertical on November 2, 2011, on the subsequent day a cover of dense haze is quite visible in the valley region and just above the site which is spreading over the southern valley and foothills of the Himalayas, shown by the image captured from MODIS Terra satellite on November 3, 2011 (~05:15 hours local time), in Figure 7.4c.

Besides the reduction in visibility, the detrimental effects of transport of smoke plumes are mainly the increase in the production of greenhouse gases, changes in the air quality, and the impact on health and environment at regional and continental scales (e.g., Sarangi et al., 2014; Kumar et al., 2018). Among the by-products of agricultural and forest fires, carbon monoxide (CO) is the most active component to contribute in both air pollution and climate change (Sarangi et al., 2014) and particularly plays a critical role in reducing the hydroxide ion (OH⁻) concentrations. It has the residence time of even more than 60 days and gets easily transported to long distances. It also affects methane (CH_4), carbon dioxide (CO_2), and tropospheric ozone (O_3). The transported CO reaching the surrounding regions and high-altitude locations in the Himalayas may also affect human health by creating respiratory problems, help developing heart diseases, and reduce the efficiency of vital organs (e.g., Pope and Dockery, 2006; Ding et al., 2015).

Considering the harmful effects of crop residue and forest burning, and to preserve the environment for future generations, it becomes essential that the community, especially the villagers and farmers, should be made aware of the negative consequences. At the same time, the technocrats and scientists should join a common platform to promote the sustainable technologies or alternate clean mechanisms to mitigate this issue. Through awareness campaigns, government initiatives, viable schemes, and training programs, the people concerned must be trained on the sustainable agricultural practices by which they can reduce their environmental footprints.

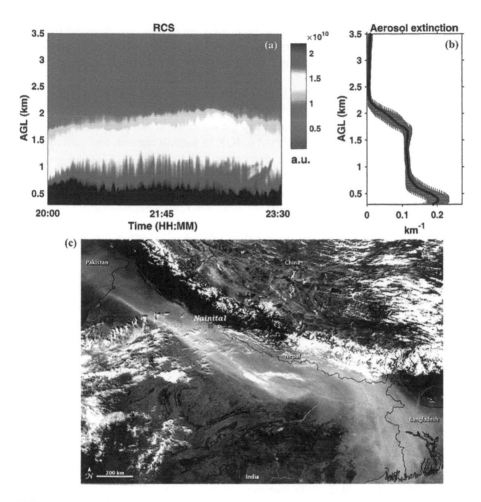

FIGURE 7.4 (a) Height–time contour plot of the LiDAR range-corrected signal (RCS) of November 2, 2011 (20:00–23:30 hours. local time), from ARIES, Nainital. Vertical and temporal resolutions are 15 m and 120 seconds, respectively. (b) Corresponding aerosol extinction profile showing mean (solid black) and standard deviation (dark blue bars) for the profiles captured between 20:00 and 23:30 hours on November 2, 2011, and (c) MODIS Terra satellite image of November 3, 2011, at ~05:15 hours local time showing the haze stretching from Pakistan southeastward to Bangladesh (NASA image courtesy MODIS Rapid Response Team at NASA GSFC, caption by Michon Scott).

CONCLUSION

The agricultural and forest fires in the north Indian states, originating mainly from Punjab, Haryana, and Uttarakhand, are the major sources for the atmospheric gaseous and particulate pollutants during summer (April and May) and post-monsoon (October and November). The analysis on the seasonal hotspots and their intensity in terms of FRP (MW) over a decade confirmed the recurrent occurrence of such fires in these states. During summer, the highest active fires are observed in entire Punjab and the subregions of Haryana and Uttarakhand, while post-monsoon, the hotspots are observed to be widespread in Punjab and over a portion of northwestern Haryana. The analysis revealed that the highest fire occurrences took place in the north Indian region during 2016 since 2010. A rising trend in active fire detection is observed during the month of May in Punjab, while a decreasing trend is seen during October. The pollutants resulting from the agricultural and forest fires, under the prevailing winds, often get transported to IGP, to foothill regions, and subsequently to the higher altitudes in the Himalayan region and significantly impact visibility and impose substantial health and environmental costs. Taking into account the fire emission characteristics and meteorological conditions, more extensive studies in this direction are further required for effective mitigation strategies in order to achieve sustainable development.

ACKNOWLEDGMENTS

The authors gratefully acknowledge the use of data and Rapid Response imagery from the Land, Atmosphere Near real-time Capability for EOS (LANCE) FIRMS operated by NASA's Earth Science Data and Information System (ESDIS) with funding provided by NASA Headquarters. The authors are thankful to those responsible for the creation of the TRMM and MERRA-2 data used in the study. Thanks are also due to the Director and colleagues at ARIES, Nainital, for their constant support and encouragement.

REFERENCES

Calle, A., Casanova, J. L., Romo, A., 2006. Fire detection and monitoring using MSG Spinning Enhanced Visible and Infrared Imager (SEVIRI) data. *J. Geophys. Res.* 111: G04S06. DOI: 10.1029/2005JG000116.

Crutzen, P.J., Andreae, M.O., 1990. Biomass burning in the tropics: impact on atmospheric chemistry and biogeochemical cycles. *Science.* 250(4988): 1669–1678. DOI: 10.1126/science.250.4988.1669.

Devisscher, T., Anderson, L. O., Aragão, L. E., Galván, L., Malhi, Y., 2016. Increased wildfire risk driven by climate and development interactions in the Bolivian Chiquitania, Southern Amazonia. *PLoS One.* 11(9): e0161323. DOI: 10.1371/journal.pone.0161323.

Ding, K., Liu, J., Ding, A., Liu, Q., Zhao, T. L., Shi, J., Han, Y., Wang, H., Jiang, F., 2015. Uplifting of carbon monoxide from biomass burning and anthropogenic sources to the free troposphere in East Asia. *Atmos. Chem. Phys.* 15: 2843–2866. DOI: 10.5194/acp-15-2843-2015.

Escobar, H., 2019. Amazon fires clearly linked to deforestation, scientists say. *Science.* 365(6456): 853. DOI: 10.1126/science.365.6456.853.

Giglio, L., Kendall, J.D., Tucker, C.J., 2000. Remote sensing of fires with the TRMM VIRS. *Int. J. Remote Sens.* 21: 203–207. DOI: 10.1080/014311600211109.

Giglio, L., Csiszar, I., Justice, C.O., 2006. Global distribution and seasonality of active fires as observed with the Terra and Aqua Moderate Resolution Imaging Spectroradiometer (MODIS) sensors. *J. Geophys. Res.* 111: 1–12. DOI: 10.1029/2005JG000142.

Giglio, L., Loboda, T., Roy, D.P., Quayle, B., Justice, C.O., 2009. An active-fire based burned area mapping algorithm for the MODIS sensor. *Remote Sens. Environ.* 113: 408–420. DOI: 10.1016/j.rse.2008.10.006.

Giglio, L., Schroeder, W., Justice, C.O., 2016. The collection 6 MODIS active fire detection algorithm and fire products. *Remote Sens. Environ.* 178: 31–41. DOI: 10.1016/j.rse.2016.02.054.

Kasischke, E.S., Hewson, J.H., Stocks, B., van der Werf, G., Randerson, J., 2003. The use of ATSR active fire counts for estimating relative patterns of biomass burning – a study from the boreal forest region. *Geophys. Res. Lett.* 30. DOI: 10.1029/2003GL017859.

Kumar, A., Singh, N., Anshumali, Solanki, R., 2018. Evaluation and utilization of MODIS and CALIPSO aerosol retrievals over a complex terrain in Himalaya. *Remote Sens. Environ.* 206: 139–155. DOI: 10.1016/j.rse.2017.12.019.

Negi, G.C.S., Rawal, R.S., 2019. Himalayan biodiversity in the face of climate change. In: *Tropical Ecosystems: Structure, Functions and Challenges in the Face of Global Change.* Springer. pp. 263–277. DOI: 10.1007/978-981-13-8249-9_14.

Pant, G.B., Kumar, P.P., Revadekar, J.V., Singh, N., 2018. *Climate Change in the Himalayas.* Springer, Cham. DOI: 10.1007/978-3-319-61654-4.

Pausas, J.G., Keeley, J.E., 2009. A burning story: The role of fire in the history of life. *BioScience.* 59(7): 593–601. DOI: 10.1525/bio.2009.59.7.10.

Pope III, C.A., Dockery, D.W., 2006. Health effects of fine particulate air pollution: Lines that connect. *J. Air Waste Manage.* 56: 709–742. DOI: 10.1080/10473289.2006.10464485.

Ramanathan, V., Ramana, M.V., 2005. Persistent, widespread, and strongly absorbing haze over the Himalayan foothills and the Indo-Gangetic Plains. *Pure Appl. Geophys.* 162: 1609–1626. DOI: 10.1007/s00024-005-2685-8.

Sarangi, T., Naja, M., Ojha, N., Kumar, R., Lal, S., Venkataramani, S., Kumar, A., Sagar, R., Chandola, H.C., 2014. First simultaneous measurements of ozone, CO, and NO_y at a high-altitude regional representative site in the central Himalayas. *J. Geophys. Res. Atmos.* 119: 1592–1611. DOI: 10.1002/2013JD020631.

Singh, C. P., Panigrahy, S., 2011. Characterisation of residue burning from agricultural system in India using space based observations. *J. Indian Soc. Remote Sens.* 39(3): 423–429. DOI: 10.1007/s12524-011-0119-x.

Singh, R. D., Gumber, S., Tewari, P., Singh, S. P., 2016. Nature of forest fires in Uttarakhand: Frequency, size and seasonal patterns in relation to pre-monsoonal environment. *Curr. Sci.* 111(2): 398–403. DOI: 10.18520/cs/v111/i2/398-403.

Vadrevu, K.P., Lasko, K., Giglio, L., Schroeder, W., Biswas, S., Justice, C., 2019. Trends in Vegetation fires in South and Southeast Asian Countries. *Sci. Rep.* 9: 7422. DOI: 10.1038/s41598-019-43940-x.

Vijayakumar, K., Safai, P.D., Devara, P.C.S., Rao, S.V.B., Jayasankar, C.K., 2016. Effects of agriculture crop residue burning on aerosol properties and long-range transport over northern India: A study using satellite data and model simulations. *Atmos. Res.* 178–179: 155–163. DOI: 10.1016/j.atmosres.2016.04.003.

Yadav, I.C., Devi, N.L., 2018. *Biomass Burning, Regional Air Quality, and Climate Change. Earth Systems and Environmental Sciences.* Edition: Encyclopedia of Environmental Health. Elsevier. DOI: 10.1016/B978-0-12-409548-9.11022-X.

8 Spatial Point Patterns and Scale Analysis of Vegetation Fires in Laos and Cambodia

Krishna Prasad Vadrevu
NASA Marshall Space Flight Center, USA

Sumalika Biswas
Smithsonian Conservation Biology Institute, USA

Aditya Eaturu
University of Alabama Huntsville, USA

CONTENTS

INTRODUCTION

The impact of fires on landscape structure and composition in different ecosystems of the world is well documented (Goldammer and Price, 1998). In particular, the tropical ecosystems are highly fire-sensitive; thus, with the increasing anthropogenic pressure on these systems, the vegetation is continuously degraded (Goldammer and Seibert, 1990; Hayasaka et al., 2014; Hayasaka and Sepriando, 2018; Inoue, 2018; Biswas et al., 2015a, b; Vadrevu et al., 2019). The drivers of fires can include both climate and anthropogenic factors (Vadrevu et al., 2006; Prasad and Badarinth, 2004; Prasad et al., 2005, 2008a, b; Israr et al., 2018; Saharjo and Yungan, 2018; Oanh et al., 2018; Tariq and Ul-Haq, 2018). In the literature, both positive and negative effects of fires have been highlighted by different researchers. Specific to the negative effects, fires result in the loss of forests and their attendant ecosystem services, such as timber, recreation, nutrients, water retention including land degradation, and bioenergy loss (Myers, 1996; Justice et al., 2015; Vadrevu and Lasko, 2015). Repeated burning also modifies the nutrient balance of soils, especially through the process of pyrodenitrification (Crutzen and Andreae, 1990; Prasad et al., 2004). Specifically, in Asia, fires are used as a management tool for clearing forests, for example, through slash and burn, and clearing of agricultural residues after harvest (Prasad et al., 2001; Badarinath et al., 2008a, b; Lasko et al., 2017; Lasko and Vadrevu, 2018). The biomass burning resulting from such activities is an important source of greenhouse gas emissions and aerosols (Goldammer and Seibert, 1990; Kant et al., 2000a, b; Prasad et al., 2000; 2002; 2003; Lasko et al., 2018a, b; Vadrevu et al., 2013; 2014a, b; 2015; 2017; 2018; Zhang et al., 2018; Ramachandran, 2018). Outdoor fires, such as wildfires and slash-and-burn agriculture can emit substantial amounts of particulate matter (PM) and other pollutants into the atmosphere. These emissions may significantly impact air quality on both local and regional scales (Crutzen and Andreae, 1990; Gupta et al., 2001). In contrast to the above negative effects of fires, the positive effects include facilitating the uptake and availability of nutrients by plants, increasing the availability and palatability of food for herbivores, and promoting the vegetation growth such as grass cover in some ecosystems (Bond and Wilgen, 2012). Fires can also determine the type of vegetation modifying both the structure and composition and altering the structure of the landscape (Dansereau and Bergeron, 1993) and can affect ecological processes (Turner, 1989). Both Laos and Cambodia have recurrent fire events every year.

Quantifying where fires occur and their spatial and geographic gradients can help in understanding their ecological and environmental impacts (Vazquez and Moreno, 2001; Vadrevu, 2008; Vadrevu et al., 2008; Vadrevu and Justice, 2011). Over the past decades, spatial information technologies have been widely used in fire detection, mapping, and monitoring studies. In particular, remote sensing technology with its multitemporal, multispectral, synoptic, and repetitive coverage capabilities can provide valuable information on the fire counts, the amount of area burned, and the type of ecosystem burned (Vadrevu et al., 2012; Petropoulos et al., 2013). Fires, because of their high temperature, emit thermal radiation with a peak in the middle-infrared region, in accordance with Planck's theory of blackbody radiation. Therefore, active fire sensing is often done using middle-infrared and also thermal infrared (usually around 3.7–11 mm) information from satellites (Matson et al., 1987; Kant et al., 2000b). The most commonly available remote sensing fire datasets are from the MODIS and Visible Infrared Imaging Radiometer Suite (VIIRS) satellite datasets that have these channels. The products are available as point datasets with x, y location information at 1 km and 375 m resolution, respectively, at a global scale (Schroeder et al., 2014; Giglio et al., 2016).

Integrating fire datasets with the spatial point pattern statistics can provide useful information for fire management and mitigation options at different scales. The aim of spatial point statistics is to analyze the geometrical structure of patterns formed by objects that are distributed randomly in one-, two-, or three-dimensional space (Palmer and McGlinn, 2017). The spatial point pattern statistics thus can help to understand and describe the interactions among the points (e.g., fire datasets in our case). Most often, the patterns relate to the degree of clustering or randomness among points and the spatial scale at which they operate and the patterns can provide information on the underlying processes (Cressie, 1991; Diggle, 2003). In particular, the stochastic and spatial nature of fires poses challenges related to fire risk, mitigation, and management. In this study, we compare and contrast the spatial point patters and scale variability related to vegetation fires in Laos and Cambodia using the robust VIIRS 375 m satellite datasets. The results identify spatial configuration, variations, patterns, and scale in fire events useful for fire management and mitigation efforts in these countries.

DATASETS

We used the Near real-time (NRT) Suomi National Polar-orbiting Partnership (Suomi NPP) VIIRS I-Band 375 m active fire product (VNP14IMGTDL_NRT) for the study. In contrast to other coarser-resolution (≥ 1 km) satellite fire detection products such as MODIS, the improved 375 m data provide greater response over fires of relatively small areas, as well as improved mapping of large fire perimeters (Schroeder et al., 2014). Thus, the data are well suited for use in support of fire management including other science applications (Vadrevu and Lasko, 2018). The data are available from various formats including the TXT, SHP, KML, and WMS from the NASA FIRMS website (https://earthdata.nasa.gov/active-fire-data). We used these VIIRS 375 m location (x, y) datasets to infer the spatial point patterns and scale in Laos and Cambodia. To infer the spatial patterns, we first gridded the fire datasets for each month from 2012 to 2019 at a resolution of 5 minutes. We then averaged the datasets for different months and used only August month data for both Laos (Figure 8.1) and Cambodia (Figure 8.2) as it had most fires in any given year.

METHODS

Spatial Pattern Analysis of Fire Events

Fire ignitions detected from VIIRS satellite data with the event location information (x- and y-coordinates) have been assessed using spatial statistical tools. One of the simplest theoretical models for analyzing the spatial point patterns is that of complete

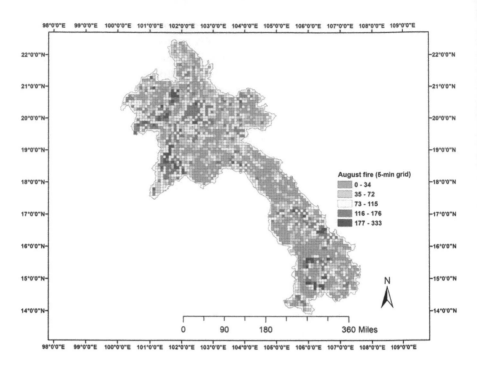

FIGURE 8.1　VIIRS 375 m August fires aggregated at 5-arc-minute grid cells (~9×9 km) for Laos. The data have been averaged from 2012 to 2019. Relatively more fires can be seen in the northern Laos provinces than in the southern provinces.

spatial randomness (CSR) (Cressie, 1991; Diggle, 2003; Rosenberg and Anderson, 2011; Vadrevu et al., 2012). A general definition of CSR is that the events in any region of observation "A" represent a homogeneous, planar Poisson process (Diggle, 2003; Wong and Lee, 2005). This process incorporates a single parameter, "l," or the intensity, or the mean number of events per unit area. The actual number of events in region "A," "n" say, is an observation from a Poisson distribution with mean l $|A|$, where $|A|$ denotes the area of the region "A." If we consider "n" as fixed, we arrive at the following definition of CSR: (1) Each of the events (n) is equally likely to occur at any point within A; and (2) the "n" events are located independently of each other. Rejection of CSR is a minimal prerequisite to model an observed spatial pattern; tests are used to explore a set of data and to assist in the formulation of plausible alternatives to CSR. CSR thus acts as a dividing hypothesis to distinguish between patterns, which are broadly classifiable as "regular" or "aggregated" (Diggle, 2003). Thus, if the CSR hypothesis is rejected, then the distribution may be categorized as clumped or uniform. In this study, to identify the spatial patterns of fire events, we used different indices of dispersion (Ludwig and Reynolds, 1988; Rosenberg and Anderson, 2011). We also attempted scale analysis using semivariograms (Isaaks and Srivastava, 1989) and correlogram analysis using Moran's I (Moran, 1950). The individual methods are described below.

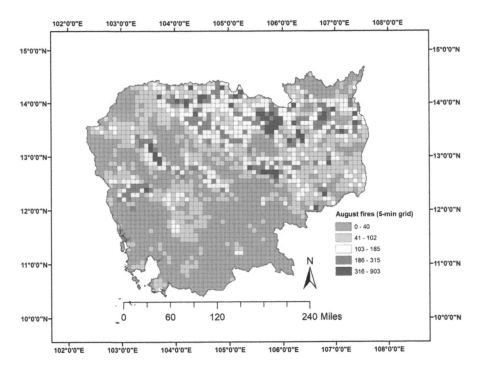

FIGURE 8.2 VIIRS 375 m August fires aggregated at 5-arc-minute grid cells (~9 × 9 km) for Cambodia. The data have been averaged from 2012 to 2019. Relatively more fires can be seen in the northeastern provinces than in the southern provinces.

DISPERSION INDICES

Most of the dispersion indices are based on the mean ($\bar{x}$) and variance (s^2) of the quadrat counts, and the number of quadrats (n) and indices primarily examine the deviation from a random (Poisson) distribution.

Index of Dispersion (ID)

It is also called variance-to-mean ratio and is calculated as (Ludwig and Reynolds, 1988; Wong and Lee, 2005) (Equation 8.1)

$$\text{ID} = s^2 / \bar{x} \tag{8.1}$$

where $\bar{x}$ and s^2 are sample mean and variance, respectively. If the sample is in agreement with the theoretical Poisson series, we would expect this ratio to be equal to 1.0. Therefore, we can test for significant departures of ID from 1.0 using the chi-square test statistic, as (Equation 8.2)

$$\chi^2 = \left(\sum_{i=1}^{N} (x_i - \bar{x})^2 \right) / \bar{x} \tag{8.2}$$

and ID $(N-1)$, where "x_i" is the number of individuals in the ith sampling unit and "N" is the total number of spatial units. For small sample sizes $(N<30)$, χ^2 is a good approximation to chi-square with $N-1$ degrees of freedom. If the value for χ^2 falls between the chi-square table values at the 0.975 and 0.025 probability levels $(P>0.05)$, agreement with a Poisson (random) distribution is accepted (i.e., $s^2=\bar{x}$). In contrast, values for χ^2 less than the 0.975 probability level suggest a regular pattern, whereas χ^2 values greater than the 0.025 probability level suggest a clumped pattern.

Index of Clumping (IC) or Index of Cluster Size (ICS)

This index is measure of dispersion (David and Moore, 1954). Under a random distribution of points, ICS is expected to equal 0 (Equation 8.3). The ID in the equation represents the Index of Dispersion. The positive values indicate a clumped distribution; the negative values indicate a regular distribution; and under a random distribution of points, the index is equal to 0.

$$ICS = \frac{s^2}{\bar{x}} - 1 = ID - 1 \tag{8.3}$$

Index of Mean Crowding (IMC)

The IMC (Lloyd, 1967) is the average number of other points contained in the quadrat that contains a randomly chosen point. It is related to the ICS (Equation 8.4).

$$IMC = \bar{x} + \frac{s^2}{\bar{x}} - 1 = \bar{x} + ICS \tag{8.4}$$

Index of Cluster Frequency (ICF)

The ICF (Douglas, 1975) is a measure of aggregation and is equal to k of the negative binomial distribution. The ICF is proportional to the quadrat area and is related to the ICS (Equation 8.5).

$$ICF = \frac{\bar{x}}{\dfrac{s^2}{\bar{x}} - 1} = \frac{\bar{x}}{ICS} \tag{8.5}$$

Green's Index

Green (1966) proposed a modification of the ICS that is independent of n. Green's index (GI) is computed as (Ludwig and Reynolds, 1988, Equation 8.6)

$$GI = \frac{\dfrac{s^2}{\bar{x}} - 1}{n-1} = \frac{ICS}{n-1} \tag{8.6}$$

where GI is Green's index, s^2 is the variance, $\bar{x}$ is the mean number of individuals per sampling unit, and n is the total number of fire spots in a sample. GI varies between 0 (for random) and 1 (for maximum clumping). Green's index can be used to compare samples that vary in the total number of individuals, their sample means, and the number of spatial units in the sample. Consequently, for the numerous variations of ID that have been proposed to measure the degree of clumping, GI seems most recommendable (Ludwig and Reynolds, 1988).

Index of Patchiness (IP)

The IP (Lloyd, 1967) is related to the ICF and the IMC and is similar to Morisita's index. It is a measure of pattern intensity that is unaffected by thinning (the random removal of points). The index is given as (Equation 8.7)

$$IP = \frac{\bar{x} + \frac{s^2}{\bar{x}} - 1}{\bar{x}} = \frac{IMC}{\bar{x}} = 1 + \frac{1}{ICF} \tag{8.7}$$

Morisita's Index (I_M)

Morisita's index (Morisita, 1959) is related to the IP. It is the scaled probability that two points chosen at random from the whole population are in the same quadrat. The higher the value, the more clumped the distribution, and the index is given as (Equation 8.8)

$$I_M = \frac{n \sum x(x-1)}{n\bar{x} \ (\ n\bar{x} - 1)} = \frac{n\bar{x}IP}{(n\bar{x} - 1)} \tag{8.8}$$

Semivariogram Analysis

Variograms are one of the primary forms of spatial analysis used in geostatistics (Isaaks and Srivastava, 1989). Specifically, the semivariograms are useful to characterize the spread or variation of data, in a regionalized variable, z (fire events in our case), based on distance (Rossi et al., 1992). The geographic distance between two samples is termed the spatial lag. The semivariance is computed as (Equation 8.9)

$$\gamma(h) = \left\{ \sum [z(i) - z(i+h)]^2 \right\} \Big/ 2N(h) \tag{8.9}$$

where $\gamma(h)$ is the semivariance of a lag of distance h, $z(i)$ is the value of a regionalized variable z at location i, $z(i+h)$ is the value of z at a location separated from I by lag h, and $N(h)$ is the number of pairs of points separated by lag h. The summation is over all pairs of points separated by distance h. In plain English, the semivariance is half of the average squared difference of all pairs of points separated by a given distance. A semivariogram is a plot of semivariance versus the lag distance. As with the variance, the semivariance cannot be less than zero, but it is not bounded on the top (Palmer and McGlinn, 2017).

Correlogram Analysis

Correlograms are most commonly used to compute autocorrelation in the datasets (Cliff and Ord, 1981). A typical correlogram consists of a series of estimated autocorrelation coefficients calculated for different spatial relationships. The most common indices of autocorrelation are Moran's I and Geary's c. In this study, we used Moran's I to assess autocorrelation in the fire location datasets.

Moran's I (Moran, 1950) is calculated as (Equation 8.10)

$$I = n \; \frac{\sum_{ij} w_{ij}\left(y_i - \bar{y}\right)\left(y_j - \bar{y}\right)}{W \sum_{i=1}^{n}\left(y_i - \bar{y}\right)^2} \tag{8.10}$$

where y_i is the value of the variable at the ith location, n is the number of points, w_{ij} is a weight indicating something about the spatial relationship of points "i" and "j," $\sum_{ij}$ indicates the double sum over all i and all j where $i \neq j$, and $W = \sum_{ij} w_{ij}$, the sum of the values in the weight matrix. Moran's I is similar to a correlation coefficient, and the values range from 1 to –1, with an expected value of $-\frac{1}{n-1}$; for a large n, this is approximately zero. Positive values of I indicate positive spatial autocorrelation, and negative values of I indicate negative spatial autocorrelation. We assumed randomness in the observed fire data at the observed locations; thus, the variance of I is estimated as (Equations 8.11–8.14)

$$V(I) = \frac{n\left[\left(n^2 - 3n + 3\right)S_1 - nS_2 + 3W^2\right] - b_2\left[\left(n^2 - n\right)S_1 - 2nS_2 + 6W^2\right]}{(n-1)(n-2)(n-3)W^2}$$

$$- \frac{1}{(n-1)^2} \tag{8.11}$$

where

$$S_1 = \frac{1}{2}\sum_{ij}\left(w_{ij} + w_{ji}\right)^2 \tag{8.12}$$

$$S_2 = \frac{1}{2}\sum_{i=1}^{n}\left(w_i + w \cdot i\right)^2 \tag{8.13}$$

w_i and $w \cdot i$ are the sums of the ith row and ith column of the weight matrix, respectively, and

$$b_2 = \frac{n\sum_{i=1}^{n}\left(y_i - \bar{y}\right)^4}{\left(\sum_{i=1}^{n}\left(\left(y_i - \bar{y}\right)^2\right)\right)^2} \tag{8.14}$$

The significance of the correlogram is calculated using a Bonferroni procedure (Oden, 1984).

RESULTS AND DISCUSSION

Results on the spatial point pattern analysis of fires suggested a clumped pattern in both Cambodia and Laos. The ID results suggested a relatively higher clumping of fires in Cambodia than in Laos, as reflected in their values. Similarly, the IMC, IC, MI, IP, and GI too were relatively higher for Cambodia than for Laos (Figure 8.3a and b). In contrast to these indices, the ICF was higher for Laos than for Cambodia (Figure 8.3b). To infer these spatial patterns, we explored fire data variations at a provincial level in Cambodia and Laos.

Results from the VIIRS 385 m gridded fires for August 2019 at 5-minute intervals showed variation in fires from 0 to 333 fires per grid cell with relatively more fires in northern Laos compared to the south (Figure 8.1). The northern provinces in Laos which had more fires were Louangphabang, Oudomxay, Xaignabouri, and Viangchan; in the south, the provinces with dominant fires were Savannakhet, Salavan, and Champasak. In the case of Cambodia, fires varied from 0 to 903 fires per grid cell with a relatively higher number of fires in northern Cambodia (Figure 8.2), more specifically in the provinces of Preah Vihear, Stung Treng, and Ratanakiri. Also, the central provinces of Kratie and Kampong Thom and the western provinces including Battambang and Pursat had more fires.

An idealized semivariogram is given in Figure 8.4 (Palmer and McGlinn, 2017). At distances less than the range, R, there is a spatial dependence between the datasets; *i.e., the* closer samples are more similar than the distant samples. In general, as the distance between the points increases, semivariance increases; i.e., the values become increasingly dissimilar. The increase continues until the points are so far

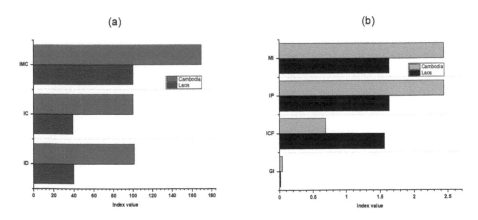

FIGURE 8.3 Dispersion indices for fire datasets in Laos and Cambodia. (a) Index of Dispersion (ID), Index of Clumping (IC), and Index of Mean Crowding (IMC) with higher values in Cambodia. (b) Except the Index of Cluster Frequency (ICF), other indices, i.e., Green's index (GI), Index of Patchiness (IP), and Morisita's index (MI), were higher for Cambodia than for Laos.

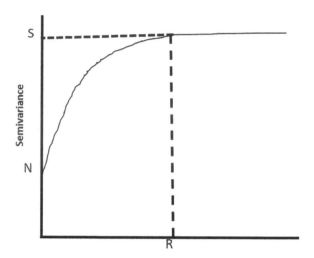

FIGURE 8.4   An ideal semivariogram. N = nugget; R = range; S = sill.

apart that they are not related to each other and their squared difference becomes equal to the average variance of the samples (the flat region called the sill). In contrast, at distances of at least h, the data are spatially independent, and therefore, samples separated by longer distances are dissimilar without any autocorrelation; in such cases, conventional statistics can be used; if not, spatial autocorrelation needs to be addressed. The sill indicates the amount of "background" variation (Palmer and McGlinn, 2017). The maximum distance within which the samples are still autocorrelated is called the range. The unresolved variation at a very fine scale is termed the nugget effect, denoted by N. The semivariogram specifically addresses how variance increases as a function of scale, which might be linked to various underlying processes governing the shape of the semivariograms.

The semivariograms for Laos and Cambodia are given in Figure 8.5a and b. In the case of Laos fire datasets, the spherical model fitted well (Figure 8.5a), whereas for Cambodia, the model was exponential (Figure 8.5b). Further, a relative comparison of semivariance of fire points suggested higher values for Cambodia (14,000)

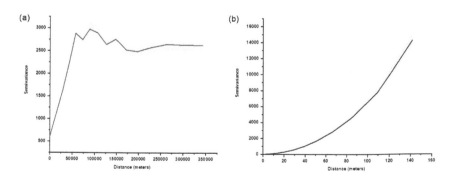

FIGURE 8.5   Omnidirectional semivariograms of VIIRS 375 m fires in Laos and Cambodia. Spherical and exponential models for Laos (a) and Cambodia (b) can be noted.

than for Laos (2800). The range (R) or the influence of correlation (or autocorrelation) between the fire points for Laos was found to be around 55,000 m, whereas in Cambodia, no such range could be found as the distance increased. In addition, for the Laos dataset, a nugget effect was found in contrast to Cambodia where no such effect was noticed.

Results from the correlogram analysis using Moran's I autocorrelation function for Laos and Cambodia are in shown in Figure 8.6a and b). The interpretation of autocorrelogram is almost similar to of semivariance; however, with a distance decay, the autocorrelograms often look like upside-down semivariance. Specifically, Moran's I value of 0 means there is no relationship between the datasets, −1 suggests a strong negative correlation, and 1 suggests a strong positive relationship. Thus, a close observation of Moran's I in the case of Laos (Figure 8.6a) suggests that the fire points showed positive autocorrelation at less than 50,000 m in contrast to less than 120 m in Cambodia (Figure 8.6b). The autocorrelation was a much smaller distance in Cambodia, unlike Laos.

A review of literature related to fire characteristics in different regions of the world suggests that our understanding of the spatial processes at which fires occur is very limited. It is often difficult to control and manage fires without knowing their spatial scale, extent, and fire regime characteristics. In this study, to evaluate the spatial structure in fire point patterns, we used a variety of statistics in conjunction with scale analysis. The computed dispersion indices, semivariograms, and Moran's I for fire points revealed clear differences in spatial pattern of fires in Laos and Cambodia. Specifically, although the fires were clustered in both the countries, the fires showed spatial dependence up to distances of 50,000 m in Laos and almost none in Cambodia. These patterns suggest differences in both the forest structural properties and anthropogenic activities operating at the ground level. Although the methods used in the study can only describe patterns, there might be underlying processes that might have affected these spatial patterns. For example, the larger scale variation of fires in Laos may be attributed to forest patch sizes and fires that are larger in size related to slash-and-burn agriculture in remote areas. In contrast, the smaller spatial structure

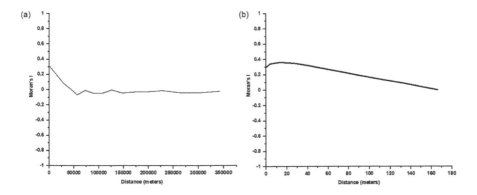

FIGURE 8.6 Moran's I for VIIRS 375 m fire datasets showing autocorrelation for Laos (a) and Cambodia (b). Fire datasets showed a positive Moran's I at 50,000 m in Laos compared to less than 120 m in Cambodia.

of fire characteristics observed in Cambodia might be related to more agricultural fires where the field size is small. Also, the results from the various indices, i.e., ID, IC, GI, IMC, IP, and MI, support these findings, wherein relatively higher clumping of fire events was observed in Cambodia than in Laos. High semivariance observed in Laos in conjunction with Moran's I too confirms these inferences. However, these inferences need to be substantiated with the ground-based information, which unfortunately in our case was not available. Nevertheless, the results highlighted in the study can serve as a baseline to infer any such pattern–process relationships and validation from field studies. We infer that the use of these spatial statistics is only the first step in exploring spatial patterns, which can be effectively used to build models and hypothesis to explain the underlying fire events. The results specific to fire characteristics in different provinces and scale characteristics highlighted in both Laos and Cambodia should help resource managers and environmental scientists to focus on fire management and mitigation efforts.

ACKNOWLEDGMENTS

The authors gratefully appreciate the VIIRS 375 m product developers for sharing the fire datasets.

REFERENCES

Badarinath, K.V.S., Kharol, S.K., Krishna Prasad, V., Kaskaoutis, D.G. and Kambezidis, H.D. 2008a. Variation in aerosol properties over Hyderabad, India during intense cyclonic conditions. *International Journal of Remote Sensing*. 29(15), 4575–4597.

Badarinath, K.V.S., Kharol, S.K., Prasad, V.K., Sharma, A.R., Reddi, E.U.B., Kambezidis, H.D. and Kaskaoutis, D.G. 2008b. Influence of natural and anthropogenic activities on UV Index variations–a study over tropical urban region using ground based observations and satellite data. *Journal of Atmospheric Chemistry*. 59(3), 219–236.

Biswas, S., Lasko, K.D. and Vadrevu, K.P. 2015a. Fire disturbance in tropical forests of Myanmar—Analysis using MODIS satellite datasets. *IEEE Journal of Selected Topics in Applied Earth Observations and Remote Sensing*. 8(5), 2273–2281.

Biswas, S., Vadrevu, K.P., Lwin, Z.M., Lasko, K. and Justice, C.O. 2015b. Factors controlling vegetation fires in protected and non-protected areas of Myanmar. *PLoS One*. 10(4), e0124346.

Bond, W.J. and Van Wilgen, B.W., 2012. *Fire and Plants* (Vol. 14). Dordrecht: Springer Science & Business Media.

Cliff, A.D. and Ord. J.K., 1981. *Spatial Processes*. London: Pion.

Cressie, N.A.C. 1991. *Statistics for Spatial Data*. New York: John Wiley & Sons, 900p.

Crutzen, P.J. and Andreae, M. O. 1990. Biomass burning in the tropics impact on atmospheric chemical and biogeochemical cycles. *Science*. 250, 1669–1678.

Dansereau, P.R. and Bergeron, Y., 1993. Fire history in the southern boreal forest of northwestern Quebec. *Canadian Journal of Forest Research*. 23(1), 25–32.

David, F.N. and Moore, P.G. 1954. Notes on contagious distributions in plant populations. *Annals of Botany of London* 18, 47–53.

Diggle, P. 2003. *Statistical Analysis of Spatial Point Patterns*. London: Arnold Publishing

Douglas, J.B. 1975. Clustering and aggregation. *Sankhya, Series B*. 37, 398–417.

Giglio, L., Schroeder, W. and Justice, C.O., 2016. The collection 6 MODIS active fire detection algorithm and fire products. *Remote Sensing of Environment*. 178, 31–41.

Goldammer, J. G. and Price, C. 1998. Potential impacts of climate change on fire regimes in the tropics based on MAGICC and a GISS GCM-derived lightning model. *Climate Change.* 39, 273–296.

Goldammer, J.G. and Seibert, B., 1990. The impact of droughts and forest fires on tropical lowland rain forest of East Kalimantan. In: *Fire in the Tropical Biota.* Goldammar, J.G. (Ed). Berlin, Heidelberg: Springer. pp. 11–31.

Green, R.H. 1966. Measurement of non-randomness in spatial distributions. *Researches in Population Ecology.* 8, 1–7.

Gupta, P.K., Prasad, V.K., Sharma, C., Sarkar, A.K., Kant, Y., Badarinath, K.V.S. and Mitra, A.P. 2001. CH_4 emissions from biomass burning of shifting cultivation areas of tropical deciduous forests–experimental results from ground-based measurements. *Chemosphere-Global Change Science.* 3(2), 133–143.

Hayasaka, H. and Sepriando, A. 2018. Severe air pollution due to peat fires during 2015 super El Niño in Central Kalimantan, Indonesia. In: *Land-Atmospheric Research Applications in South/Southeast Asia.* Vadrevu, K.P., Ohara, T., and Justice, C. (Eds). Cham: Springer. pp. 129–142.

Hayasaka, H., Noguchi, I., Putra, E.I., Yulianti, N. and Vadrevu, K. 2014. Peat-fire-related air pollution in Central Kalimantan, Indonesia. *Environmental Pollution.* 195, 257–266. doi: 10.1016/j.envpol.2014.06.031.

Inoue, Y. 2018. Ecosystem carbon stock, Atmosphere and Food security in Slash-and-Burn Land Use: A Geospatial Study in Mountainous Region of Laos. In: *Land-Atmospheric Research Applications in South/Southeast Asia.* Vadrevu, K.P., Ohara, T., and Justice, C. (Eds). Cham: Springer. pp. 641–666.

Isaaks, E.H. and Srivastava, R.M. 1989. *An Introduction to Applied Geostatistics.* Oxford: Oxford University Press.

Israr, I., Jaya, S.N.I, Saharjo, H.S., Kuncahyo, B. and Vadrevu, K.P. 2018. Spatio-temporal analysis of land and forest fires in Indonesia using MODIS active fire dataset. In: *Land-Atmospheric Research Applications in South/Southeast Asia.* Vadrevu, K.P., Ohara, T., and Justice, C. (Eds). Cham: Springer. pp. 105–128.

Justice, C., Gutman, G. and Vadrevu, K.P. 2015. NASA land cover and land use change (LCLUC): An interdisciplinary research program. *Journal of Environmental Management.* 148(15), 4–9.

Kant, Y., Ghosh, A.B., Sharma, M.C., Gupta, P.K., Prasad, V.K., Badarinath, K.V.S. and Mitra, A.P. 2000a. Studies on aerosol optical depth in biomass burning areas using satellite and ground-based observations. *Infrared Physics & Technology.* 41(1), 21–28.

Kant, Y., Prasad, V.K. and Badarinath, K.V.S. 2000b. Algorithm for detection of active fire zones using NOAA AVHRR data. *Infrared Physics & Technology.* 41(1), 29–34.

Lasko, K. and Vadrevu, K.P. 2018. Improved rice residue burning emissions estimates: Accounting for practice-specific emission factors in air pollution assessments of Vietnam. *Environmental Pollution.* 236(5), 795–806.

Lasko, K., Vadrevu, K.P., Tran, V.T., Ellicott, E., Nguyen, T.T., Bui, H.Q. and Justice, C. 2017. Satellites may underestimate rice residue and associated burning emissions in Vietnam. *Environmental Research Letters.* 12(8), 085006.

Lasko, K., Vadrevu, K.P. and Nguyen, T.T.N. 2018a. Analysis of air pollution over Hanoi, Vietnam using multi-satellite and MERRA reanalysis datasets. *PLoS One.* 13(5), e0196629.

Lasko, K., Vadrevu, K.P., Tran, V.T. and Justice, C. 2018b. Mapping double and single crop paddy rice with Sentinel-1A at varying spatial scales and polarizations in Hanoi, Vietnam. *IEEE Journal of Selected Topics in Applied Earth Observations and Remote Sensing.* 11(2), 498–512.

Lloyd, M. 1967. Mean crowding. *Journal of Animal Ecology.* 36, 1–30.

Ludwig, J.A. and Reynolds, J.F. 1988. *Statistical Ecology. A Primer on Methods and Computing.* New York: Wiley.

Matson, M., Stephens, G. and Robinson, J., 1987. Fire detection using data from the NOAA-N satellites. *International Journal of Remote Sensing*. 8(7), 961–970.

Moran, P.A.P. 1950. Notes on continuous stochastic phenomena. *Biometrika*. 37, 17–23.

Morisita, M. 1959. Measuring the dispersion and the analysis of distribution patterns. *Memoires of the Faculty of Science, Kyushu University, Series E. Biology*. 2, 215–235.

Myers, N. 1996. The world's forests: Problems and potentials. *Environmental Conservation*. 23, 156–169.

Oanh, N.T.K., Permadi, D.A., Dong, N.P. and Nguyet, D.A. 2018. Emission of toxic air pollutants and greenhouse gases from crop residue open burning in Southeast Asia. In: *Land-Atmospheric Research Applications in South/Southeast Asia*. Vadrevu, K.P., Ohara, T., and Justice, C. (Eds). Cham: Springer. pp. 47–68.

Oden, N.L. 1984. Assessing the significance of a spatial correlogram. *Geographical Analysis* 16, 1–16.

Palmer, M.W. and McGlinn, D.J. 2017. Scale detection using Semivariograms and Autocorrelograms. In: *Learning Landscape Ecology*. Gergel, S.E., and Turner, M.G. (Eds). New York: Springer-Verlag. doi: 10.1007/978-1-4939-6374-4_5.

Petropoulos, G.P., Vadrevu, K.P. and Kalaitzidis, C. 2013. Spectral angle mapper and object-based classification combined with hyperspectral remote sensing imagery for obtaining land use/cover mapping in a Mediterranean region. *Geocarto International*. 28(2), 114–129.

Prasad, V.K. and Badarinth, K.V.S., 2004. Land use changes and trends in human appropriation of above ground net primary production (HANPP) in India (1961–98). *Geographical Journal*. 170(1), 51–63.

Prasad, V.K., Gupta, P.K., Sharma, C., Sarkar, A.K., Kant, Y., Badarinath, K.V.S., Rajagopal, T. and Mitra, A.P. 2000. NO$_x$ emissions from biomass burning of shifting cultivation areas from tropical deciduous forests of India–estimates from ground-based measurements. *Atmospheric Environment*. 34(20), 3271–3280.

Prasad, V.K., Kant, Y. and Badarinath, K.V.S. 2001. CENTURY ecosystem model application for quantifying vegetation dynamics in shifting cultivation areas: A case study from Rampa Forests, Eastern Ghats (India). *Ecological Research*. 16(3), 497–507.

Prasad, V.K., Kant, Y., Gupta, P.K., Elvidge, C. and Badarinath, K.V.S. 2002. Biomass burning and related trace gas emissions from tropical dry deciduous forests of India: A study using DMSP-OLS data and ground-based measurements. *International Journal of Remote Sensing*. 23(14), 2837–2851.

Prasad, V.K., Lata, M. and Badarinath, K.V.S. 2003. Trace gas emissions from biomass burning from northeast region in India—estimates from satellite remote sensing data and GIS. *Environmentalist*. 23(3), 229–236.

Prasad, V.K., Badarinath, K.V.S., Yonemura, S. and Tsuruta, H. 2004. Regional inventory of soil surface nitrogen balances in Indian agriculture (2000–2001). *Journal of Environmental Management*. 73(3), 209–218.

Prasad, V.K., Anuradha, E. and Badarinath, K.V.S. 2005. Climatic controls of vegetation vigor in four contrasting forest types of India—Evaluation from national oceanic and atmospheric administration's advanced very high resolution radiometer datasets (1990–2000). *International Journal of Biometeorology*. 50(1), 6–16.

Prasad, V.K., Badarinath, K.V.S. and Eaturu, A. 2008a. Biophysical and anthropogenic controls of forest fires in the Deccan Plateau, India. *Journal of Environmental Management*. 86(1), 1–13.w

Prasad, V.K., Badarinath, K.V.S. and Eaturu, A. 2008b. Effects of precipitation, temperature and topographic parameters on evergreen vegetation greenery in the Western Ghats, India. *International Journal of Climatology: A Journal of the Royal Meteorological Society*. 28(13), 1807–1819.

Ramachandran, S. 2018. Aerosols and climate change: Present understanding, challenges and future outlook. In: *Land-Atmospheric Research Applications in South/Southeast Asia.* Vadrevu, K.P., Ohara, T. and Justice, C. (Eds). Cham: Springer. pp. 341–378.

Rosenberg, M.S. and Anderson, C.D. 2011. PASSaGE: Pattern analysis, spatial statistics and geographic exegesis. Version 2. *Methods in Ecology and Evolution.* 2(3), 229–232.

Rossi, R.E., Mulla, D.J., Journel, A.G. and Franz, E.H. 1992. Geostatistical tools for modeling and interpreting ecological spatial dependence. *Ecological Monographs.* 62(2), 277–314.

Saharjo, B.H. and Yungan, A. 2018. Forest and land fires in Riau province; A Case study in fire prevention, policy implementation with local concession holders. In: *Land-Atmospheric Research Applications in South/Southeast Asia.* Vadrevu, K.P., Ohara, T., and Justice, C. (Eds). Cham: Springer. pp. 143–170

Schroeder, W., Oliva, P., Giglio, L. and Csiszar, I.A. 2014. The New VIIRS 375 m active fire detection data product: Algorithm description and initial assessment. *Remote Sensing of Environment,* 143, 85–96.

Tariq, S. and Ul-Haq, Z. 2018. Satellite remote sensing of aerosols and gaseous pollution over Pakistan. 2018. In: *Land-Atmospheric Research Applications in South/Southeast Asia.* Vadrevu, K.P., Ohara, T., and Justice, C. (Eds). Cham: Springer. pp. 523–552.

Turner, M.G., 1989. Landscape ecology: The effect of pattern on process. *Annual Review of Ecology and Systematics,* 20(1), 171–197.

Vadrevu, K.P. 2008. Analysis of fire events and controlling factors in eastern India using spatial scan and multivariate statistics. *Geografiska Annaler: Series A, Physical Geography.* 90(4), 315–328.

Vadrevu, K.P. and Justice, C.O. 2011. Vegetation fires in the Asian region: Satellite observational needs and priorities. *Global Environmental Research.* 15(1), 65–76.

Vadrevu, K.P. and Lasko, K.P., 2015. Fire regimes and potential bioenergy loss from agricultural lands in the Indo-Gangetic Plains. *Journal of Environmental Management.* 148, 10–20.

Vadrevu, K.P. and Lasko, K. 2018. Intercomparison of MODIS AQUA and VIIRS I-Band fires and emissions in an agricultural landscape—Implications for air pollution research. *Remote Sensing.* 10(7), 978. doi:10.3390/rs10070978.

Vadrevu, K.P., Eaturu, A. and Badarinath, K.V.S. 2006. Spatial distribution of forest fires and controlling factors in Andhra Pradesh, India using spot satellite datasets. *Environmental Monitoring and Assessment.* 123(1–3), 75–96.

Vadrevu, K.P., Badarinath, K.V.S. and Anuradha, E. 2008. Spatial patterns in vegetation fires in the Indian region. *Environmental Monitoring and Assessment.* 147(1–3), 1. doi:10.1007/s10661-007-0092-6.

Vadrevu, K.P., Csiszar, I., Ellicott, E., Giglio, L., Badarinath, K.V.S., Vermote, E. and Justice, C. 2012. Hotspot analysis of vegetation fires and intensity in the Indian region. *IEEE Journal of Selected Topics in Applied Earth Observations and Remote Sensing.* 6(1), 224–238.

Vadrevu, K.P., Giglio, L. and Justice, C. 2013. Satellite based analysis of fire–carbon monoxide relationships from forest and agricultural residue burning (2003–2011). *Atmospheric Environment.* 64, 179–191.

Vadrevu, K.P., Ohara T. and Justice, C. 2014a. Air pollution in Asia. *Environmental Pollution.* 12, 233–235.

Vadrevu, K.P., Lasko, K., Giglio, L. and Justice, C. 2014b. Analysis of Southeast Asian pollution episode during June 2013 using satellite remote sensing datasets. *Environmental Pollution.* 12, 245–256.

Vadrevu, K.P., Lasko, K., Giglio, L. and Justice, C. 2015. Vegetation fires, absorbing aerosols and smoke plume characteristics in diverse biomass burning regions of Asia. *Environmental Research Letters.* 10(10), 105003.

Vadrevu, K.P., Ohara, T. and Justice, C. 2017. Land cover, land use changes and air pollution in Asia: A synthesis. *Environmental Research Letters.* 12(12), 120201.

Vadrevu, K.P., Ohara, T. and Justice, C. (Eds). 2018. *Land-Atmospheric Research Applications in South and Southeast Asia.* Cham: Springer.

Vadrevu, K.P., Lasko, K., Giglio, L., Schroeder, W., Biswas, S. and Justice, C. 2019. Trends in vegetation fires in south and Southeast Asian countries. *Scientific Reports,* 9(1), 7422. doi:10.1038/s41598-019-43940-x.

Vazquez, A., & Moreno, J. M. (2001). Spatial distribution of forest fires in Sierra de Gredos (Central Spain). *Forest Ecology and Management,* 147, 55–65.

Wong, D. W. S. and Lee, J. (2005). *Statistical Analysis of Geographic Information with ArcView GIS and ArcGIS.* New York: Wiley.

Zhang, Y.Z., Wong, M.S. and Campbell, J.R. 2018. Conceptualizing how severe haze events are impacting long-term satellite based trend studies of aerosol optical thickness in Asia. In: *Land-Atmospheric Research Applications in South/Southeast Asia.* Vadrevu, K.P., Ohara, T. and Justice, C. (Eds). Cham: Springer. pp. 425–446.

Section II

Land Use, Forests, and Biomass Burning

9 Vegetation Fire Status and Management in Bhutan

Pankaj Thapa
Sherubtse College, Royal University of Bhutan, Bhutan

Krishna Prasad Vadrevu
NASA Marshall Space Flight Center, USA

Aditya Eaturu
University of Alabama Huntsville, USA

Sumalika Biswas
Smithsonian Conservation Biology Institute, USA

CONTENTS

INTRODUCTION

Vegetation fires in the Asian region receive increasing attention because of a wide range of ecological, economic, social, and political impacts. Several researchers discussed the role of fire in creating and maintaining landscape structure, composition, function, and ecological integrity (Goldammer and Seibert, 1990; Prasad et al., 2001; 2003; 2005; Prasad and Badarinath, 2004). At the local scale, fire can stimulate soil microbial processes and combust vegetation, ultimately altering the soil's structure and composition (Kennard and Gholz, 2001). Important drivers of fires in Asia include slash-and-burn agriculture, clearing of forested lands for oil palm expansion, and agricultural residue or waste burning for clearing of land for the next crop (Badarinath et al., 2008a,b; Biswas et al., 2015a, b; Hayasaka et al., 2014; Israr et al., 2018; Inoue, 2018; Oanh et al., 2018; Saharjo and Yungan, 2018; Lasko et al., 2017; 2018a, b; Vadrevu et al., 2006; 2008; 2012; 2013; 2014a, b; 2019; Vadrevu and Lasko, 2015; 2018). At regional and local scales, the burning of biomass from these activities can result in the release of large amounts of radiatively active gases, aerosols, and other chemically active species that significantly alter the Earth's radiation balance and atmospheric chemistry (Andreae and Merlet, 2001; Kant et al., 2000; Gupta et al., 2001; Vadrevu and Choi, 2011; Vadrevu and Justice, 2011; Justice et al., 2015; Lasko and Vadrevu, 2018; Ramachandran, 2018). Fires can result in the loss of biodiversity and result in adverse health effects due to the smoke released during the biomass burning. The use of remote sensing data for mapping of forests and fire occurrences in various ecosystems, including emissions estimation, has been demonstrated successfully by different researchers in Asia (Prasad et al., 2000; 2002; 2004; 2005; 2008a, b; Gupta et al., 2001; Hayasaka and Sepriando, 2018; Vadrevu, 2008; Vadrevu et al., 2015, 2017; 2018; Chowdhary and Gupta, 2018; Zhang et al., 2018; Tariq and Ul-Haq, 2018). This study provides a detailed review of fires' current status, their causative factors, and strategies adopted for fire prevention and mitigation in Bhutan.

Bhutan is one of the top ten global biodiversity hotspots and is well known for maintaining high environmental standards by currently preserving about 70% of its geographical area under forest cover. Bhutan also passed a constitutional mandate to maintain 60% of its area under forest cover at all times and pledges to remain carbon neutral (RGoB, 2009). Almost 99.8% of the country's forest cover is natural forests, found in three distinct eco-floristic precincts, namely alpine (> 4000 MSL), temperate (~ 2000–4000 m), and subtropical (< 2000 m) zones (RGoB 2009).

The abundance of virgin forests and richly intact biodiversity, which contains more than 5500 species of vascular plants and over 770 bird and 200 mammal species, rightly qualify the country to be known as the "naturalist's paradise" (Tharchen, 2013). Most of the studies indicate that Bhutan's forest cover has increased over the past few decades. The Food and Agriculture Organization (FAO, 2010) estimates a constant annual increase of about 0.34% from 1990 to 2010, while Gilani et al. (2015) estimate that the forest's average yearly growth rate was 59 km^2 (0.22%) per year between 1990 and 2010. However, forest cover's overall quality and density could be compromised mainly due to an increase in forest fragmentation, leading to an increase in patch number and a decrease in patch size (Sharma et al., 2016).

This can be mainly attributed to the collection of firewood and forest fire, which together contribute to more than 50% of annual forest degradation and greenhouse gas (GHG) emission (MoA, 2017a).

A detailed forest inventory, including forest growing stock assessment, is the necessary prerequisite to estimate forest biomass and carbon stocks. Advancement in geospatial techniques and field data collection has dramatically enhanced the inventory, assessment, monitoring, and forest cover management, including estimation of forest biomass and carbon stocks at various spatial and temporal scales. Evaluation of forest resource potentials and randomized branch sampling are some of the Department of Forest's ongoing Bhutan activities. Being a member of the UN-REDD Programme, Bhutan plans to follow the National Forest Monitoring System (NFMS) and satellite land monitoring system (SLMS) for monitoring REDD+ activities and generate data for national GHG inventory (Phuntsho et al., 2015). Geospatial techniques offer a wide range of data for biomass estimation, assessment, and monitoring (Murthy et al., 2015), and measurement of biomass using spaceborne SAR (Qazi and Gilani, 2015; Watanabe et al., 2015), LiDAR (Peuhkurinen et al., 2015), Landsat ETM+, Advanced Land Observation System (ALOS) PALSAR data, and other optical sensors and spectral models (Rahman, 2015; Couteron et al., 2015; Mon and Myint, 2015).

At a national level, the application of geospatial techniques in Bhutan dates back to the early 1990s when the first LULC mapping at 1:50,000 was carried out by the Ministry of Agriculture (MoA) in 1994 using SPOT satellite imagery of December 1989 (panchromatic) and color images of June, July, and August 1990. The ground-truthing was done between 1993 and 1994, and a report was published in 1997 (MoA, 1997). This was followed by LULC mapping in 2010 using multispectral ALOS AVNIR-2 satellite images of 2006–2009 (MoAF, 2010). Few studies focus on forest fires and emissions from biomass burning in Bhutan compared to these studies. The present study is one of the first attempts to document fire ecology and emissions from biomass burning in Bhutan.

GEOGRAPHY

The Himalayan kingdom of Bhutan is a landlocked country located in the Eastern Himalayas, sharing borders with China in the north and India in the south (Figure 9.1). The land extends from 26° 40′ to 28° 15′ latitude and 88° 40′ to 92° 15′ longitude, spanning over 38,394 km² of geographical area. Compared to the other South Asian countries, Bhutan is one of the least populated (735,553 persons), with a very low population density of only 19 persons/km² (PHCB, 2017). Even though almost 62% of the country's population live in rural areas and are dependent upon subsistence farming and animal husbandry for their livelihood, the area under agriculture is reported to have reduced from 7.7% in 1994 (MoA 1997) to 2.75% in 2010 (FRMD, 2017).

Bhutan's relief features are marked by rugged terrain, steep slopes, and elevation ranging from 100 to 7500 MSL within a narrow north–south aerial distance of about 150 km. This has resulted in highly diverse geographical conditions favoring high biodiversity and climatic conditions ranging from subtropical to alpine types. About

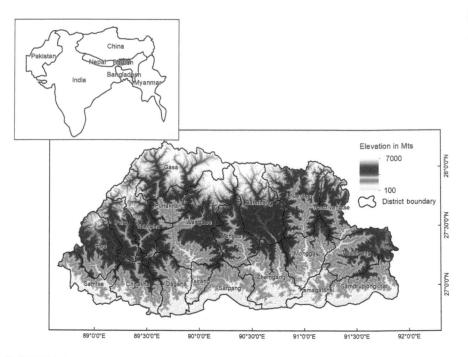

FIGURE 9.1   Bhutan country depicting physiography.

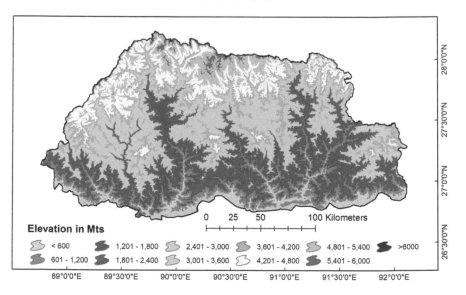

FIGURE 9.2   Bhutan elevation gradients.

57% of the geographical area lies below 3000 m, beyond which there are very few
human settlements (Figure 9.2). Since most of the landforms in Bhutan are very rugged and demarcated as protected areas, land availability for agriculture and urban
expansion is highly dispersed and limited (Figure 9.2).

LAND USE/COVER ASSESSMENT

The Royal Government of Bhutan carried Bhutan's land use/land cover assessment for 2015 (Forest Resources Management Division (FRMD), 2016). The land cover assessment was done using the Landsat data and rule-based classification. A total of 20 classes were derived. The dominance of forest and preservation of the country's rich biodiversity can be realized by the fact that forest, which is defined as land with trees spanning more than 0.5 ha with trees higher than 5 m and a canopy cover of more than 10% (National Forest Policy of Bhutan, 2011), is the largest land cover in Bhutan occupying 70.77% of the country's geographical area (FRMD, 2017). Shrubs (13.13%), snow and glaciers (5.35%), and rocky outcrops (4.15%) are the other three main land cover types. Agriculture (2.75%) and built-up areas (0.19%) constitute less than 3% of the total geographical area (MoAF, 2016). The distribution of area under different LULC is provided in Table 9.1 and Figure 9.3. The dominance of blue pine, chir pine, mixed conifer, and broadleaf forest with conifer, which comprises almost 45% of the total forest area, increases its susceptibility to frequent forest fires (UNISDR, 1994). Conifer forests, especially chir pine and blue pine, are the most common forests affected by forest fires in Bhutan. The high resin content in this species, including the dry and windy climatic conditions and the distribution of settlements and agricultural practices, aggravates the cause for large-scale destruction of vegetation in the temperate belts.

PROTECTED AREAS

More than 51% of the country's total geographical area is under the protected areas (42.7%), biological corridors (8.6%), and a Royal Botanical Park (0.1%). The protected areas include 5 national parks, 4 wildlife sanctuaries, a strict nature reserve, and a Royal Botanical Park (RGoB, 2017) distributed in all three types of ecological regions spread over various landscapes (Figure 9.4). The uniqueness of Bhutan's protected areas is that these areas are also inhabited by local communities, which are a part of the Integrated Conservation and Development Programme (ICDP) (Tharchen 2013).

VEGETATION TYPES

The most recent and first-ever National Forest Inventory (NFI) estimates 27,309 km^2 or 71% of the country's total geographical area as forest cover (FRMD 2016). The inventory also confirms that Bhutan has one of the highest forest (stem) densities (213 trees per hectare) in the world, which mainly comprise broadleaf (50%) and coniferous (20%) forests. Details of forest type, area, characteristics, and distribution are provided in Table 9.2.

REVIEW OF FIRES IN BHUTAN

SLASH-AND-BURN AGRICULTURE

There are very limited and sporadic details of time-series data or written records on forest fire in Bhutan. Shifting cultivation, locally known as *Tseri*, was an age-old tradition among the local inhabitants clearing and burning biomass for agriculture

TABLE 9.1
LULC of Bhutan

Land Cover Class	Sub Class	Area (ha)	Area (%)
Alpine scrubs		130097.72	3.39
Built-up		7457.03	0.19
Cultivated agriculture		105682.4	2.75
	Chhuzhing	31891.87	0.83
	Kamzhing	68260.64	1.78
	Orchards	5529.92	0.14
Forests		2717161.64	70.77
	Blue pine	101155.06	2.64
	Broadleaf	1763899.46	45.94
	Chir pine	101537.45	2.64
	Fir	230983.99	6.02
	Mixed conifer	519585.68	13.53
Landslides		3730.22	0.10
Meadows		96273.61	2.51
Moraines		14393.94	0.37
Non-built-up		595.89	0.02
Rocky outcrops		159455.55	4.15
	Rocky outcrops	119754.16	3.12
	Scree	39701.39	1.03
Shrubs		374032.56	9.74
Snow and glacier		205343.63	5.35
Water bodies		25175.78	0.65
	Lakes	6252.58	0.16
	Rivers	18923.20	0.49
Grand total		3839400.00	100.00

Source: FRMD (2017)

as well as to reduce pests and diseases. Shifting cultivation was the second-largest agricultural practice, accounting for almost 6.7% of cultivated land in 1978, next to dryland (9.6%) agriculture (IFAD 1981). It was most prevalent in elevations ranging from 2500 to 3800 m in areas dominated by grasses, and short shrubs interspersed with blue pine (Pinus wallichiana), in slopes ranging from 20% to 40% (Roder et al., 1993). Though the land was kept fallow for almost 15–20 years, continuous grazing reduced the aboveground biomass; therefore, farmers had to depend on litter and roots in soil mounds and add blue pine biomass collected from nearby forests to add fuel while burning the debris in preparation for cultivation. Since *Tseri* was also the main reason for forest fires, the Royal Government of Bhutan (RGoB, 1969) passed the "Bhutan Forest Act of 1969," which prohibited fresh clearance of land for shifting

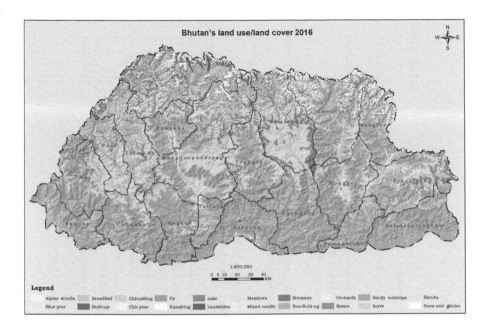

FIGURE 9.3 Bhutan land use/cover (2016) map. (Forest Resource Management Division.)

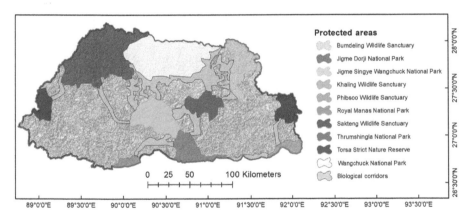

FIGURE 9.4 Protected areas in Bhutan. (Forest Resource Management Division.)

cultivation, and reserved the right to withdraw the practice of shifting cultivation in areas where it was practiced prior to the issue of this act if it endangered the safety of the highways and public property. In 1974 (RGoB, 1974), National Forest Policy was formulated, which clearly states that shifting cultivation needs to be abolished if forests have to be conserved, as the composition and growth of natural vegetation were gradually declining after each cycle of shifting cultivation. This policy raised specific problems to the shifting cultivators, many of whom were solely dependent on it for their livelihood, and phasing it out was not actively pursued (FAO, 1995). Dryland agriculture, which was mostly under shifting cultivation, occupied almost

TABLE 9.2
Forest Types and Distribution

Forest Type	Forest Area (ha)	Forest Cover (%)	Characteristics	Altitude Range (in meters a.s.l.)
Subtropical forest	241,804	6	• Contain many tropical genera and species, forming dense jungle with scattered sal trees in some areas	200–1000
Warm broadleaf forest	693,683	18	• Type of subtropical forest, but occurs at higher altitude with lower rainfall • Contains mixture of evergreen and deciduous broadleaf species • Many of the tropical genera, e.g., Duabanga, Pterospermum, and Tetrameles, are absent	1000–2000
Chir pine forest	98,563	3	• Low-altitude xerophytic forest occurring in the deeper dry valleys of Bhutan • Almost no other tree species occur in such forest other than chir pine	900–1800
Cool broadleaf forest	986,765	26	• Found on moist exposed slopes • Mixed forest in which oaks are less common. Both deciduous and evergreen, e.g., Lauraceae, and Exbucklandia are abundant together with dense shrubs, climbers, and epiphytes	2000–2900
Evergreen oak forest	31,464	1	• Characteristic feature of some parts of central Bhutan • Composition varies according to altitude and rainfall • At lower levels, Castanopsis hystrix and C. tribuloides are often dominant. Quercus lamellosa are more common in higher elevations • With increasing dryness, more xerophytic Quercus species, e.g., Q. lanata, Q. griffithii, and Q. semecarpifolia, and Pinus wallichiana are seen • Not much of shrub layer. Shady humid floors are dominated by small herbs	1800–2000
Blue pine forest	137,230	4	• Temperate equivalent of chir pine forest occupies the dry valleys of Bhutan • Blue pine dominant with Quercus species in some places • Xerophytic shrubs occur and herbs mostly appear during the monsoon season	2100–3000

(Continued)

TABLE 9.2 (Continued)

Forest Types and Distribution

Forest Type	Forest Area (ha)	Forest Cover (%)	Characteristics	Altitude Range (in meters a.s.l.)
Spruce forest	40,183	1	• Spruce forest with hemlock and fir forests occupies the montane cloud forest zone of Bhutan • Often mixed with each other, but separate forests can frequently be recognized • Spruce are found at lower altitude than hemlock and fir	2700–3100
Hemlock forest	88,327	2	• Appears at higher altitude than spruce where Tsuga dumosa is the dominant species mixed with spruce and fir • Shrubby and arborescent rhododendrons are frequent with dense growth of ferns, lichens, and bryophytes	2800–3100
Fir forest	352,552	9	• Occurs in the highest ridges of Bhutan below tree line, where huge tracts are covered by no other tree species than fir (Abies densa) and some hemlock and birch • Luxuriant undergrowth of rhododendrons and other shrubs with many small herbs on mossy ground layer is found • As tree lines approach, the firs become stunted and are mixed with junipers and smaller rhododendron species	3300–3800
Juniper–rhododendron scrub	57,242	1	• Moist scrub vegetation occurs above tree line throughout northern and central Bhutan • Consists of scattered shrubs of junipers, rhododendron, and Potentilla arbuscula but with rich herb layer appearing during the monsoon • Damp grassy meadow commonly found in this zone	3700–4200
Dry alpine scrub	2654	0	• Xerophytic vegetation including juniper, rhododendron, and scrub	4000–4600

Source: Compiled from FRMD (2016)

60% (54,900 ha) of agricultural land, while the remaining land (32,800 ha) was utilized for wetland agriculture (RGOB, 1990). Thus, the practice of shifting cultivation at a very local level was reported till 2010 (MoAF, 2010), and the imposition of the ban on shifting cultivation was reiterated in the National Biodiversity Strategies and Action Plan of Bhutan in 2014 (RGoB, 2014). Though the practice was formerly widespread even in national parks, its intermediate disturbance to biodiversity is considered to have ceased, and forest regenerated in most of the abandoned sites (Siebert and Belsky, 2014). However, adding forest litter in the manure and burning it and burning biomass at the initial stage to clear the fields even for permanent cultivation is a common practice even now. In the present form of agricultural practices, residues from both wetlands dominated by paddy cultivation and dryland agriculture where maize, potato, and other vegetables are cultivated are mostly used as fodder for the cattle. However, farmers still prefer to burn the vegetation in the vicinity of their agricultural land to clear the dry forest litter and prevent their crops from pests and diseases. Such practices often lead to forest fires.

FIRE INCIDENCES AND AREA BURNT

Long-term and detailed data and information on forest fire in Bhutan are almost nonexistent. International Forest Fire News (IFFN) was one of the only sources of data that provided some information about a forest fire in Bhutan between 1981 and 1985 (UNISDR 1994) and between 1988 and 2004 (UNISDR 2006). Information such as location and forest type is damaged, and fire drivers are collected by the forestry officials at the local level, but these are not readily available. However, from 2008 onward, the Department of Forests and Park Services (DoFPS), Bhutan, has been compiling details about forest fire, including damage and loss of life and property (RGoB, 2017). The Forest Fire Rules of Bhutan (2012) include prescribed reporting forms for firsthand information and detailed forest fire reports, including a GPS survey of burnt areas. The information is meant to be used to determine the future course of action and rehabilitation. In the following sections, we review the details on the number of fire incidences and burnt areas in Bhutan.

The IFFN of UNISDR (2006) documents the number of forest fire incidents from 1988 to 2004 and the area burned between 1993 and 2004. Almost 61 fire incidents and almost 59,000 acres of forested area burnt per annum within this period. Maximum damage was caused in 1993, as it destroyed more than 180,000 acres of forest in 84 forest incidents. However, in some years like 1999 and 2000, though there were more than 100 fire incidents, the area burnt was around 83,000 acres. Fires in Wangdue destroyed almost 37,000 acres of forest in 2006. It was estimated that on average, almost 50 fire incidents occur annually (FAO, 2001), leading to forest degradation and loss of biodiversity. The number of forest fires between 1981 and 1985, including the damage caused, is provided in Table 9.3.

The Department of Forestry (DoF) recorded 526 incidents of a forest fire, affecting nearly 70,000 ha of forest between 1999/2000 and 2007/2008. Forest fires are more widespread and recurrent in the eastern region, accounting for almost two-thirds of the forest area burnt during this period. The May 2007 fire in *Athang, Wangdue*

TABLE 9.3
Fire Incidence and Damage Caused between 1981 and 1985 in Bhutan

Year	Number of Fires	Area Burnt (ha)	Damage Nugultrun (@1 US$ = 12 NU)
1981–1982	74	12,843	111,104,400
1982–1983	64	5487	54,531,700
1983–1984	47	7243	42,337,000
1984–1985	47	3943	22,041,000

Source: UNISDR (1994)

Phodrang Dzongkhag, destroyed more than 15,000 ha of forest and killed numerous wildlife and few critically endangered white-bellied herons (RGOB, 2009).

The amount of area burned between 2008 and 2013 destroyed 47,500 acres of forest land (RGoB 2014). There were 66 forest fire incidences in 2013–2014, which affected 46,694 acres of forests. Even though there were more incidences (72 cases) and as many as seven fire incidences in 5 days in 2016 (Dorji, 2016), the damage (21,057 acres) was almost half of the area destroyed by a forest fire in 2013–2014. The incidence of forest fire (37 cases) as well as the area (16,214 acres) destroyed by forest fire further reduced in 2017–2018. It is also estimated that almost 10,000 acres of blue pine forest and 6000 acres of chir pine were destroyed, while most of the forest fires (21 cases) were reported in the chir pine forest. But, nearly half of the damage (8000 acres) was caused by a single forest fire in Paro (Nima, 2018) due to a short circuit from power transmission lines. The fire lasted for over a week and took more than 500 men to contain it (Gyelmo, 2016) finally.

Data received through personal communication with Kinley Tshering (Forest Fire Management Program, Department of Forests and Park Services, Ministry of Agriculture and Forests) record 510 incidences of a forest fire from 2008 to 2017. On average, about 50 fire incidences take place per year. During 2008 and 2009, the number of fire incidences was 38 and 79, respectively; however, the damage caused has been drastically reduced from the previous years. For example, in 2009, 79 fire incidences spread over 3895 acres of forest, whereas in 2014, there were only 55 incidences which burnt almost 35,573 acres of forest (Figure 9.5).

SATELLITE FIRE DATA AND TRENDS

In this study, we used both the MODIS (collection 6.1) and VIIRS datasets from 2002 to 2019 to characterize Bhutan's fires. Our analysis suggested a range of 134–503 fire counts with a mean of 308 fires per year (averaged across 2002–2009) with a peak during March. Further, of different years, 2010 had the highest with 504 fire counts and 2015 the least with 134 fire counts (Figure 9.6; Table 9.4). MODIS-derived burnt areas for different years (2002–2016) derived from the Global Fire Emissions Database suggested an annual average area of 136.3 km² burnt every year with the highest 2018.14 km² during 2012 and the least area of 59.33 km² burnt during 2003 (Figure 9.7).

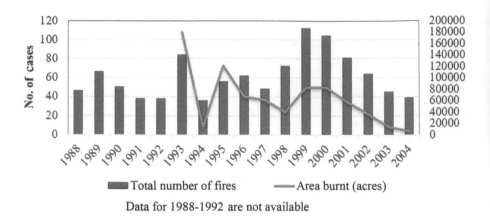

FIGURE 9.5   Incidence and area damaged by forest fire (1988–2004) (UNISDR, 2006).

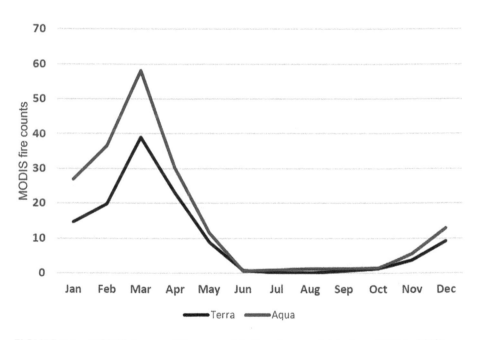

FIGURE 9.6   MODIS Aqua and Terra monthly fires (averaged data from 2002 to 2019).

VIIRS 1-Band 375 m product suggested a range of 522–2166 fire counts with a mean of 1125 fires per year (averaged across 2012–2019) with a peak during March. Also, 2012 showed the highest fire counts with 923 and 585 fire counts during March and April of that year and the least fires during 2019 (Table 9.5). Overall, VIIRS data due to higher resolution showed 3.7 times more fires than MODIS datasets. Further, both MODIS and VIIRS data showed a decreasing trend of fires in Bhutan, suggesting strong policy and regulatory measures.

TABLE 9.4

MODIS (Aqua and Terra Combined Fires) for Bhutan (2002–2019)

Year	Jan	Feb	Mar	Apr	May	Jun	Jul	Aug	Sep	Oct	Nov	Dec	Total
2002	5	34	46	6	8	0	0	0	1	4	11	29	144
2003	66	25	39	36	14	4	0	0	0	2	9	5	200
2004	14	49	102	13	19	0	0	4	1	2	23	37	264
2005	27	25	66	99	2	0	0	0	10	0	6	28	263
2006	129	60	167	54	3	0	4	5	2	2	1	11	438
2007	33	8	81	12	200	1	3	1	0	0	6	29	374
2008	50	75	42	61	9	0	0	1	4	0	8	25	275
2009	59	74	187	73	12	3	4	0	5	2	6	14	439
2010	88	87	233	41	14	2	0	1	0	0	0	38	504
2011	22	73	141	36	1	1	0	2	5	17	15	17	330
2012	14	59	157	159	53	0	0	0	0	2	27	32	503
2013	62	75	59	20	12	0	0	0	0	0	13	31	272
2014	15	59	84	204	7	2	2	0	4	10	8	23	418
2015	12	60	215	5	4	0	0	0	0	0	2	3	301
2016	11	63	59	59	5	2	0	1	0	4	13	47	264
2017	52	121	23	59	2	0	0	3	0	1	3	10	274
2018	31	42	31	19	1	8	4	3	0	3	7	19	168
2019	56	43	16	4	0	0	1	4	0	0	10	0	134

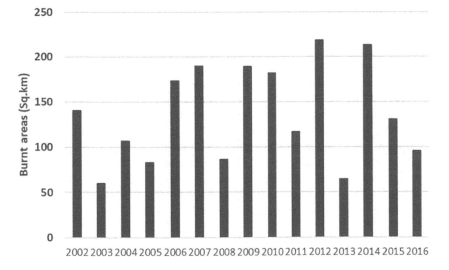

FIGURE 9.7   MODIS-derived burnt areas (in km²).

Drivers of Forest Fires

This section provides a review of various studies that highlighted the drivers of fires in Bhutan. The National Soil Services Centre considers forest fires to be a recurrent and widespread phenomenon leading to gradual degeneration of forest,

TABLE 9.5

VIIRS 375 m I-Band Active Fires Retrieved from 2012 to 2019

Year	Jan	Feb	Mar	Apr	May	Jun	Jul	Aug	Sep	Oct	Nov	Dec	Total
2012	56	238	923	585	147	0	0	5	2	2	33	175	2166
2013	214	222	154	114	30	3	0	7	1	2	27	75	849
2014	66	179	319	979	42	6	3	2	17	42	52	67	1774
2015	41	224	1008	20	8	0	13	3	1	3	3	37	1361
2016	48	274	154	218	20	1	0	8	0	7	24	147	901
2017	201	364	101	96	8	1	0	2	2	5	5	63	848
2018	96	144	81	87	3	21	12	9	2	19	47	61	582
2019	200	182	71	52	1	4	0	12	0	0	0	0	522

land degradation, and change in the ecosystem (NSSC, 2014). Clearing of land for agriculture and horticulture, regeneration of lemongrass and forage for cattle, and prevention against wildlife incursion are the main reasons for forest fires' deliberate causes. Moreover, a forest fire is also reported to be an effective means to control pests such as wild boar, deer, monkey, and bear, which are known to depredate crops and livestock, especially where settlements are located near a dense forest. Gyelmo (2016) points out that a vast range of causes such as the burning of agricultural waste, children playing with flammable materials, smokers disposing of burning matches or cigarettes, picnickers, cattle herders, roadside workers and hikers making campfires, and even electrical short circuits make it difficult for officials to identify the culprit behind a particular fire. Natural causes such as a fire ignited by rolling stones and lightning, and those originating from a construction site, beehive collection, and fire originating from across the border are also reported to cause a forest fire.

According to UNISDR (1994), 70% of forest fires recorded during 1981–1985 were caused by fires escaping from agricultural fields while burning debris (40%) and burning forest to regenerate the growth of grass for cattle grazing (30%), and the remaining 30% were caused by uncontrolled fire from camping, cooking, warming, and road maintenance activities. However, between 1993 and 2005, 60% of the forest fires were due to intentional burning to aggravate the growth of lemongrass and new grass for cattle, including the burning of agriculture debris. Smokers, children playing with an ignition source, roadside workers, picnickers, campfires, and other accidental causes contributed to 40% of forest fires (UNISDR 2006). A case study conducted by Wangmo (2019) suggests that the drivers of 77% of the forest fires between 2001 and 2007 were unknown but speculated to be due to anthropogenic causes, whereas 21% were accidental and 2% intentional, consuming a total of 154,692 ha of the forest. Thus, a forest fire is a deliberate, recurrent, and widespread phenomenon in Bhutan, leading to gradual degeneration, land degradation, and ecosystem change (NSSC, 2014).

The MoA ranks timber harvesting, firewood, and forest fires among the top three forest degradation drivers, resulting in total annual degradation of 359,914 m³/

TABLE 9.6

Main Drivers and Emission of GHGs

Driver	Annual Degradation (m³/ha)	% of Annual Degradation	Annual GHG Emission (tCO₂e/ha)	% Annual GHG Emission	Ranking
Timber harvesting	163,009	45	117,394	44	1st
Firewood	84,936	24	61,168	23	2nd
Forest fires	111,969	31	88,560	33	3rd
Total	359,914		267,122		

Source: Adapted from MoA (2017a).

ha and GHG emission of 267,122 tCO_2e/ha (MoA, 2017b). Together, forest fire and firewood contribute to almost 55% of annual degradation and GHG emission (Table 9.6).

Lemongrass Cultivation and Fires

Anthropogenic activities cause almost all forest fires in Bhutan. Most of the forest fires are deliberately set to invigorate the growth of pastures or commercially valuable grasses such as lemongrass or occur due to general public carelessness (MoA, 2004). Lemongrass grows at elevations ranging from 600 to 1700 MSL and is associated with chir pine as it grows best under open conditions or stands with low canopy cover. Studies conducted by Darabant (2017) reveal that forest fire does not affect biomass yield but increases the yield of essential oil from lemongrass. Its high value and demand as raw material for synthetic vitamin A and ingredient in perfumes, soaps, and cosmetics have led to the forest's intentional burning.

Climate Change and Fires

Climate change is likely to aggravate forest fire as a rise in temperature and windy conditions including drought combined with more frequent lightning is expected to cause a greater risk of forest fires, such as two consecutive forest fires in Trashigang in 2016 caused by lightning (Tshering, 2016). Higher temperatures decrease soil moistures, and extended periods of drought owing to climate change are likely to enhance the frequency and area affected by wildfires (NEC, 2016). Though the impacts of climate change on Bhutan's biodiversity and ecosystems *per se* are yet to be ascertained (National Biodiversity Strategies and Action Plan of Bhutan (RGoB), 2014), there are observations of blue pine encroaching into spruce/maple/birch forests and decline in certain species of forests (Gratzer et al., 1997) and montane cloud forests due to moisture stress (Wangda and Ohsawa, 2010). Climate change will accelerate the incidence and intensity of forest fire and further exacerbate pests and diseases and accelerate invasive alien species' growth.

DISCUSSION

NEEDS AND PRIORITIES OF FIRE MITIGATION AND MANAGEMENT

The conservation of Bhutan's rich biodiversity, success in achieving the mandate to preserve 60% of the area under forest cover, pledge to remaining carbon neutral, and attaining sustainable development goals will rely heavily on mitigation and forest fire management. The country's reliance on hydropower, agriculture, and forestry sector for economic development will largely depend directly or indirectly on forest resources. Bhutan's vulnerability to climate change impacts makes the country even more vulnerable to a forest fires and vice versa. The second National Adaptation Programme of Action (NAPA) includes forest fire management to address the risks of climate-induced disasters and increase their resilience to climate change (NEC, 2016). However, currently available geospatial techniques, including satellite data in detecting, monitoring, assessing, and studying the impacts of a forest fire, have not been explored and applied to their full potentials.

Strict rules and regulations such as the "Forest and Nature Conservation Act 1995" that prohibits setting forest fires and imposes fines and penalties and even imprisonment have not successfully mitigated forest fire in Bhutan. Enforcing the existing fines and penalties is too severe on offenders, who are generally the poor peasants or children. The lack of ad hoc implementation of research on forest fire management and public education and awareness, and the absence of adequate resources, including firefighting training and equipment, are other necessities for managing forest fire (NSSC, 2014).

The DoFPS has prepared detailed forest and nature conservation rules and regulations, which include acquiring a permit from the local administration for burning agricultural debris, which is decided upon weather condition, construction of adequate fire lines, regular maintenance around substations, and other power line installations, including activities involving roadsides, campfire, bonfire, logging, and manufacturing (RGoB, 2017). These rules and regulations along with forest fire prevention campaign conducted by the FFMS (MoAF, 2015), an adaptation of appropriate preventive and controlled burning techniques for pasture, agriculture, and horticulture, and awareness about the dangers and effects of a forest fire, including quick response supported by various stakeholders and local communities, are primary steps to reduce and manage forest fire at the grassroot level.

The Regional Database System (http://rds.icimod.org) developed by ICIMOD provides data on "Active Fire Incidents in Bhutan" based on MODIS images from March 2012 to June 2016, which can be downloaded as point shapefiles. The interactive mapping application for visualization of active fire locations, including *Dzongkhag*-level fire count statistics, is a useful tool for identifying and analyzing spatiotemporal distribution of forest fire (http://geoapps.icimod.org/Bhutanforestfire/#). However, many factors such as cloud cover, dense tree canopy, the small size of the fire, satellite overpass, the confidence level of fire pixels, and lack of ground validation are impediments in using such data. Obtaining accurate and timely information, including early warning of a forest fire, fire's strength, and probable causes; reaching out to areas at risk; and improving preparedness are major challenges to stopping forest fire (ICIMOD, 2015). It is also reported that the Ministry of Agriculture and Forests (MoAF) has carried out fire hazard mapping based on vulnerability, past forest fire incidences, proximity to forest, roads, settlements, aspects, and several other factors (MoAF, 2015). This can be further improved to map fire-prone areas, predict the behavior and impacts of a forest fire, and estimate the actual loss in the quality and quantity of forest species, which are some of the immediate priorities for mitigating and managing forest fire.

INVOLVEMENT OF LOCAL GOVERNMENT AND COMMUNITIES IN MANAGING FIRES

Forest fire programs are decentralized to the *Dzongkhag* Administration (DA), Department of Forest Services (DFS), and local communities (MoA, 2004). The local government is mainly responsible for fire prevention and suppression. Stakeholders such as NGOs, private sectors, and municipal and local communities are mostly involved in fire suppression (UNISDR, 2006). The functions and responsibilities of the MoAF, including local governments, schools and institutions, and individuals for prevention and control of forest fire, are specified in the Forest Fire Rules of Bhutan (2012).

As of December 2017, 733 community forests (CFs) involving 30,352 households spread over 20 *Dzongkhags* aim to promote sustainable forest management through people's participation (RGoB, 2017) and they can play a crucial role in mitigating and managing forest fires. The geographical conditions such as rugged terrain, steep slopes, dense vegetation, and weather conditions pose severe fire control challenges in Bhutan. However, a participatory approach involving local people can offer practical solutions, as most fires are anthropogenic.

CONCLUSIONS

Forest fire is a significant problem and concern for preserving pristine biodiversity and a healthy environment of Bhutan. The country's socioeconomic development, which is heavily dependent on hydropower, forestry, and agriculture, can be severely affected if the areas destroyed by forest fire continue to rise even though the incidence has reduced. In this study, we reviewed the land cover, forestry, fires, forest conservation, fire management, and mitigation aspects. Remotely sensed data from MODIS and VIIRS helped characterize the temporal fire events, including the intensity, timing of burning, and the area burnt. These statistics can be effectively used for fire management and mitigation efforts. Further, we also infer the need to integrate field data, involve relevant stakeholders, invest in building robust spatial data infrastructure, and implement viable management practices to address fire problems in Bhutan.

REFERENCES

Andreae, M.O., and Merlet, P. 2001. Emission of trace gases and aerosols from biomass burning. *Global Biogeochemical Cycles*, 15(4), 955–966.

Badarinath, K.V.S., Kharol, S.K., Krishna Prasad, V., Kaskaoutis, D.G., and Kambezidis, H.D. 2008a. Variation in aerosol properties over Hyderabad, India during intense cyclonic conditions. *International Journal of Remote Sensing*. 29(15), 4575–4597.

Badarinath, K.V.S., Kharol, S.K., Prasad, V.K., Sharma, A.R., Reddi, E.U.B., Kambezidis, H.D., and Kaskaoutis, D.G. 2008b. Influence of natural and anthropogenic activities on UV Index variations–A study over tropical urban region using ground-based observations and satellite data. *Journal of Atmospheric Chemistry*. 59(3), 219–236.

Biswas, S., Lasko, K.D., and Vadrevu, K.P. 2015a. Fire disturbance in tropical forests of Myanmar—Analysis using MODIS satellite datasets. *IEEE Journal of Selected Topics in Applied Earth Observations and Remote Sensing*. 8(5), 2273–2281.

Biswas, S., Vadrevu, K.P., Lwin, Z.M., Lasko, K., and Justice, C.O. 2015b. Factors controlling vegetation fires in protected and non-protected areas of Myanmar. *PLoS One*. 10(4), e0124346.

Chowdary, V., and Gupta, M.K. 2018. Automated forest fire detection and monitoring techniques: A survey. In: R. Singh et al. (Eds). *Intelligent Communication, Control and Devices, Advances in Intelligent Systems and Computing*. Springer Nature Singapore Pvt Ltd. pp. 1111–1117. doi: 10.1007/978-981-10-5903-2_116.

Couteron, P., Barbier, N., Deblauwe, V., Pélissier, R., and Ploton, P. 2015. Texture analysis of very high spatial resolution optical images as a way to monitor vegetation and forest biomass in the tropics. In: Murthy, M.S.R., Wesselman, S., and Gilani, H. (Eds). *Multi-Scale Forest Biomass Assessment and Monitoring in the Hindu Kush Himalayan Region: A Geospatial Perspective*. ICIMOD, Kathmandu. pp. 157–164.

Darabant, A. 2017. Fire as a land management tool in Chir Pine forests with Lemon Grass understory. *Research Gate*. doi: 10.13140/RG.2.2.18308.94085.

Dorji, J.K. 2016. Forest Fires in Chuzom and Wangdue Yet to be Contained. Kuensel, February 17, 2016. Accessed from http://www.kuenselonline.com/forest-fires-in-chuzom-and-wangdue-yet-to-be-contained/

FAO. 1995. Shifting Cultivation in Bhutan: A Gradual Approach to Modifying Land Use Patterns, A Case Study from Pema Gatshel District, Bhutan. Community forestry Case Study, Series 11. Accessed from http://www.fao.org/3/V8380E/V8380E00.HTM#TopOfPage; http://www.fao.org/3/V8380E/V8380E05.htm.

FAO. 2010. *Global Forest Resources Assessment 2010: Country Report Bhutan*. Food and Agriculture Organization of the United Nations, Rome.

FAO. 2001. *Global Forest Fire Assessment, 1990–2000. The Forest Resource Assessment Programme*, Food and Agriculture Organization of the United Nations, Rome, p. 170.

Forest Fire Rules of Bhutan. 2012. *Royal Government of Bhutan*. Ministry of Agriculture and Forests, Department of Forest and Park Services.

FRMD. 2016. *National Forest Inventory Report Volume-I*. Forest Resource Management Division, Department of Forests and Park Services, ISBN: 978-99936-743-1-3.

FRMD. 2017. *Land Use and Land Cover of Bhutan 2016. Maps and Statistics*. Forest Resources Management Division Department of Forests & Park Services Ministry of Agriculture and Forests, Thimphu, ISBN: 978-99936-743-2-0.

Gilani, H., Shrestha, H.L., Murthy, M., Phuntso, P., Pradhan, S., Bajracharya, B., and Shrestha, B. 2015. Decadal land cover change dynamics in Bhutan. *Journal of Environmental Management*. 148, 91–100.

Goldammer, J.G., and Seibert, B., 1990. The impact of droughts and forest fires on tropical lowland rain forest of East Kalimantan. In: Goldammer, J.G. (Ed). *Fire in the Tropical Biota*. Springer, Berlin, Heidelberg: pp. 11–31.

Gratzer, G., Rai, P.B., and Glatzel, G. 1997. Ecology of the Abies densa forests in IFMP Ura, Bhutan. In RGoB 2014. National Biodiversity Strategies and Action Plan of Bhutan. National Biodiversity Centre, Ministry of Agriculture and Forests, Royal Government of Bhutan.

Gupta, P.K., Prasad, V.K., Sharma, C., Sarkar, A.K., Kant, Y., Badarinath, K.V.S., and Mitra, A.P. 2001. CH_4 emissions from biomass burning of shifting cultivation areas of tropical deciduous forests–experimental results from ground-based measurements. *Chemosphere-Global Change Science*. 3(2), 133–143.

Gyelmo, D. 2016. Forest Fires Burning up Bhutan's Wilderness, thethirdpole.net. Accessed from https://www.thethirdpole.net/en/2016/03/21/forest-fires-burning-up-bhutans-wilderness/

Hayasaka, H., and Sepriando, A. 2018. Severe air pollution due to peat fires during 2015 super El Niño in Central Kalimantan, Indonesia. In: Vadrevu, KP, Ohara, T., and Justice, C. (Eds). *Land-Atmospheric Research Applications in South/Southeast Asia*. Springer, Cham. pp. 129–142.

Hayasaka, H., Noguchi, I., Putra, E.I., Yulianti, N., and Vadrevu, K. 2014. Peat-fire-related air pollution in Central Kalimantan, Indonesia. *Environmental Pollution*. 195, 257–266. doi: 10.1016/j.envpol.2014.06.031.

ICIMOD. 2015. Enhancing Community Preparedness in Forest Fire Management in Nepal, ICIMOD, Nepal. Accessed from http://geoportal.icimod.org/?q=19001.

IFAD. 1981. *Appraisal of Small Farmers Development Irrigation Rehabilitation Project*. IFAD, Thimpu.

Inoue, Y. 2018. Ecosystem carbon stock, atmosphere and food security in slash-and-burn land use: A geospatial study in mountainous region of Laos. In: Vadrevu, K.P., Ohara, T., and Justice, C. (Eds). *Land-Atmospheric Research applications in South/Southeast Asia*. Springer, Cham. pp. 641–666.

Israr, I., Jaya, S.N.I., Saharjo, H.S., Kuncahyo, B., and Vadrevu, K.P. 2018. Spatio-temporal analysis of land and forest fires in Indonesia using MODIS active fire dataset. In: Vadrevu, K.P., Ohara, T., and Justice, C. (Eds). *Land-Atmospheric Research Applications in South/Southeast Asia*. Springer, Cham. pp. 105–128.

Justice, C., Gutman, G., and Vadrevu, K.P. 2015. NASA land cover and land use change (LCLUC): An interdisciplinary research program. *Journal of Environmental Management*. 148(15), 4–9.

Kant, Y., Ghosh, A.B., Sharma, M.C., Gupta, P.K., Prasad, V.K., Badarinath, K.V.S., and Mitra, A.P. 2000. Studies on aerosol optical depth in biomass burning areas using satellite and ground-based observations. *Infrared Physics & Technology*. 41(1), 21–28.

Kennard, D.K., and Gholz, H.L. 2001. Effects of high-and low-intensity fires on soil properties and plant growth in a Bolivian dry forest. Plant and Soil, 234(1), 119–129.

Lasko, K., and Vadrevu, K.P. 2018. Improved rice residue burning emissions estimates: Accounting for practice-specific emission factors in air pollution assessments of Vietnam. *Environmental Pollution*. 236(5), 795–806.

Lasko, K., Vadrevu, K.P., Tran, V.T., Ellicott, E., Nguyen, T.T., Bui, H.Q., and Justice, C. 2017. Satellites may underestimate rice residue and associated burning emissions in Vietnam. *Environmental Research Letters*. 12(8), 085006.

Lasko, K., Vadrevu, K.P., and Nguyen, T.T.N. 2018a. Analysis of air pollution over Hanoi, Vietnam using multi-satellite and MERRA reanalysis datasets. *PLoS One*. 13(5), e0196629.

Lasko, K., Vadrevu, K.P., Tran, V.T., and Justice, C. 2018b. Mapping double and single crop paddy rice with Sentinel-1A at varying spatial scales and polarizations in Hanoi, Vietnam. *IEEE Journal of Selected Topics in Applied Earth Observations and Remote Sensing*. 11(2), 498–512.

MoA. 1997. *Atlas of Bhutan, Land Cover and Area Statistics of 20 Dzongkhags*. Policy and Planning Division, MoA, Thimphu.

MoA. 2004. *Bhutan Biological Conservation Complex*. Nature Conservation Division, Department of Forestry Services, Ministry of Agriculture.

MoA. 2017a. Drivers of Deforestation and Forest Degradation in Bhutan Final Report, Watershed Management Division Department of Forests and Park Services, Ministry of Agriculture and Forests Royal Government of Bhutan.

MoA. 2017b. *Atlas of Bhutan, Land use Land Cover 2016 Statistics*. Forest Resources Management Division, Department of Forests and Park Services, Ministry of Agriculture and Forests, Thimphu.

MoAF. 2010. *National Forest Policy of Bhutan*. Ministry of Agriculture and Forests, Royal Government of Bhutan, Thimphu.

MoAF. 2015. Door-to-door forest fire prevention campaign conducted. Ministry of Agriculture and Forests, RGoB, Accessed from http://www.moaf.gov.bt/door-to-door-forest-fire-prevention-campaign-conducted/

Mon, M.S., and Myint, A.A. 2015. Estimating above ground biomass of tropical mixed deciduous forests using Landsat ETM+ imagery for two reserved forests in Bago Yoma Region, Myanmar. In: Murthy, M.S.R., Wesselman, S., and Gilani, H. (Eds). (2015) *Multi-Scale Forest Biomass Assessment and Monitoring in the Hindu Kush Himalayan region: A Geospatial Perspective*. ICIMOD, Kathmandu. pp. 165–177.

Murthy, M.S.R., Wesselman, S., and Gilani, H. 2015. *Multi-Scale Forest Biomass Assessment and Monitoring in the Hindu Kush Himalayan Region: A Geospatial Perspective*. International Centre for Integrated Mountain Development, Nepal. pp. 3–19. Retrieved from http://lib.icimod.org/record/30997/files/Biomass%20book.pdf.

National Forest Policy of Bhutan. 2011. http://www.dofps.gov.bt/wp-content/uploads/2020/06/Forest-Policy-of-Bhutan-2011.pdf

NEC. 2016. Bhutan Environment Outlook 2016 (Draft). National Environment Commission Secretariat, Royal Government of Bhutan, Thimphu. Accessed www.nec.gov.bt/nec1/wp-content/uploads/2016/05/BEO-2016-final.doc.

Nima. 2018. Fire Destroyed about 16,000 acres of Forest Reserve in 2017-18. Kuensel, August 8, 2018. Accessed from http://www.kuenselonline.com/fire-destroyed-about-16000-acres-of-forest-reserve-in-2017-18/).

NSSC. 2014. The National Action Program to Combat Land Degradation. National Soil Services Centre (NSSC), Department of Agriculture, Ministry of Agriculture and Forests, Royal Government of Bhutan, Thimphu. http://www.nssc.gov.bt/wp-content/uploads/2014/08/Aligned-NAP-FINAL-REPORT.pdf.

Oanh, N.T.K., Permadi, D.A., Dong, N.P., and Nguyet, D.A. 2018. Emission of toxic air pollutants and greenhouse gases from crop residue open burning in Southeast Asia. In: Vadrevu, K.P., Ohara, T., and Justice, C. (Eds). *Land-Atmospheric Research applications in South/Southeast Asia*. Springer, Cham. pp. 47–68.

Peuhkurinen, J., Kauranne, T., Hämäläinen, J., and Gautam, B. 2015. Estimation of forest biomass using the Lidar-assisted multi-source programme. In: Murthy, M.S.R., Wesselman, S., and Gilani, H. (Eds). *Multi-Scale Forest Biomass Assessment and Monitoring in the Hindu Kush Himalayan Region: A Geospatial Perspective*. ICIMOD, Kathmandu. pp. 112–123.

PHCB. 2017. *Population & Housing Census of Bhutan*, National Report. National Statistics Bureau, Royal Government of Bhutan, Thimphu, ISBN: 978-99936-28-50-7.

Phuntsho, Tshering, K., Rai, A., et al. 2015. Bhutan's geospatial information system for forest biomass assessment. In: Murthy, M.S.R., Wesselman, S., and Gilani, H. (Eds). *Multi-Scale Forest Biomass Assessment and Monitoring in the Hindu Kush Himalayan Region: A Geospatial Perspective*. International Centre for Integrated Mountain Development, Nepal. pp. 30–35. Retrieved from http://lib.icimod.org/record/30997/files/Biomass%20book.pdf.

Prasad, V.K., and Badarinth, K.V.S., 2004. Land use changes and trends in human appropriation of above ground net primary production (HANPP) in India (1961–98). *Geographical Journal*. 170(1), 51–63.

Prasad, V.K., Gupta, P.K., Sharma, C., Sarkar, A.K., Kant, Y., Badarinath, K.V.S., Rajagopal, T., and Mitra, A.P. 2000. NO_x emissions from biomass burning of shifting cultivation areas from tropical deciduous forests of India–estimates from ground-based measurements. *Atmospheric Environment*. 34(20), 3271–3280.

Prasad, V.K., Kant, Y., and Badarinath, K.V.S. 2001. CENTURY ecosystem model application for quantifying vegetation dynamics in shifting cultivation areas: A case study from Rampa Forests, Eastern Ghats (India). *Ecological Research.* 16(3), 497–507.

Prasad, V.K., Kant, Y., Gupta, P.K., Elvidge, C., and Badarinath, K.V.S. 2002. Biomass burning and related trace gas emissions from tropical dry deciduous forests of India: A study using DMSP-OLS data and ground-based measurements. *International Journal of Remote Sensing.* 23(14), 2837–2851.

Prasad, V.K., Lata, M., and Badarinath, K.V.S. 2003. Trace gas emissions from biomass burning from northeast region in India—estimates from satellite remote sensing data and GIS. *Environmentalist.* 23(3), 229–236.

Prasad, V.K., Badarinath, K.V.S., Yonemura, S., and Tsuruta, H. 2004. Regional inventory of soil surface nitrogen balances in Indian agriculture (2000–2001). *Journal of Environmental Management.* 73(3), 209–218.

Prasad, V.K., Anuradha, E., and Badarinath, K.V.S. 2005. Climatic controls of vegetation vigor in four contrasting forest types of India—Evaluation from national oceanic and atmospheric administration's advanced very high resolution radiometer datasets (1990–2000). *International Journal of Biometeorology.* 50(1), 6–16.

Prasad, V.K., Badarinath, K.V.S., and Eaturu, A. 2008a. Biophysical and anthropogenic controls of forest fires in the Deccan Plateau, India. *Journal of Environmental Management.* 86(1), 1–13.

Prasad, V.K., Badarinath, K.V.S., and Eaturu, A. 2008b. Effects of precipitation, temperature and topographic parameters on evergreen vegetation greenery in the Western Ghats, India. *International Journal of Climatology: A Journal of the Royal Meteorological Society.* 28(13), 1807–1819.

Qazi, W.A., and Gilani. H. 2015. Exploring the use of spaceborne SAR for above ground biomass measurements in the Hindu Kush Himalayan Region and Pakistan. In: Murthy, M.S.R., Wesselman, S., and Gilani, H. (Eds). *Multi-Scale Forest Biomass Assessment and Monitoring in the Hindu Kush Himalayan Region: A Geospatial Perspective.* ICIMOD, Kathmandu. pp. 102–111.

Rahman, M.M. 2015. Forest biomass assessment in Southeastern Bangladesh using Landsat ETM+ and ALOS PALSAR data. In: Murthy, M.S.R., Wesselman, S., and Gilani, H. (Eds). 2015. *Multi-Scale Forest Biomass Assessment and Monitoring in the Hindu Kush Himalayan Region: A Geospatial Perspective.* ICIMOD, Kathmandu. pp. 124–135.

Ramachandran, S. 2018. Aerosols and climate change: Present understanding, challenges and future outlook. In: Vadrevu, K.P., Ohara, T., and Justice, C. (Eds). *Land-Atmospheric Research Applications in South/Southeast Asia.* Springer, Cham. pp. 341–378.

RgoB. 1969. The Bhutan Forest Act, Royal Government of Bhutan. Ministry of Trade, Industries and Forests. Accessed from http://www.biodiv.be/bhutan/legal-instruments/acts/bhutan-forest-act-1969/download/en/1/Bhutan%20Forest%20Act%201969.pdf?action=view

RgoB. 1974. National Forest Policy 1974. Department of Forest, Ministry of Trade, Industry and Forests, Royal Government of Bhutan. Accessed from file:///C:/Users/user/Downloads/National%20Forest%20Policy%201974.pdf.

RGOB. 1990. *Statistical Year Book of Bhutan, 1990.* Central Statistical Organization, Planning Commission, Royal Government of Bhutan, Thimphu.

RGoB. 2009. *Biodiversity Action Plan 2009.* National Biodiversity Centre, Ministry of Agriculture, Royal Government of Bhutan, Thimphu.

RGoB. 2014. *National Biodiversity Strategies and Action Plan of Bhutan.* National Biodiversity Centre, Ministry of Agriculture and Forests, Royal Government of Bhutan, p. 57. Accessed from http://nbc.gov.bt/wp-content/uploads/2010/06/2015-02-12-FINAL_NBSAP2014_web.pdf.

RGoB. 2017. *Forestry Facts and Figures, 2017*. Department of Forests and Park Services, Ministry of Agriculture and Forests, Royal Government of Bhutan, Thimphu.

Roder, W., Calvert, O., and Dorji, Y. 1993. Effect of burning on selected soil parameters in a grass fallow shifting cultivation system in Bhutan. *Plant and Soil*, 149, 51–58, Springer, Kluwer Academic Publishers, Netherlands. Accessed from https://www.jstor.org/stable/pdf/42938808.pdf?refreqid=search%3Af1b3d1dc00a91e612cc7a33921e589d3.

Saharjo, B.H., and Yungan, A. 2018. Forest and land fires in Riau province; A Case study in fire prevention, policy implementation with local concession holders. In: Vadrevu, K.P., Ohara, T., and Justice, C. (Eds). *Land-Atmospheric Research Applications in South/Southeast Asia*. Springer, Cham. pp. 143–170.

Sharma, K., Robeson, S.M., Thapa, P., and Saikia, A. 2016. *Land-Use/Land-Cover Change and Forest Fragmentation in the Jigme Dorji National Park, Bhutan, Physical Geography*, Taylor & Francis, UK, pp. 1–18, doi: 10.1080/02723646.2016.1248212.

Siebert, S.F., and Belsky, J. M. 2014. Historic livelihoods and land uses as ecological disturbances and their role in enhancing biodiversity: An example from Bhutan. *Biological Conservation*. 177, 82–89. doi: 10.1016/j.biocon.2014.06.015.

Tariq, S., and Ul-Haq, Z. 2018. Satellite remote sensing of aerosols and gaseous pollution over Pakistan. In: Vadrevu, K.P., Ohara, T., and Justice, C. (Eds). *Land- Atmospheric Research applications in South/Southeast Asia*. Springer, Cham. pp. 523–552.

Tharchen, L. 2013. *Protected Areas and Biodiversity of Bhutan*. CDC Printers Pvt. Ltd., Kolkata. ISBN 978-99936-651-6-8.

Tshering, U. 2016. Lightning Sparks Two Consecutive Forest Fires in Trashigang Earlier This Month. Ministry of Agriculture and Forests, Royal Government of Bhutan. Accessed from http://www.moaf.gov.bt/lightning-sparks-two-consecutive-forest-fires-in-trashigang-earlier-this-month/#more-6561.

UNISDR. 1994. Seasonality of Forest Fires in Bhutan (IFFN No. 10 – January 1994, pp. 5–9) http://gfmc.online/iffn/country/bt/bt_1.html.

UNISDR. 2006. Fire Situation in Bhutan, International Forest Fire News (IFFN) No.34 (January–June 2006, pp. 55–63), ISBN 1029-0864 retrieved from http://gfmc.online/wp-content/uploads/07-IFFN-34-Bhutan-3.pdf.

Vadrevu, K.P. 2008. Analysis of fire events and controlling factors in eastern India using spatial scan and multivariate statistics. *Geografiska Annaler: Series A, Physical Geography*. 90(4), 315–328.

Vadrevu, K.P., and Choi, Y. 2011. Wavelet analysis of airborne CO_2 measurements and related meteorological parameters over heterogeneous landscapes. *Atmospheric Research*. 102(1–2), 77–90.

Vadrevu, K.P., and Justice, C.O. 2011. Vegetation fires in the Asian region: Satellite observational needs and priorities. *Global Environmental Research*. 15(1), 65–76.

Vadrevu, K.P., and Lasko, K.P., 2015. Fire regimes and potential bioenergy loss from agricultural lands in the Indo-Gangetic Plains. *Journal of Environmental Management*. 148, 10–20.

Vadrevu, K.P., and Lasko, K. 2018. Intercomparison of MODIS AQUA and VIIRS I-Band fires and emissions in an agricultural landscape—Implications for air pollution research. *Remote Sensing*. 10(7), 978. doi:10.3390/rs10070978.

Vadrevu, K.P., Eaturu, A., and Badarinath, K.V.S. 2006. Spatial distribution of forest fires and controlling factors in Andhra Pradesh, India using spot satellite datasets. *Environmental Monitoring and Assessment*. 123(1–3), 75–96.

Vadrevu, K.P., Badarinath, K.V.S., and Anuradha, E. 2008. Spatial patterns in vegetation fires in the Indian region. *Environmental Monitoring and Assessment*. 147(1–3), 1. doi:10.1007/s10661-007-0092-6.

Vadrevu, K.P., Csiszar, I., Ellicott, E., Giglio, L., Badarinath, K.V.S., Vermote, E., and Justice, C. 2012. Hotspot analysis of vegetation fires and intensity in the Indian region. *IEEE Journal of Selected Topics in Applied Earth Observations and Remote Sensing*. 6(1), 224–238.

Vadrevu, K.P., Giglio, L., and Justice, C. 2013. Satellite based analysis of fire–carbon monoxide relationships from forest and agricultural residue burning (2003–2011). *Atmospheric Environment*. 64, 179–191.

Vadrevu, K.P., Ohara T., and Justice, C. 2014a. Air pollution in Asia. *Environmental Pollution*. 12, 233–235.

Vadrevu, K.P., Lasko, K., Giglio, L., and Justice, C. 2014b. Analysis of Southeast Asian pollution episode during June 2013 using satellite remote sensing datasets. *Environmental Pollution*. 12, 245–256.

Vadrevu, K.P., Lasko, K., Giglio, L., and Justice, C. 2015. Vegetation fires, absorbing aerosols and smoke plume characteristics in diverse biomass burning regions of Asia. *Environmental Research Letters*. 10(10), 105003.

Vadrevu, K.P., Ohara, T., and Justice, C. 2017. Land cover, land use changes and air pollution in Asia: A synthesis. *Environmental Research Letters*. 12(12), 120201.

Vadrevu, K.P., Ohara, T., and Justice, C. (Eds). 2018. *Land-Atmospheric Research Applications in South and Southeast Asia*. Springer, Cham.

Vadrevu, K.P., Lasko, K., Giglio, L., Schroeder, W., Biswas, S., and Justice, C. 2019. Trends in vegetation fires in south and Southeast Asian countries. *Scientific Reports*, 9(1), 7422. doi:10.1038/s41598-019-43940-x.

Wangda, P., and Ohsawa, M. 2010. Temperature and Humidity as Determinants of the Transition from Dry Pine to Humid Cloud Forests in the Bhutan Himalaya. In RGoB 2014. National Biodiversity Strategies and Action Plan of Bhutan, National Biodiversity Centre, Ministry of Agriculture and Forests, Royal Government of Bhutan.

Wangdi, T., Norbu, N., Dendup, P., Singye, R., and Chettri, B.B. 2012. *In Search of Sacred Giants: An Assessment of Tsenden Forests in Bhutan*. UWICE Press, Bumthang. p. 33.

Wangmo, K. 2019. *A Case Study on Forest Fire Situation in Trashigang*. Research Gate, Bhutan. Accessed from https://www.researchgate.net/publication/268352552_A_Case_Study_on_Forest_Fire_Situation_in_Trashigang_Bhutan.

Watanabe, M., Thapa, R.B., Motohka, T., and Shimada, M. 2015. JAXA's activities for REDD+. In: Murthy, M.S.R., Wesselman, S., and Gilani, H. (Eds). 2015. *Multi-Scale Forest Biomass Assessment and Monitoring In the Hindu Kush Himalayan Region: A Geospatial Perspective*. ICIMOD, Kathmandu. pp. 85–92.

Zhang, Y.Z., Wong, M.S., and Campbell, J.R. 2018. Conceptualizing how severe haze events are impacting long-term satellite-based trend studies of aerosol optical thickness in Asia. In: Vadrevu, K.P., Ohara, T., and Justice, C. (Eds). *Land-Atmospheric Research Applications in South/Southeast Asia*. Springer, Cham. pp. 425–446.

10 Biomass Burning in Malaysia: Sources and Impacts

Justin Sentian, Franky Herman,
Vivian Kong Wan Yee, and Carolyn Melissa Payus
Universiti Malaysia Sabah, Malaysia

Mohd Sharul Mohd Nadzir
National University of Malaysia, Malaysia

CONTENTS

INTRODUCTION

Biomass burning is generally understood as the burning of biological material derived from both plants and animals which may be sourced from natural forests, planted forest and agricultural crop residues, or industrial waste and co-products (Hao and Liu, 1994; Taylor, 2010; Biswas et al., 2015a, b). Biomass burning occurrences, either by human-initiated or naturally induced burning, are widespread and largely concentrated in the tropical forests of South America, Western Africa, and South/Southeast Asia. It can be argued that human activities are the main contributor to biomass burning (Kant et al., 2000; Prasad et al., 2000; 2001; 2002; Gupta et al., 2001; Justice et al., 2015; Vadrevu et al., 2017; 2018; 2019; Vadrevu and

Lasko, 2018), although there are reports of natural biomass burning caused by volcanic activity (e.g., Ainsworth and Kauffman, 2009) and lightning (e.g., Larjavaara et al., 2005). The increased prevalence of severe biomass burning due to forest fires in ASEAN countries (Malaysia, Indonesia, Brunei, Singapore, and Timor-Leste), and particularly in Indonesia, is strongly linked with man-made changes in land use, largely related to agricultural expansion (Hayasaka and Sepriando, 2018; Lasko et al., 2017; 2018a, b; Lasko and Vadrevu, 2018). Incidences of biomass burning due to forest fires in Kalimantan, much like other fire in Indonesia, were caused by anthropogenic activities such as logging and agriculture development (palm oil plantations and rice fields) (Bowen et al., 2001; Aiken, 2004; Badarinath et al., 2008a, b; Field et al., 2009; Wooster et al., 2012; Yulianti and Hayasaka, 2013; Vadrevu et al., 2012;2013;2014a, b;2015; Vadrevu and Lasko, 2015; Vadrevu and Justice, 2011; Sloan et al., 2017). Similarly, in Malaysia, land clearing and burning for agriculture by local farmers during the prolonged dry seasons following the El Niño phenomenon in 1997/98 caused massive forest fires. About 98% (63,331 ha) of the country's biomass burning during this period involved large areas of peat forest in Sabah and Sarawak (Mat Isa, 2003).

Large-scale biomass burning that recurs almost annually in ASEAN plays a vital role in global environmental changes (Taylor, 2010), atmospheric composition and climate variation (Permadi and Oanh, 2013; Lee et al., 2017; Cui et al., 2018), biogeochemical materials' cycles (Liu et al., 2017), human health (Reddington et al., 2014), and economic activities (Schultz et al., 2008). Biomass burning emissions from forest fires, agricultural waste, and peatland burning are recognized as an important cause of greenhouse gases (GHGs) and aerosols to the atmosphere (Shi and Yamaguchi, 2014; Huijnen et al., 2016; Tacconi, 2016; Vadrevu et al., 2018). Combustion of biomass materials releases a variety of gases such as CO, CO_2, and CH_4; volatile and semi-volatile organic compounds; aldehydes; organic acids; and inorganic elements, including particulate matter (PM) (Yokelson et al., 2013; Shi and Yamaguchi, 2014; Gilman et al., 2015; Cui et al., 2018). The characteristics of these emissions depend on the natural condition of the biomass, wood species, burning stages (flaming or smoldering), weather (moisture, humidity, and temperature), and emission factors of the biomass (Jain et al., 2006; Taylor, 2010; Giglio et al., 2013; Goto and Suzuki, 2013). Biomass burning is also found to be the major source of reactive nitrogen compounds such as NO_x, PAN, HNO_3, and nitrate (Kondo et al., 2004). In Malaysia, high concentrations of PM are always associated with regional biomass burning. Based on the source apportionment method, Sulong et al. (2017) reported about 38.5% of the PM in ambient air from biomass burning during June 2015 and January 2016, and about 31% during July–September 2013 and January–February 2014 (Khan et al., 2016).

In this study, we highlight the extent of biomass burning in ASEAN countries (specifically, Malaysia and Indonesia) and how the spatial variability and temporal variability of important gases and particulates are influenced in Malaysia due to transboundary emissions and under extreme weather conditions. In reference to Malaysia and, to some extent, other regions in Southeast Asia and East Asia, assessment and discussion of biomass burning emissions and its potential impacts were highlighted.

BIOMASS BURNING IN MALAYSIA AND SOUTH ASEAN

Recent biomass burning and the resultant atmospheric pollution in Malaysia and ASEAN countries have thrown light on the importance of biomass fire management practices. Biomass burning occurs throughout the year in the ASEAN region and is becoming a major concern because of its adverse impacts on regional air quality (Chuang et al., 2016; Huang et al., 2016; Pani et al., 2016a,b). Haze episodes that are occasionally related to biomass burning and transboundary pollution can impair the Malaysian economy and the daily lives of the general population (Othman et al., 2014). Regardless of the intensity of fires and emissions, biomass burning releases atmospheric pollutants which harm human health in this region (Betha et al., 2014; Behere et al., 2015a; Behera et al., 2015b; Khan et al., 2016; Sulong et al., 2017). Previous studies have shown that air quality issues in Malaysia were not caused by transboundary and external causative factors alone but are exacerbated by local anthropogenic drivers such as agriculture practices, vehicular exhaust, and industrial emissions (Dominick et al., 2012; Sulong et al., 2017; Latif et al., 2018). In some cases, persistent, prolonged, and severe air quality issues, especially prolonged haze episodes, are associated with El Niño–Southern Oscillation (ENSO) events in the region (Latif et al., 2018; Sentian et al., 2018).

The intensity of biomass burning in Malaysia and other ASEAN regions is reflected (though not necessarily in its entirety) by fire hotspot counts as detected by satellite observations (NOAA-18/19). In the years between 2006 and 2017, based on the hotspot counts, the intensity of fires and emissions from the four main subregions in Southeast ASEAN were found the highest in Sumatera (46%), followed by Kalimantan (42%), Sabah/Sarawak (7%), and Peninsular Malaysia (5%) (Figure 10.1). During the same period, annually, the highest fire hotspot counts were observed in 2012 (13.33%), followed by 2006 (12.50%), 2014 (12.48%), 2009 (12.36%), and 2011 (10.48%). Seasonally, hotspot counts were higher during the southeast monsoon (July–October), with the highest fires in September (26%) and August (24%) (Figure 10.2).

SPATIAL AND TEMPORAL VARIATIONS OF LONG-TERM AIR QUALITY IN MALAYSIA

Malaysia's air quality is greatly influenced by the seasonal monsoon winds and transboundary pollution (Ab Manan et al., 2016, 2018; Othman et al., 2014; Khan et al., 2016). It is becoming more apparent that seasonal open biomass burning in Kalimantan and Sumatera Indonesia contributes to air quality deterioration in Malaysia (Latif et al., 2018; Sentian et al., 2018). During 2006 and 2016, monthly concentrations of the pollutants PM_{10}, CO, NO_x, and O_3 over Malaysia showed large variations (Figure 10.3).

In the northern region (Figure 10.4), long-term average concentrations of CO (0.58 ppm), NO_x (0.017 ppm), O_3 (0.019 ppm), and PM_{10} (44.55 µg/m³) revealed this region to be the second-most polluted area in Malaysia after the central subregion except PM_{10}, which was slightly lower than the central and southern subregions. Pollutants such as CO and NO_2 and PM_{10} recorded the highest values at Perai station compared to others. Mobile sources and industrial emissions are the main sources

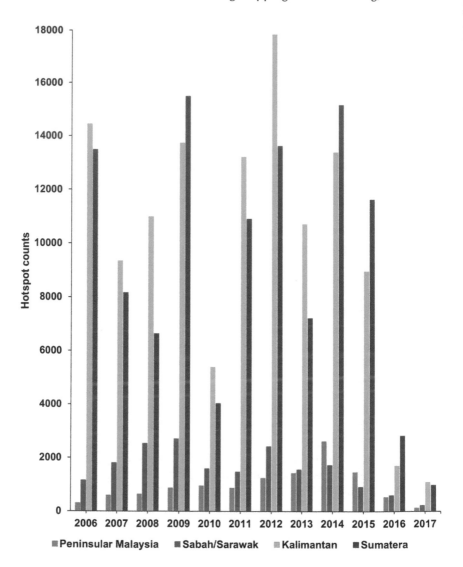

FIGURE 10.1 Annual hotspot counts in Malaysia (Peninsular Malaysia and Sabah/ Sarawak) and Indonesia (Kalimantan and Sumatera). (Raw data source: http://asmc.asean. org/asmc-hotspot/.)

for this station as Perai is an urban area with large industrial activities. However, for the O_3 pollution, the highest concentration was observed at USM station, of about 0.023 ppm, higher than the national average concentration of 0.017 ppm. Perai station also showed the highest coefficient of variation (CV) for all pollutants in the northern region. There were no notable trends of NO_2 and O_3 across all stations in the north of the subregion. For CO, there were slight decreasing trends at Perai and Kangar stations. However, small increasing trends of CO were observed at USM and Ipoh stations, which are located in larger urban areas than Perai and Kangar stations. For

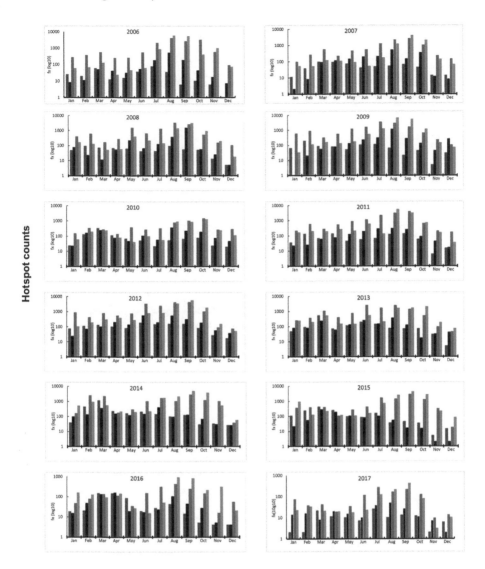

FIGURE 10.2 Monthly hotspot counts in Peninsular Malaysia (brown bar), Sabah/Sarawak (black bar), Sumatera (red bar), and Kalimantan (blue bar) for the period between 2006 and 2016. (Raw data source: http://asmc.asean.org/asmc-hotspot/.)

PM_{10}, moderate increasing trends were observed at Kangar, USM, and Ipoh stations, while a significant moderate decreasing trend was observed at Perai station.

In more urbanized, industrialized, and highly populated areas such as the central subregion, all pollutants' levels were highest compared to all other subregions in Malaysia. This subregion is also considered as a hotspot area for high levels of pollution. Between 2006 and 2016, the monthly average concentrations of pollutants in this subregion are relatively higher than the national averages: CO (0.98 ppm), NO_2 (0.04 ppm), O_3 (0.019 ppm), and PM_{10} (53.80 μg/m³). Spatially, Petaling Jaya station

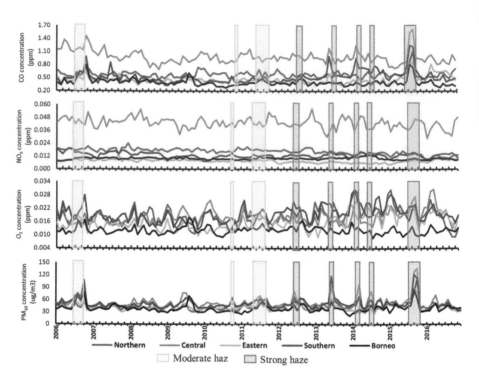

FIGURE 10.3 Monthly mean concentration of CO, NO$_x$, O$_3$, and PM$_{10}$ for each region in Malaysia over the period 2006–2016.

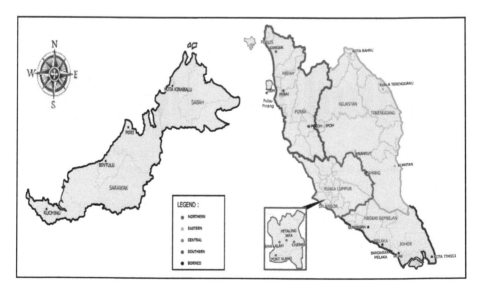

FIGURE 10.4 Location of air quality monitoring stations for each region (northern – red; eastern – green; central – blue; southern – purple; Borneo – black) in Malaysia.

recorded the highest level of primary pollutants such as CO and NO_x, which largely originate from mobile sources, while secondary pollutants such as O_3 were highest at Shah Alam station. PM_{10} was highest at the Port Klang station's coastal area, which is geographically opposite Sumatera, Indonesia, and historically known as the main source for PM due to biomass burning. The pollutants of NO_2 and O_3 at all stations showed no notable trends. However, for CO, small but significant decreasing trends were observed at Petaling Jaya station and Shah Alam station. A small decreasing trend of PM_{10} was observed at Shah Alam station while other stations observed a small increasing trend at the 95% confidence interval.

Similar to the northern subregion, the southern subregion is also experiencing rapid urbanization and industrialization. In the southern subregion, the long-term average concentrations of CO (0.51 ppm), NO_x (0.013 ppm), O_3 (0.020 ppm), and PM_{10} (50.30 µg/m³) were comparable to those in the northern subregion. Long-term monthly average concentrations of pollutants at four monitoring stations were largely varied. Among the four monitoring stations, Melaka station showed the highest pollution level, particularly NO_2 (0.019 ppm) and PM_{10} (61.53 µg/m³). Melaka station, which is located in the coastal area and is relatively closer to Sumatera, Indonesia, showed the second-highest PM_{10} levels in Malaysia after Port Klang station. Long-term pollution trends were not significant for NO_2 and O_3 at all stations. However, there was a small but significant increasing trend for CO at all stations. For PM_{10}, significant small and decreasing trends were observed at all stations except at the Seremban station.

The eastern subregion is considered the least polluted area as urbanization and industrialization are relatively low compared to other subregions in Peninsular Malaysia. The long-term average concentrations of CO (0.45 ppm), NO_x (0.008 ppm), O_3 (0.016 ppm), and PM_{10} (41.79 µg/m³) were observed to be lower than the national average concentrations. Similar to the other subregions in Malaysia, there were no notable trends for NO_2 and O_3 pollution in the eastern subregion. Small but significant decreasing trends for CO pollution were observed at Jerantut and Kota Bharu stations while increasing trends were observed at Kuantan and Kuala Terengganu stations. Notable increasing trends for PM_{10} were observed at Kuantan and Kota Bharu stations while decreasing trends were observed at Jerantut and Kuala Terengganu stations.

Over the Malaysian Borneo subregion, the pollution levels were comparatively lower, and the air is considered the cleanest in Malaysia. This is attributed to the low urbanization and industrialization rates and large forested areas. The long-term monthly average concentrations of CO (0.40 ppm), NO_x (0.009 ppm), O_3 (0.012 ppm), and PM_{10} (39.05 µg/m³) were the lowest in Malaysia and well below the national average concentrations. No notable long-term trends for NO_2 and O_3 were seen at any of the monitoring stations. For CO, no notable trends were observed at Bintulu and Miri stations. Small but significant decreasing trends were observed for PM_{10} at Kuching and Bintulu stations while increasing trends were observed at Miri and Kota Kinabalu stations.

Primary Climate and Chemically Active Compounds' Emissions

Biomass burning emissions such as CO_2, black carbon (BC), and primary organic aerosols (POA), including the other emitted emissions such as CH_4, tropospheric

O_3, sulfate, and secondary organic aerosols (SOA), are chemically active and have an important role in atmospheric chemistry (Wang et al., 2012). Biomass emissions are also important sources of tropospheric NO_x, CO, volatile organic compounds (VOCs), and SO_2, which play an essential role in atmospheric oxidation processes. NO_x and VOCs are essential precursors of SOA (Majdi et al., 2019) and nitrate aerosols, which have a similar climate effect on a global scale as sulfate aerosols of smaller magnitude (Chen et al., 2017). The increase in sulfate aerosols and SOA is responsible for increasing the Earth's albedo through solar radiation scattering and consequently causes a direct cooling effect (Zhang et al., 2018). Indirectly, these aerosols also act as cloud condensation nuclei (CCN) and eventually affect cloud and precipitation (Zhang et al., 2018).

Intense biomass burning in Indonesia (Kalimantan and Sumatera) during September–October 2015 had an estimated total carbon release of 227 TgC, of which 83% was in the form of CO_2 (692 $TgCO_2$), 16% CO (84 TgCO), and 1% CH_4 (3.2 $TgCH_4$) (Huijnen et al., 2016). These emissions are comparatively higher than the previous intense biomass burning episodes in September–October 1997. Compared with other types of carbon, CH_4 constitutes a relatively small fraction of the carbon from biomass burning with less than 10% (Dlugokencky and Houweling, 2015).

Apart from being a highly oxidizing secondary pollutant, tropospheric O_3 is also responsible for warming the troposphere–surface climate system through an increase in the absorption of both solar and longwave radiation (Chen et al., 2017). Therefore, understanding the changes in atmospheric chemical composition due to biomass burning emissions and the effect on climate–chemistry interactions is vital for understanding the present and future climate. Tropospheric O_3 is not directly produced from biomass burning; however, biomass burning is known as a significant source of O_3 precursors such as NO_x and CO. Due to the short lifetime of NO_x, emissions of NO_x from biomass burning have the most significant impact on the source region, and therefore, O_3 production is also limited to the source region, which is estimated to be about 15% of the total O_3 concentrations in the tropics (Galanter et al., 2000). Several previous studies (see Chan et al., 2000, 2003; Galanter et al., 2000; Andreae and Marlet, 2001; Thompson et al., 2001; Streets et al., 2003; Takegawa et al., 2003) have found that the emissions from biomass burning have enhanced O_3 concentrations over large regions. Enhancement of O_3 over Hong Kong, for example, was linked to the biomass burning in Indochina through chemical mechanisms such as the CO–OH–CH_4 cycle and photochemistry involving NO_x and VOCs (Chan et al., 2000). However, this was not the case in Guangzhou, where there was no significant effect of biomass burning on O_3 concentration (Deng et al., 2008). In Malaysia, the high-ozone episode on August 7, 2006 (burning season during a southwest monsoon), was due to transboundary pollution from Sumatra's biomass burning. Long-range transport from Indochina is thought to have caused the high-ozone episode on February 24, 2008 (Ying et al., 2013).

CLIMATIC FACTORS OF BIOMASS BURNING OCCURRENCES

The tropical rainforest generally has a highly humid microclimate and is less susceptible to ignition sources (Laurance and Williamson 2001; Taylor, 2010). However, canopy disturbance corresponding with changing climate and human activity increases the susceptibility of rainforest to ignition sources. As suggested by Cochrane et al. (1999) and Stott (2000), the addition of dead, combustible organic matter and the perturbation of forest microclimates can provide suitable environments for the growth and then drying-out of herbaceous and shrubby plants. This is particularly true when disturbance coincides with a prolonged drought event, during which vegetation, litter, and surface layers of the substrate are pre-dried and therefore are prone to burn, and the extent of burning intensity once started may be increased (Stott, 2000).

Extreme weather conditions in Malaysia are always associated with long periods of a dry atmosphere, a precursor to severe weather such as drought. This situation frequently occurs during the El Niño events, which are characterized by prolonged dry conditions. Biomass burning in Sumatera and Kalimantan is strongly linked with El Niño events (Chen et al., 2016a,b; Field et al., 2016; Yin et al., 2016; Fanin and van der Werf, 2017) as well as the Indian Ocean Dipole (IOD) (Pan et al., 2018). El Niño phenomena frequently occur along the equatorial line in the Pacific Ocean. The increase in sea surface temperature (SST) can weaken the Walker circulation that induces drought conditions in the region and, in turn, leads to severe fires (Chen et al., 2016a,b; Yin et al., 2016). El Niño events in the region can be classified into two types, namely Eastern Pacific (EP) and Central Pacific (CP), the names referring to the central location of the SST anomalies (Ashok et al., 2007; Kao and Yu, 2009). Within the period 1979–2016, twelve El Niño occurrences were identified, where four events were classified as EP and eight events as CP (Pan et al., 2018). El Niño events and fire occurrences are mostly associated with the SST anomalies' location rather than the El Niño strength (Yeh et al., 2009; Chen et al., 2016a,b; Pan et al., 2018). The four EP El Niño events occurred in 1982, 1987, 1997, and 2006, and during these years, huge biomass burning events were detected in the southern Kalimantan and Sumatera. In contrast, during the CP El Niño events (1976, 1977, 1979, 1991, 2002, 2004, 2009, and 2015), the intensities of biomass burning occurrences were consistently less (Pan et al., 2018). The higher ability of the EP El Niño in weakening the Walker circulation compared to the CP El Niño is the main cause for the prolonged drought conditions in the region from July to October and potential increase of fires (Yeh et al., 2009; Field et al., 2016; Yin et al., 2016). For example, the 2015 El Niño phenomenon was categorized as an EP El Niño. Huge forest fires occurred in Kalimantan and Sumatera extending into November and were responsible for the transboundary pollution across Malaysia (Figure 10.5). The strength of the 2015 El Niño was not the major factor for the large and intense biomass burning in Kalimantan but rather the central location of the SST anomalies associated with this El Niño (Pan et al., 2018).

Pan et al. well addressed the role of climate variability on the Indonesian fire activity (2018) by considering (1) the presence of different types of El Niño and (2) the interaction between El Niño and the IOD. They concluded that intense and prolonged Indonesian drought and fires occur in the EP-type El Niño, during which the amounts of emitted carbon are almost double those during the CP type. By further separating the CP-type El Niño according to the IOD phase, they show that fire

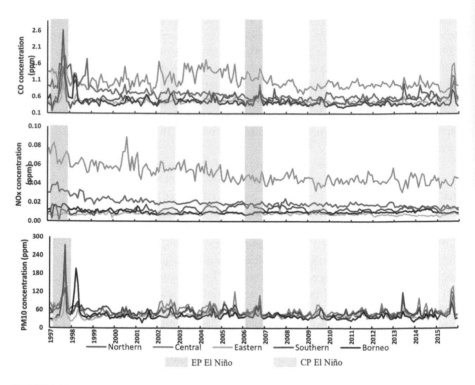

FIGURE 10.5   Air pollution concentration during El Niño phenomena over the period 1997–2015 (EP: Eastern Pacific; CP: Central Pacific).

seasons with less burning intensity and shorter duration are predominantly associated with weakly positive or even negative phases of the IOD phenomena. Moreover, fire intensity exhibits geographic diversity, whereby fires are always more intense in southern Kalimantan than southern Sumatra in all El Niño events, although it is also less dry in the southern Kalimantan region. This study's results can be applied to drought early warning, fire management, and air quality forecast in Indonesia and adjacent areas by identifying the type of El Niño and the phase of the IOD in advance (Pan et al., 2018).

Annually, large biomass burning events in Southeast ASEAN countries, mostly between July and October, are favored by persistent droughts during the dry season, coincident with the southwest monsoon (Lestari et al., 2014). The weakening of the Walker circulation associated with tropical ocean warming has increased the probability of severe droughts, despite increasing tropical-mean precipitation in the region by about 25% from 1951–2000 to 2001–2050 (Field et al., 2009). In the future, this region will experience drier conditions during the boreal summer as an effect of the anthropogenic warming (Field et al., 2009; Lestari et al., 2014), particularly in southern Sumatera and southern Kalimantan, where warming of between 3.0°C and 3.1°C has been projected under climate scenarios (SRES B2 and A2) (Figure 10.6) (Sentian and Kong, 2013). This suggests that there will be more severe fires in these

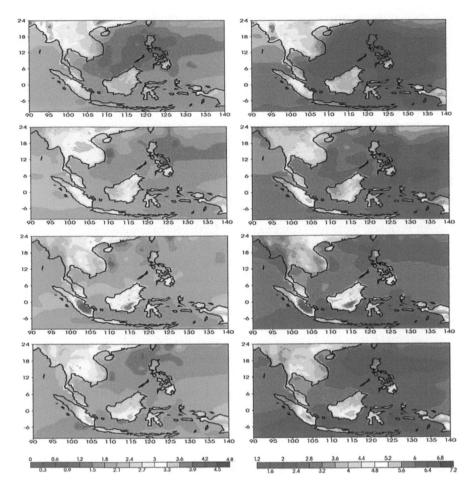

FIGURE 10.6 Projected seasonal surface temperature over the Southeast Asian region under climate scenarios SRES B2 and SRES A2 (right). Note: DJF – Dec–Jan–Feb; MAM – Mar–Apr–May; JJA – Jun–Jul–Aug; SON – Sept–Oct–Nov.

subregions, which will become additional important sources of CO_2, NO_x, and CO emissions to the atmosphere.

ASSOCIATED IMPACTS OF BIOMASS BURNING

AIR POLLUTION AND HAZE

Open biomass burning in Indonesia produces transboundary smoke haze, an annual phenomenon that causes deterioration in Malaysia's local air quality. The southwest-erly wind during the southwest monsoon plays a significant role in the transport of transboundary pollution from intense biomass burning in Sumatera and Kalimantan, Indonesia, which contributes to high concentrations of fine particles and atmo-spheric gases to the Malaysian regional atmosphere (Sulong et al., 2017; Latif et al.,

2018; Sentian et al., 2018). Emissions from the local biomass burning contributing to Malaysia's haze were comparatively smaller than the emissions from Indonesia (Afroz et al., 2003). Based on the biomass burning hotspots, most of Malaysia's burning areas are paddy fields and palm oil plantations (Mahmud, 2005). However, annually a substantial number of fire hotspots are observed in large peat swamp areas in the Sarawak and Klias Peninsular in Sabah, which emits smoke for relatively longer periods of up to 2–4 months, thus exacerbating the local haze episodes.

The rising trend of transboundary pollution and severe episodic haze in Malaysia originates from Indonesia and is associated with seasonal biomass burning and monsoon seasons (Sentian et al., 2018). Haze episode in Malaysia has been recorded since the 1970s; however, due to the haze's intensity during the 1990s and by the wide media coverage, they have been constantly highlighted. The haze episode in 1997, which occurred from September to November and coincided with the El Niño event, resulted in an Air Pollution Index (API) exceeding the 500 level (hazardous) in Miri, Sarawak. In August 2005, the haze episode in Peninsular Malaysia was considered more severe than the 1997's episode, particularly in the Klang Valley area, where API readings in most monitoring stations exceeded 500. Intense haze episodes in Malaysia due to the vast and intense biomass burning in Indonesia (Sumatra and Kalimantan) were also recorded in October 2010, June 2013, March and July 2014, and September–October 2015. Most of the haze episodes occurred during the southwest monsoon and, in some cases, coincided with the El Niño event. During the haze episodes, PM_{10} in most regions (see Figure 10.5) was relatively higher than non-haze periods, particularly in 2006, 2013, 2014, and 2015. PM_{10} was also observed as the primary pollutant that contributed to the higher API in Malaysia. Other gaseous pollutants such as CO and O_3 were also notably higher concentrations during these periods.

Respiratory and Health

Haze from biomass smoke contains a large and diverse number of chemical components, of which many are considered to have health implications. These include PM, sulfur dioxide (SO_2), nitrogen dioxide (NO_2), carbon monoxide (CO), and ozone (O_3) (Ab Manan et al., 2018). For most studies, PM is the major component of the haze (Cheng et al., 2013; Huang et al., 2014) and is significantly associated with adverse health effects (Mabahwi et al., 2014; Khan et al., 2016; Sulong et al., 2017). Haze episodes have contributed to increasing hospital visits for treatments relating to chronic obstructive pulmonary diseases, upper respiratory infections, asthma, and rhinitis. A previous study conducted by Nasir et al. (2000) claimed that 285,227 asthma attacks, 3889 cases of chronic bronchitis in adults, and 118,804 cases related to children, 2003 respiratory admissions and 26,864 emergency issues were due to the 1997 haze episode in Malaysia. In comparison with Indonesia for the same haze episode, an estimated 20 million people suffered from respiratory problems, with 19,800–48,100 premature mortalities (Heil, 2007). In severely affected fire areas in Indonesia, more than 90% of people had a respiratory illness, and elderly individuals suffered a serious deterioration in overall health (Kunii et al., 2002). In the neighboring country of Singapore, health impact surveillance during the 1997 haze episode in Southeast Asia indicated a 30% increase in hospital attendances for haze-related conditions, though there was no significant increase in hospital admissions or mortality. During the same period, there were a 12% increase for upper respiratory tract illness, 19% for asthma, and 26% for rhinitis (Emmanuel, 2000).

In a recent study, Sahani et al. (2014) have shown that haze events were found to be significantly associated with natural and respiratory mortality in Malaysia. Respiratory mortality increased by 19% due to haze episodes, where children and senior citizens were more likely to suffer. A total of 126,822 deaths were recorded for the period between 2006 and 2007 associated with the haze event originating from biomass burning in this region. Although the mortality impact of haze varied according to gender, age, and location, PM's regional transboundary pollution was the major underlying reason for health impacts and the accelerating inpatient cost, estimated at around USD 91,000 per year in Malaysia. In the same period, Hayasaka et al. (2014) found that in Palangkaraya, located in the central Kalimantan area of Indonesia, the air quality was rated very unhealthy or dangerous for 60 days during the fire season of 2006.

ATMOSPHERIC ACIDITY AND NEUTRALIZATION

Tropical biomass burning emits large amounts of cationic species such as K, Ca, and Mg, whereas the anionic species are mainly sulfate, chloride, nitrate, carbonate, oxalate, and acetate (Sentian, 2002). The presence of mineral acids, such as sulfuric acid and nitric acid, may contribute to the atmospheric acidity. However, large oxides of alkali and alkaline earth elements produced during the biomass burning process may play a role in neutralizing the atmospheric acidity near the burning site. Away from the burning areas, the neutralization effects are uncertain due to the difficulty in assessing the alkaline particles' loss from air masses. An investigation on the impact of large biomass burning in Indonesia in 1997 on atmospheric acidity in Singapore found the mean pH values of rainwater varying from 3.79 to 6.20 with a volume-weighted mean of 4.35 (Balasubramanian et al., 1999). Also, higher volume-weighted mean concentrations of Cl^-, NO_3^-, SO_4^{2-}, $HCOO^-$, Ca^{2+}, NH_4^+, and K^+ were observed. There were a large number of rain events with elevated concentrations of these ions during the biomass burning period. High concentrations of sulfate, nitrate, and ammonium during these periods are attributed to air masses' long residence time leading to progressive gas to particle conversion of biomass burning emission components. The decrease in pH of precipitation in response to the increased concentrations of acid is only marginal, which is ascribed to the neutralization of the acidity by NH_3 and $CaCO_3$.

In Brunei's neighboring country, the monthly contribution of H^+ during forest fires and haze in 1998 varied between 0.0% and 5.35% (Radojevic, 2003), suggesting the biomass burning emissions as a minor source of atmospheric acidity through a wet deposition. Despite the high conductivity and ionic content of rainwater samples during the haze episodes compared to the non-haze periods in Singapore and Brunei (see Balasubramanian et al., 1999; Radojevic,2003), atmospheric acidity was less pronounced due to the higher deposition rates during non-haze periods (due to higher precipitation). Similar results were observed at two rural monitoring sites (Danum Valley, Sabah, and Tanah Rata Cameron Highland) between 2006 and 2015 but slightly acidic rainwater in Petaling Jaya, located in urban areas. Emissions from other local anthropogenic sources such as industries and motor vehicles potentially contribute to the local atmospheric acidity. Earlier study by Ayers et al. (2003) in Petaling Jaya found that annual deposition fluxes of acidic sulfur and nitrogen species during 1993 and 1998 have shown high acid deposition, in the range of 277–480 $m_{eq}/m^2/year$.

TROPICAL PLANTS' PHYSIOLOGY AND PRODUCTIVITY

The increasing trend of haze occurrences in Southeast Asia over the last four decades and the related air quality degradation has impacted the general population and economy of Malaysia and the tropical forest ecosystems. Recent investigations have indicated that haze has critical negative effects on plant ecology corresponding to the radiation scattering caused by the aerosol radiative effect (Wang et al., 2015; Pani et al., 2016a, b). During the haze events, there can be a reduction in solar radiation reaching the Earth's surface; thus, it can lower the photosynthesis rate reducing the plant production (Tang et al., 1996; Yoneda et al., 2000). During the 1994 haze episode, two varieties of hybrid rice in Malaysia, MR151 and MR123, displayed a 50% reduction in growth rate and abnormal ripening, while in Indonesia, there was a 2%–3% reduction in the paddy rice yield. In a similar investigation, Chameides et al. (1999) found a 5%–30% reduction of solar irradiance in China, contributing to a reduction of crop yield of at least 5%–30%.

The sulfate and organic carbon particles emitted during the biomass burning can modulate optical properties as much as 73%–92% total light extinction (Fermen et al., 1981; Wolff et al., 1986). An earlier study by Tang et al. (1996) conducted on the effect of haze on photon flux density (PFD) in a tropical forest in Malaysia (Pasoh Forest Reserve) conducted during clear days and hazy days, both under and on top of the tree canopy concluded that more than 90% of the PFD readings were below 1000 pmol photons mm^2/s on a hazy day, while over 60% of PFD readings were over 1000 pmol photons mm^2/s on a clear day (Figure 10.7). The total PFD and mean PFD on the hazy days were only about 50% of those on the clear days. The reduction of PFD due to haze under the canopy was much higher than that on the top of the canopy (Tang et al., 1996). Under the hazy conditions, almost all PFD values were lower than 50 pmol photons mm^2/s, which indicates a few sun flecks were available for understory plants (Chazdon, 1988). The study also indicated that leaf carbon gain decreased for both under- and top-canopy leaves during hazy days. The decrease in leaf carbon gain indicates that PPD reduction during hazy days may result in reduced production in the understory plants in tropical forests. Under hazy conditions, canopy trees can still maintain a fairly high positive carbon gain; thus, PFD reduction due to haze would have more important impacts on the growth and survival of understory plants than the canopy trees (Kitijima, 1994; Tang et al., 1996). In the long term, frequent and intense haze occurrences in the region may have considerable ecological effects on the regeneration of tree species in a tropical forest.

PALM OIL PRODUCTIVITY

Oil palm is one of the world's major vegetable crops, and Malaysia is the second-largest producer in the world after Indonesia. Oil palm can live ~25 years with the

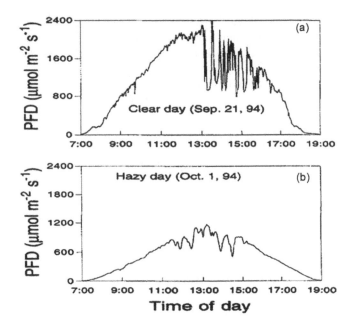

FIGURE 10.7 Diurnal variations of photon flux density (PFD) during a clear day (a; September 21, 1994) and a hazy day (b; October 1, 1994) at Pasoh Forest Reserve Malaysia (Tang et al., 1996.)

peak oil yields varying from 12 to 18 t/ha/year (Woittiez et al., 2017). The potential yield as defined by the fruit bunch yield and oil content is determined by the active photosynthetic radiation (PAR), temperature, ambient CO_2 concentration, and crop genetic characteristics, under perfect crop management (van Ittersum and Rabbinge, 1997). Oil palm productivity is linked with its ability to intercept radiation throughout the year by its permanent leaf canopy. Apart from the clouds, the incoming solar radiation is also influenced by other environmental conditions such as thick haze due to particle emissions from the intense biomass burning. Increasing incidences of atmospheric pollution in the Southeast Asian region, particularly of high particle concentrations, will lead to substantial solar radiation reduction, which can affect oil palm yields in Malaysia. Typical average radiation for optimum oil palm growth is between 15 and 23 MJ/m²/day (Paramananthan et al., 2000; Henson, 2000). Reduction in the available PAR due to haze caused by forest burning is a common issue in Indonesia. Forest burning occurs mostly during the dry season when available radiation is at its peak and is likely to reduce yields significantly (Woittiez et al., 2017). However, earlier modeling results by Henson (2000) have shown little indication of the negative effects of haze on oil palm yields in Malaysia. The model results have shown no immediate apparent effect of reduced radiation on the oil palm yield due to the long time required for bunch morphogenesis, the complexity of the process, and the presence of assimilating stores which serve to buffer the palm against periods of adverse conditions (Henson, 2000). It was also inferred that, apart from radiation, other climatic factors are equally important in influencing the physiological processes of oil palm productivity, such as temperature, atmospheric vapor pressure deficit, and soil water availability. A lower atmospheric pressure deficit, lower temperature, and improved soil water supply tended to offset yield reductions due to lower light intensity. In some circumstances, it was found that predicted yields were higher under low or moderate radiation than under high radiation since high radiation was closely linked with high evapotranspiration rates and lower rainfall, which led to an increased likelihood of soil water deficits and drought, thus reducing the yields (Henson, 2000).

CONCLUSION AND FINAL REMARKS

The effects of biomass burning emissions on the environment, including temporal and spatial variations of air quality due to local and transboundary pollution in Southeast Asia and, more specifically, Malaysia, were reviewed. Long-term air pollution analysis in Malaysia revealed that intense and large biomass burning in Indonesia (Sumatera and Kalimantan) has significantly contributed to higher pollution levels, mainly PM during the southwest monsoons and the El Niño events. The probability of severe fires is likely to increase in Malaysia and neighboring countries due to the tropical ocean warming and the weakening of the Walker circulation. Emissions from the biomass burning contain a large and diverse number of chemical compounds, and therefore can cause atmospheric acidity and neutralization, and respiratory and human health issues, including the reduction in plant productivity, which were reviewed in the study.

REFERENCES

Ab Manan, N., Hod, R., Sahani, M., Mohd Yusoff, H., Ismail, R. and Wan Mahiyuddin, W.R. 2016. The impact of air pollution and haze on hospital admission for cardiovascular and respiratory diseases. *International Journal of Public Health Research*. 6(1), 707–712.

Ab Manan, N., Abdul Manaf, M.R. and Hod, R. 2018. The Malaysia haze and its health economic impact: A literature review. *Malaysian Journal of Public Health Medicine*. 18(1), 38–45.

Afroz, R., Hassan, M.N. and Ibrahim, N.A. 2003. Review of air pollution and health impacts in Malaysia. *Environmental Research*. 92(2), 71–77.

Aiken, S.R. 2004. Runaway fires, smoke-haze pollution, and unnatural disasters in Indonesia. *Geographical Review*. 94, 55–79.

Ainsworth, A.A. and Kauffman, J.B. 2009. Response of native Hawaiian woody species to lava-ignited wildfires in tropical forests and shrub lands. *Plant Ecology*. 201, 197–209.

Andreae, M.O. and Marlet, P. 2001. Emissions of trace gases and aerosols from biomass burning. *Global Biogeochemical Cycles*. 15, 955–966.

Ashok, K., Behera, S.K., Rao, S.A., Weng, H. and Yamagata, T. 2007. El Niño Modoki and its possible teleconnection. *Journal of Geophysical Research*. 112, C11007. doi:10.1029/2006JC003798.

Ayers, G.P. Leong, C.P., Lim, S.F., Cheah, W.K., Gillett, R.W. and Manins, P.C. 2003. Atmospheric concentrations and deposition of oxidized sulfur and nitrogen Species at Petaling Jaya, Malaysia 1993–1998. *Tellus*. 52, 60–73.

Badarinath, K.V.S., Kharol, S.K., Krishna Prasad, V., Kaskaoutis, D.G. and Kambezidis, H.D. 2008a. Variation in aerosol properties over Hyderabad, India during intense cyclonic conditions. *International Journal of Remote Sensing*. 29(15), 4575–4597.

Badarinath, K.V.S., Kharol, S.K., Prasad, V.K., Sharma, A.R., Reddi, E.U.B., Kambezidis, H.D. and Kaskaoutis, D.G. 2008b. Influence of natural and anthropogenic activities on UV Index variations–a study over tropical urban region using ground based observations and satellite data. *Journal of Atmospheric Chemistry*. 59(3), 219–236.

Balasubramanian, R, Victor, T. and Begum, R. 1999. Impact of biomass burning on rainwater acidity and composition in Singapore. *Journal of Geophysical Research*. 104(D21), 26881–26890. doi:10.1029/1999JD900247.

Behere, S.N., Betha, R., Huang, X. and Balasubramanian, R. 2015a. Characterization and estimation of human airway deposition of size-resolved particulate-bound trace elements during a recent haze episode in Southeast Asia. *Environment Science Pollution Research*. 22, 4265–4280.

Behera, S.N., Cheng, J., Huang, X., Zhu, Q., Liu, P. and Balasubramanian, R. 2015b. Chemical composition and acidity of size-fractionated inorganic aerosols of 2013–14 winter haze in Shanghai and associated health risk of toxic elements. *Atmospheric Environment*. 122, 259–271.

Betha, R., Behera, S.N. and Balasubramanian. 2014. 2013 Southeast Asian smoke haze: Fractionation of particulate-bound elements and associated health risk. *Environment Science Technology*. 48, 4327–4335.

Biswas, S., Lasko, K.D. and Vadrevu, K.P. 2015a. Fire disturbance in tropical forests of Myanmar—Analysis using MODIS satellite datasets. *IEEE Journal of Selected Topics in Applied Earth Observations and Remote Sensing*. 8(5), 2273–2281.

Biswas, S., Vadrevu, K.P., Lwin, Z.M., Lasko, K. and Justice, C.O. 2015b. Factors controlling vegetation fires in protected and non-protected areas of Myanmar. *PLoS One* 10(4), e0124346.

Bowen, M.R., Bompard, J.M., Anderson, I.P., Guizol, P. and Gouyon, A. 2001. Anthropogenic fires in Indonesia: A view from Sumatra. In: Peter, E. and Radojevic, M. (Eds). *Forest Fires and Regional Haze in Southeast Asia*. Nova Science, Huntington, NY. pp. 41–66.

Chameides, W.L., Yu, H., Liu, S.C., Bergin, M., Zhou, X., Mearns, L., Wang, G., Kiang, C.S., Saylor, R.D., Luo, C., Huang, Y., Steiner, A. and Giorgi, F. 1999. Case study of the effects of atmospheric aerosols and regional haze on agriculture: An opportunity to enhance crop yields in China through emission controls? *Proceedings of the National Academy of Sciences of the United States of America*. 96, 13626–13633.

Chan, C.Y., Chan, L.Y., Liu, H.Y., Christopher, S., Oltmans, S.J. and Harris, J.M. 2000. A case study on the biomass burning in Southeast Asia and enhancement of tropospheric ozone over Hong Kong. *Geophysical. Research Letter*. 27, 1479–1482.

Chan, C.Y., Chan, L.Y., Harris, J.M., Oltmans, S.J., Blake, D.R., Qin, Y., Zheng, Y.G. and Zheng, X.D. 2003. Characteristics of biomass burning emission sources, transport, and chemical speciation in enhanced springtime tropospheric ozone profile over Hong Kong. *Journal of Geophysics Research*. 108(D1), 4015. doi:10.1029/2001JD001555.

Chazdon, R.L. 1988. Sunflecks and their importance to forest understory plants. *Advance Ecology Research*. 18, 1–63.

Chen, C.C., Lin, H.W., Yu, J.Y. and Lo, M.H. 2016a. The 2015 Borneo fires: What have we learned from the 1997 and 2006 El Niños? *Environmental Research Letters*. 11(10), 10400. doi:10.1088/1748-9326/11/10/104003.

Chen, Y., Morton, D.C., Andela, N., Giglio, L. and Randerson, J.T. 2016b. How much global burned area can be forecast on seasonal timescales using sea surface temperatures? *Environmental Research Letters*. 11(4), 045001. doi:10.1088/1748-9326/11/4/045001.

Chen, J.M., Li, C.L., Ristovski, Z., Milic, A., Gu, Y.T., Islam, M.S., Wang, S.X., Hao, J.M., Zhang, H.F., He, C.G., Guo, H., Fu, H.B., Miljevic, B., Morawska, L., Thai, P., Lam, Y.F., Pereira, G., Aijun, D. and Dumka, U.C. 2017. A review of biomass burning: Emission and impacts on air quality, health and climate in China. *Science of the Total Environment.* 579, 1000–1034.

Cheng, Z., Wang, S., Jiang, J., Fu, Q., Chen, C., Xu, B., et al. 2013. Long-term trend of haze pollution and impact of particulate matter in the Yangtze River Delta, China. *Environmental Pollution.* 182, 101–110.

Chuang, H.C., Hsiao, T.C., Wang, S.H., Tsay, S.C., Lin, N.H. 2016. Characteristic of particulate matter profiling and alveolar deposition from biomass burning in Northern Thailand: The 7-SEAS study. *Aerosol and Air Quality Research.* 16, 2897–2906.

Cochrane, M.A., Alencar, A., Schulze, M.D., Souza, C.M., Nepstad, D.C., Lefebvre, P. and Davidson, E.A. 1999. Positive feedbacks in the fire dynamics of closed canopy tropical forests. *Science.* 284, 1832–1835.

Cui, M., Chen, Y., Zheng, M., Li, J., Tang, J., Han, Y., Song, D., Yan, C., Zhang, F., Tian, C., and Zhang, G. 2018. Emissions and characteristics of particulate matter from rainforest burning in the Southeast Asia. *Atmospheric Environment.* 191, 194–204.

Deng, X.J., Tie, X.X., Zhou, X.J., Wu, D., Zhong, L.J., Tan, H.B., Li, F., Huang, X.Y., Bi, X.Y. and Deng, T. 2008. Effects of Southeast Asia biomass burning on aerosols and ozone concentration over Pearl River Delta (PRD) region. *Atmospheric Environment.* 42(36), 8493–8501.

Dlugokencky, E. and Houweling, S. 2015. Chemistry of the atmosphere – Methane. In *Encyclopedia of Atmospheric Sciences (Second Edition) Reference Module in Earth Systems and Environmental Sciences.* pp. 363–371. doi:10.1016/B978-0-12-382225-3.00223-1.

Dominick, D., Juahir, H., Latif, M.T., Zain, S.M. and Aris, A.Z. 2012. Spatial assessment of air quality patterns in Malaysia using Multivariate analysis. *Atmospheric Environment.* 60, 172–181.

Emmanuel, S.C. 2000. Impact to lung function of haze from forest fires: Singapore's experience. *Respirology.* 5, 175–182.

Fanin, T. and van der Werf, G.R. 2017. Precipitation–Fire linkages in Indonesia (1997–2015). *Biogeosciences.* 14(18), 3995–4008. doi:10.5194/bg-14-3995-2017.

Fermen, M.A., Wolff, G.T. and Kelly, N.A. 1981. The nature and sources of haze in the Shenandoah Valley blue ridge. *Journal of Air Pollution Control Association.* 31, 1074–1082.

Field, R.D., van der Werf, G.R. and Shen, S.S. 2009. Human amplification of drought-induced biomass burning in Indonesia since 1960. *Nature Geoscience.* 2(3), 185. doi:10.1038/ngeo443.

Field, R.D., van der Werf, G.R., Fanin, T., Fetzer, E.J., Fuller, R., Jethva, H., Levy, R., Nathaniel, J., Livesey, J., Luo, M., Torres, O. and Worden, H.M. 2016. Indonesian fire activity and smoke pollution in 2015 show persistent non-linear sensitivity to El Niño-induced drought. *Proceedings of the National Academy of Sciences.* 113(33), 9204–9209. doi:10.1073/pnas.1524888113.

Galanter, M., Levy II, H. and Carmichael, G.R. 2000. Impacts of biomass burning on tropospheric CO, NO_x and O_3. *Journal of Geophysical Research.* 105(D5), 6633–6653.

Giglio, L., Randerson, J.T. and van der Werf, G.R. 2013. Analysis of daily, monthly, and annual burned area using fourth-generation global fire emission (GFED4). *Journal Geophysical Research: Biogeoscience.* 118(1), 317–328.

Gilman, J.B., Lerner, B.M., Kuster, W.C., Goldan, P.D., Warneke, C., Veres, P.R., Roberts, J.M., de Gouw, J.A., Burling, I.R. and Yokelson, R.J. 2015. Biomass burning emission and potential air quality impacts of volatile organic compounds and other trace gases from temperate fuels common in the United States. *Atmospheric Chemistry and Physics.* 15, 21713–21763. doi:10.5194/acpd-15-21713-2015.

Goto, Y. and Suzuki, S. 2013. Estimates of carbon emissions from forest fires in Japan, 1978–2008. *International Journal of Wildland Fire*. 22(6), 721–729.

Gupta, P.K., Prasad, V.K., Sharma, C., Sarkar, A.K., Kant, Y., Badarinath, K.V.S. and Mitra, A.P. 2001. CH_4 emissions from biomass burning of shifting cultivation areas of tropical deciduous forests–Experimental results from ground-based measurements. *Chemosphere-Global Change Science*. 3(2), 133–143.

Hao, W.M. and Liu, M.H., 1994. Spatial and temporal distribution of tropical biomass burning. *Global Biogeochemical Cycles*. 8(4), 495–503.

Hayasaka, H. and Sepriando, A. 2018. Severe air pollution due to peat fires during 2015 super El Niño in Central Kalimantan, Indonesia. In: *Land-Atmospheric Research Applications in South/Southeast Asia*. Vadrevu, K.P., Ohara, T., and Justice, C. (Eds). Springer, Cham. pp. 129–142.

Hayasaka, H., Noguchi, I., Putra, E.I., Yulianti, N. and Vadrevu, K. 2014. Peat-fire-related air pollution in Central Kalimantan, Indonesia. *Environmental Pollution*. 195, 257–266. doi:10.1016/j.envpol.2014.06.031.

Heil, A. 2007. *Indonesian Forest and Peat Fires: Emission, Air Quality and Human Health. Reports on Earth System Science*. Max Planck Institute for Meteorology, Hamburg.

Henson, I.E. 2000. Modelling the effects of 'haze' on oil palm productivity and yield. *Journal of Oil Palm Research*. 12(1), 123–134

Huang, R.J., Zhang, Y., Bozzetti, C., Ho, K.F., Cao, J.J., Han, Y., et al. 2014. High secondary aerosol contribution to particulate pollution during haze events in China. *Nature*. 514, 218–222.

Huang, X., Betha, R., Tan, L.Y. and Balasubramanian, R. 2016. Risk assessment of bioaccessible trace element in smoke haze aerosol versus urban aerosol using simulated lung fluids. *Atmospheric Environment*. 125, 505–511.

Huijnen, V., Wooster, M.J., Kaiser, J.W., Gaveau, D.I.A., Flemming, J., Parrington, M., Inness, A., Murdiyarso, D., Main, B. and van Weede, M. 2016. Fire carbon emissions over maritime Southeast Asia in 2015 largest since 1997. 2016. *Scientific Reports*. 6, 26886.

Jain, A.K., Tao, Z.N., Yang, X.J and Gillespie, C. 2006. Estimates of global biomass burning emissions for reactive greenhouse gases (CO, NMHCs and NO_x) and CO_2. *Journal of Geophysical Research Atmospheric*. 111(D6), 14.

Justice, C., Gutman, G. and Vadrevu, K.P. 2015. NASA land cover and land use change (LCLUC): An interdisciplinary research program. *Journal of Environmental Management*. 148(15), 4–9.

Kant, Y., Ghosh, A.B., Sharma, M.C., Gupta, P.K., Prasad, V.K., Badarinath, K.V.S. and Mitra, A.P. 2000. Studies on aerosol optical depth in biomass burning areas using satellite and ground-based observations. *Infrared Physics & Technology*. 41(1), 21–28.

Kao, H. and Yu, J. 2009. Contrasting eastern-Pacific and central-Pacific types of ENSO. *Journal of Climate*. 22(3), 615–632. doi:10.1175/2008JCLI2309.1.

Khan, M.F., Latif, M.T., Saw, W.H., Amil, N., Nadzir, M.S.M., Sahani, M., Tahir, N.M. and Chung, J.X. 2016. Fine particular matter in the tropical environment: Monsoonal effects, source apportionment, and health risk assessment. *Atmospheric Chemistry and Physics*. 16, 597–617.

Kitijima, K. 1994. Relative importance of the photosynthesis traits and allocation patterns as correlated of seedling shade tolerance of 13 tropical trees. *Oecologia*. 98, 419–428.

Kondo, Y., Morino, Y., Takegawa, N., Koike, M., Kita, K., Miyazaki, Y., Sachse, G.W., Vay, S.A., Avery, M.A., Flocke, F., Weinheimer, A.J., Eisele, F.L., Zondlo, M.A., Weber, R.J., Singh, H.B., Chen, G., Crawford, J., Blake, D.R., Fuelberg, H.E., Clarke, A.D., Talbot, R.W., Sandholm, R.W., Browell, E.V., Streets, D.G. and Liley, B. 2004. Impacts of biomass burning in Southeast Asia on ozone and reactive nitrogen over the western Pacific in spring. *Journal of Geophysical Research*. 109, D15S12. doi:10.1029/2003JD004203

Kunii, O., Kobayashi, S., Yajima, I., Hisamatsu, Y., Yamamura, S., Amagai, T. and Ismail, S.T. 2002. The 1997 Haze disaster in Indonesia: Its air quality and health effects. *Archives of Environmental Health an International Journal*. 57(1), 16–22. doi:10.1080/00039890209602912.

Larjavaara, M., Pennanen, J. and Tuomi, T.J. 2005. Lightening that ignites forest fires in Finland. *Agriculture for Meteorology*. 132, 171–180.

Lasko, K. and Vadrevu, K.P. 2018. Improved rice residue burning emissions estimates: Accounting for practice-specific emission factors in air pollution assessments of Vietnam. *Environmental pollution*. 236(5), 795–806.

Lasko, K., Vadrevu, K.P., Tran, V.T., Ellicott, E., Nguyen, T.T., Bui, H.Q. and Justice, C. 2017. Satellites may underestimate rice residue and associated burning emissions in Vietnam. *Environmental Research Letters*. 12(8), 085006.

Lasko, K., Vadrevu, K.P. and Nguyen, T.T.N. 2018a. Analysis of air pollution over Hanoi, Vietnam using multi-satellite and MERRA reanalysis datasets. *PLoS One*. 13(5), 0196629.

Lasko, K., Vadrevu, K.P., Tran, V.T. and Justice, C. 2018b. Mapping double and single crop paddy rice with Sentinel-1A at varying spatial scales and polarizations in Hanoi, Vietnam. *IEEE Journal of Selected Topics in Applied Earth Observations and Remote Sensing*. 11(2), 498–512.

Latif, M.T., Othman, M., Idirs, N., Juneng, L., Abdullah, A.M., Hamzah, W.P., Khan, M.F., Sulaiman, N.M.N., Jewaratnam, J., Aghamohammadi, N., Sahani, M., Xiang, C.J., Ahamad, F., Amil, N., Darus, M., Varkkey, H., Tangang, F. and Jaafar, A.B. 2018. Impact of regional haze towards air quality in Malaysia: A review. *Atmospheric Environment*. 177, 28–44.

Laurance, W.F. and Williamson, G.B. 2001 Positive feedbacks among forest fragmentation drought and climate change in the Amazon. *Conservation Biology*. 15, 1529–1535.

Lee, H.H., Bar-Or, R.Z. and Wang, C. 2017. Biomass burning aerosols and low-visibility events in Southeast Asia. *Atmospheric Chemistry and Physics*. 17(2), 965–998.

Lestari, R.K., Watanabe, M., Imada, Y., Shiogama, H., Field, R.D., Takemura, T. and Kimoto, M. 2014. Increasing potential of biomass burning over Sumatra, Indonesia induced by anthropogenic tropical warming. *Environment Research Letter*. 9, 104010.

Liu, S., Yin, Y., Liu, X., Cheng, F., Yang, J., Li, J., Dong, S. and Zhu, A. 2017. Ecosystem services and landscape change associated with plantation expansion in a tropical rainforest region of southwest China. *Ecological Modelling*. 353, 129–138.

Mabahwi, N.S., Leh, O.L.H. and Omar, D. 2014. Urban air quality and human health effects in Selangor, Malaysia. *Procedia – Social and Behavioral Sciences*. 170, 282–291.

Mahmud, M. 2005. Active fire and hotspot emissions in Peninsular Malaysia during the 2002 burning season. *Geografia-Malaysian Journal of Society and Space*. 1(1), 32–35.

Majdi, M., Turquety, S., Sartelet, K., Legorgeu, C., Menut, L. and Kim, Y.S. 2019. Euro-Mediterranean in 2007: Sensitivity to some parameterizations of emission in air quality models. *Atmospheric Chemistry and Physics Discussion*. 19, 785–812. doi:10.5194/acp-19-785-2019.

Mat Isa, A.Z. 2003. Forest Fire in Malaysia: Its Management and Impact on Biodiversity. Available from http://www.mekonginfo.org/assets/midocs/0001527-environment-forest-fire-in-malaysia-its-management-and-impact-on-biodiversity.pdf.

Nasir, M.H., Choo, W.Y., Rafia, A., Md, M.R., Theng, L.C. and Noor, M.M.H. 2000. Estimation of health damage cost for 1997-haze episode in Malaysia using Ostro model. *Proceedings of the Malaysia Science and Technology Congress (MSTC) Report*.

Othman, J., Sahani, M., Mahmud, M. and Ahmad, M.K.S. 2014. Transboundary smoke haze pollution in Malaysia: Inpatient health impacts and economic valuation. *Environment Pollution*. 189, 194–201.

Pan, X.H., Chin, M., Ichoku, C.M. and Field, R.D. 2018. Connecting Indonesian fires and drought with type of El Niño and phase of the Indian Ocean Dipole. *Journal of Geophysical Research: Atmosphere*. 123(15), 7974–7988. doi:10.1029/2018JD028402.

Pani, S.K., Wang, S.H., Lin, N.H., Lee, C.T., Tsay, S.C., Holben, B.N., Janjai, S., Hsiao, T.C., Chuang, M.T. and Chantara, S. 2016a. Radiative effect of springtime biomass-burning aerosol over Northern Indochina during 7-SEAS/BASELINE 2013 campaign. *Aerosol and Air Quality Research*. 16, 2802–2817.

Pani, S.K., Wang, S.H., Lin, N.H., Tsay, S.C., Lolli, S., Chuang, M.T., Lee, C.T., Chantara, S. and Yu, J.Y. 2016b. Assessment of aerosol optical property and radiative effect for the layer decoupling cases over the northern South China Sea during the 7 SEAS/ Dongsha Experiment. *Journal of Geophysical Research Atmosphere*. 121, 4894–4906.

Paramananthan, S., Chew, P.S. and Goh, K.J. 2000. Towards a practical framework for land evaluation for oil palm in the 21st Century. In: *Proceedings of the International Planters Conference 'Plantation Tree Crops in the New Millennium: the Way Ahead'*, 17–20 May 2000, Kuala Lumpur. pp. 869–885.

Permadi, D.A. and Kim Oanh, N.T. 2013. Assessment of biomass open burning emissions in Indonesia and potential climate forcing impact. *Atmospheric Environment*. 78, 250–258.

Prasad, V.K., Gupta, P.K., Sharma, C., Sarkar, A.K., Kant, Y., Badarinath, K.V.S., Rajagopal, T. and Mitra, A.P. 2000. NO_x emissions from biomass burning of shifting cultivation areas from tropical deciduous forests of India–estimates from ground-based measurements. *Atmospheric Environment*. 34(20), 3271–3280.

Prasad, V.K., Kant, Y. and Badarinath, K.V.S., 2001. CENTURY ecosystem model application for quantifying vegetation dynamics in shifting cultivation areas: A case study from Rampa Forests, Eastern Ghats (India). *Ecological Research*. 16(3), 497–507.

Prasad, V.K., Kant, Y., Gupta, P.K., Elvidge, C. and Badarinath, K.V.S. 2002. Biomass burning and related trace gas emissions from tropical dry deciduous forests of India: A study using DMSP-OLS data and ground-based measurements. *International Journal of Remote Sensing*. 23(14), 2837–2851.

Radojevic, M. 2003. Chemistry of forest fires and regional haze with emphasis on Southeast Asia. Pure and Applied Geophysics, 160(1): 157–187.

Reddington, C.L., Yoshioka, M., Balasubramanian, R., Ridley, D., Toh, Y.Y., Arnold, S.R. and Spracklen, D.V. 2014. Contribution on vegetation and peat fires to particulate air pollution in Southeast Asia. *Environment Research Letter*. 9(9), 12.

Sahani, M., Zainon, N.A., Wan Mahiyuddin, W.R., Latif, M.T., Hod, R., Khan, M.F., Tahir, N.M. and Chan, C.C. 2014. A case-crossover analysis of forest fire haze events and mortality in Malaysia. *Atmospheric Environment*. 96, 257–265.

Schultz, M.G., Heil, A., Hoelzemann, J.J., Spessa, A., Thonicke, K., Goldammer, J.G., Held, A.C., Pereira, J.M.C. and van het Bolsher, M. 2008. Global wildland fire emissions from 1960 to 2000. *Global Biogeochemical Cycles*. 22, GB2002.

Sentian, J. 2002. Production and neutralization of atmospheric acidity during biomass burning. *Borneo Science*. 12, 61–70.

Sentian, J. and Kong, S.S.K. 2013. High resolution climate change projection under SRESA2 scenario during summer and winter monsoon over Southeast Asia using PRECIS regional climate modelling system. *SIJ Transaction of Computer Science Engineering & Its Application (CSEA)*. 1(4), 163–173.

Sentian, J., Jemain, M.A., Gabda, D., Herman, F. and Wui, J.C.H. 2018. Long-term trends and potential associated sources of particulate matter (PM10) pollution in Malaysia. *WIT Transactions on Ecology and the Environment*. 230, 607–618.

Shi, Y. and Yamaguchi, Y. 2014. A high-resolution and multi-year emission inventory for biomass burning in Southeast Asia during 2001–2010. *Atmospheric Environment*. 98, 8–16.

Sloan, S., Locatelli, B., Wooster, M.J. and Gaeau, D.L.A. 2017. Fire activity in Borneo driven by industrial land conversion and draught during El Niño periods, 1982–2010. *Global Environmental Change*. 47, 95–109.

Stott, P. 2000. Combustion in tropical biomass fires: A critical review. *Progress in Physical Geography*. 24, 355–377.

Streets, D.G., Yarber, K.F., Woo, J.H. and Carmichael. G.R. 2003. Biomass burning in Asia: Annual and seasonal estimates and atmospheric emissions. *Global Biogeochemical Cycles*. 17(4), 1099. doi:10.1029/ 2003GB002040

Sulong, N.A., Latif, M.T., Khan, M.F., Amil, N., Ashfold, M.J., Wahab, M.I.A., Chan, K.M. and Sahani, M. 2017. Source apportionment and health risk assessment among specific age groups during haze and non-haze episodes in Kuala Lumpur, Malaysia. *Science of the Total Environment*. 601–602, 556–570.

Tacconi, I. 2016. Preventing fires and haze in Southeast Asia. *Nature Climate Change*. 6, 640–643.

Takegawa, N., Kondo, Y., Ko, M., Koike, M., Kita, K., Blakem D.R., Hu, W., Scott, C., Kawakami, S., Miyazaki, Y., Russell-Smith, J. and Ogawa, T. 2003. Photochemical production of O_3 in biomass burning plumes in the boundary layer over northern Australia. *Geophysical Research Letter*. 30(10), 1500. doi:10.1029/2003GL017017.

Tang, Y.H., Naoki, K., Akio, F. and Awang, M. 1996. Light reduction by regional haze and its effect on simulated leaf photosynthesis in a tropical forest of Malaysia. *Forest Ecology and Management*. 89, 205–211.

Taylor, D. 2010. Biomass burning, humans, and climate change in Southeast Asia. *Biodiversity Conservation*. 19, 1025–1042.

Thompson, A.M., Witte, J.C., Hudson, R.D., Guo, H., Herman, J.R. and Fujiwara, M. 2001. Tropical tropospheric ozone and biomass burning. *Science*. 291, 2128–2132.

Vadrevu, K.P. and Justice, C.O. 2011. Vegetation fires in the Asian region: Satellite observational needs and priorities. *Global Environmental Research*. 15(1), 65–76.

Vadrevu, K.P. and Lasko, K.P. 2015. Fire regimes and potential bioenergy loss from agricultural lands in the Indo-Gangetic Plains. *Journal of Environmental Management*. 148, 10–20.

Vadrevu, K.P. and Lasko, K. 2018. Intercomparison of MODIS AQUA and VIIRS I-Band fires and emissions in an agricultural landscape—Implications for air pollution research. *Remote Sensing*. 10(7), 978. doi:10.3390/rs10070978.

Vadrevu, K.P., Csiszar, I., Ellicott, E., Giglio, L., Badarinath, K.V.S., Vermote, E. and Justice, C. 2012. Hotspot analysis of vegetation fires and intensity in the Indian region. *IEEE Journal of Selected Topics in Applied Earth Observations and Remote Sensing*. 6(1), 224–238.

Vadrevu, K.P., Giglio, L. and Justice, C. 2013. Satellite based analysis of fire–carbon monoxide relationships from forest and agricultural residue burning (2003–2011). *Atmospheric Environment*. 64, 179–191.

Vadrevu, K.P., Ohara, T. and Justice, C. 2014a. Air pollution in Asia. *Environmental Pollution*. 12, 233–235.

Vadrevu, K.P., Lasko, K., Giglio, L. and Justice, C. 2014b. Analysis of Southeast Asian pollution episode during June 2013 using satellite remote sensing datasets. *Environmental Pollution*. 12, 245–256.

Vadrevu, K.P., Lasko, K., Giglio, L. and Justice, C. 2015. Vegetation fires, absorbing aerosols and smoke plume characteristics in diverse biomass burning regions of Asia. *Environmental Research Letters*. 10(10), 105003.

Vadrevu, K.P., Ohara, T. and Justice, C. 2017. Land cover, land use changes and air pollution in Asia: A synthesis. *Environmental Research Letters*. 12(12), 120201.

Vadrevu, K.P., Ohara, T. and Justice, C. (Eds). 2018. *Land-Atmospheric Research Applications in South and Southeast Asia*. Springer, Cham.

Vadrevu, K.P., Lasko, K., Giglio, L., Schroeder, W., Biswas, S. and Justice, C. 2019. Trends in vegetation fires in South and Southeast Asian countries. *Scientific Reports.* 9(1), 7422. doi:10.1038/s41598-019-43940-x.

Van Ittersum, M.K. and Rabbinge, R. 1997. Concepts in production ecology for analysis and quantification of agricultural input-output combinations. *Field Crops Research.* 52(3), 197–208. doi:10.1016/S0378-4290(97)00037-3.

Wang, W.C., Chen, J.P., Isaksen, I.S.A., Noone, K. and McGuffie, K. 2012. Climate–Chemistry Interaction: Future Tropospheric Ozone and Aerosols. In: *The Future of the World's Climate,* 2nd Edition. Henderson-Sellers, A. and McGuffie, K. (Eds.) pp. 367–399. doi:10.1016/C2010-0-67318-4.

Wang, S.H., Welton, E.J., Holben, B.N., Tsay, S.C., Lin, N.H., Giles, D., Stewart, S.A., Janjai, S., Nguyen, X.A. and Hsiao, T.C. 2015. Vertical distribution and columnar optical properties of springtime biomass-burning aerosol over Northern Indochina during 2014 7-SEAS campaign. *Aerosol Air Quality Research.* 15, 2037–2050.

Woittiez, L.S., van Wijk, M.T., Slingerland, M., van Noordwijk, M. and Giller, K.E. 2017. Yield gaps in oil palm: A quantitative review of contributing factor. *European Journal of Agronomy.* 83, 57–77. doi:10.1016/j.eja.2016.11.002.

Wolff, G.T., Kelly, N.A., Fermen, M.A., Ruthkosky, M.S., Stroup, D.P. and Korsog, P.E. 1986. Measurement of sulfur oxides, nitrogen oxides haze and fine particles at rural site on the Atlantic Coast. *Journal of Air Pollution Control Association.* 36, 585–591.

Wooster, M.J., Perry, G.L.W. and Zounas, A. 2012. Fire, draught and El Niño relationships on Borneo during the pre-Modis era (1980–2000). *Biogeosciences.* 9, 317–340.

Yeh, S.W., Kug, J.S., Dewittc, B., Kwon, M.H., Kirtman, B.P. and Jin, F.F. 2009. El Niño in a changing climate. *Nature.* 461(7263), 511–514. doi:10.1038/nature08316.

Yin, Y., Ciais, P., Chevallier, F., van der Werf, G.R., Fanin, T., Broquet, G., Boesch, H., Cozic, A., Hauglustaine, D., Szopa, S and Wang, Y.L. 2016. Variability of fire carbon emissions in equatorial Asia and its nonlinear sensitivity to El Niño. *Geophysical Research Letters.* 43, 10472–10479. doi:10.1002/2016GL070971.

Ying, Y.T., Lim, S.F. and von Glasow, R. 2013. The influence of meteorological factors and biomass burning on surface ozone concentrations at Tanah Rata, Malaysia. *Atmospheric Environment.* 70, 435–446. doi: 10.1016/j.atmosenv.2013.01.018.

Yokelson, R.J., Burling, I.R., Gilman, J.B., Warneke, C., Stockwell, C.E., de Gouw, J., Akagi, S.K., Urbanski, S.P., Veres, P., Roberts, J.M., Kuster, W.C., Reardon, J., Griffith, D.W.T., Johnson, T.J., Hosseini, S., Miller, J.W., Cocker III, D.R., Jung, H. and Weise, D.R. 2013. Coupling field and laboratory measurements to estimate the emission factors of identified and unidentified trace gases for prescribed fires. *Atmospheric Chemistry and Physics.* 13, 89–116. doi:10.5194/acp-13-89-2013.

Yoneda, T., Nishimura, S. and Chairul. 2000. Impacts of dry and hazy weather in 1997 on a tropical rainforest ecosystem in West Sumatra, Indonesia. *Ecology Research.* 15, 63–71.

Yulianti, N. and Hayasaka, H. 2013. Recent Active Fires under El Niño Conditions in Kalimantan, Indonesia. *American Journal of Plant Sciences.* 4, 3A, doi:10.4236/ajps.2013.43A087.

Zhang, H., Chen, S., Jiang, N., Wang, X., Zhang, X., Liu, J., Zang, Z., Wu, D.Y., Yuan, T.G., Luo, Y. and Zhao, D. 2018. Differences in sulfate aerosol radiative forcing between the daytime and nighttime over East Asia using the weather research and forecasting model coupled with chemistry (WRF-Chem) model. *Atmosphere.* 9, 441. doi:10.3390/atmos9110441.

11 Swidden Agriculture and Biomass Burning in the Philippines

Gay J. Perez
University of the Philippines Diliman, Philippines

Josefino C. Comiso
University of the Philippines Diliman, Philippines
NASA Goddard Space Flight Center, USA

Mylene G. Cayetano
University of the Philippines Diliman, Philippines

CONTENTS

INTRODUCTION

Swidden agriculture started about 12,000 years ago when humans stopped hunting and gathering and decided to stay put where they are and began to grow crops and domesticate animals. Among the motivations is the hesitance to leave an ideal location that, at the same time, is endowed with a perfect climate. By swidden agriculture, we mean the conversion of forest and grassland to open land suitable for farming. It is sometimes referred to as "slash-and-burns agriculture," and it is called the *kaingin* system in the Philippines. The strategy is to clean up an area by cutting down the vegetation, allowing it to dry, and then setting it on fire to produce ashes that serve as nutrients to the soil used to plant food crops. When the soil is no longer fertile, the farmer would move to another location and start the process repeatedly with the idea that they could return to the previously cultivated land after vegetation has regenerated.

The *kaingin* system was rampant in the Philippines until a few decades ago when it was discouraged by policymakers as the old-growth forest cover was reduced to just a few percent of the original, and its biodiversity became seriously compromised. Biodiversity in the Philippines has been regarded as one of the richest in the world with many endemic species (Catibog-Sinha and Heaney, 2006), but this has likely changed substantially as their habitat became considerably depleted. Records show an 86% reduction in the country's forest cover for 20 years between the 1970s and 1990s (Moya and Malayang, 2004). The process was likely facilitated by the arrival of chain saws in the 1970s and 1980s and the lack of strict control on commercial logging during the period. The designation of some areas as protected parks and the National Greening Program (NGP) introduction have helped, but it is apparent that a lot more needs to be done.

The burning of living or dead vegetation under slash and burn is part of a more general phenomenon called biomass burning, including the burning of residues from crops after harvest, including arson and the naturally occurring fires such as those caused by lightning and volcanic eruptions. It also includes the burning of grassland, which was initially considered an acceptable practice but is now regulated because it has become evident that farmers would usually allow the burning to go into adjacent forests (Predo, 1985).

Fire causes the release of large amounts of particulates and gases into the atmosphere that are expected to adversely impact humans' and other inhabitants' climate and health. Specifically, biomass burning in South/Southeast Asia is rampant, and the fire is most commonly used as a land-clearing tool for agriculture, planting commercial crops, clearing of agricultural residues, etc. (Prasad et al., 2001; 2002; Biswas et al., 2015; Justice et al., 2015; Lasko and Vadrevu, 2018; Hayasaka and Sepriando, 2018; Israr et al., 2018; Vadrevu et al., 2019). Biomass burning results in a significant release of air pollutants to the atmosphere, which can have both climate and radiative effects (Vadrevu et al., 2018; Ramachandran, 2018). In this study, we focused on the interannual variability, frequency, and extent of fire events using the historical and in situ data, including the more recent satellite data. The role of swidden agriculture and biomass burning in these events and the associated impacts of atmospheric pollution caused by fires has been elaborated.

TRENDS IN PHILIPPINE FIRE AND BIOMASS BURNING

FIRE STATISTICS AND THE ROLE OF SLASH AND BURN

The Philippines is an archipelago consisting of 7107 islands with a total land area of 300,000 km^2. Figure 11.1 shows land cover information of the entire Philippines provided by the National Mapping and Resource Information Authority (NAMRIA) using data collected primarily in 2010 or earlier (Manuel, 2014). As of this date, the data indicate that there were only 19,332.34 and 45,909.03 km^2 of closed and open forest, 62,751.73 and 61,661.77 km^2 of annual and perennial crops, and 33,530.21, 38,276.70, and 14,310.45 km^2 of shrubs, wooded grassland, and grassland,

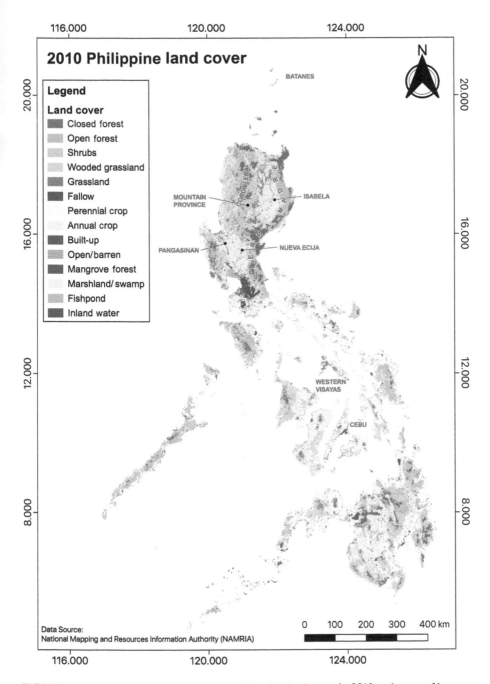

FIGURE 11.1 Location map showing the Philippine land cover in 2010 and areas of interest mentioned in the chapter. The map is constructed using the land cover data from the National Mapping and Resource Information Authority (Manuel, 2014).

respectively. The closed forest areas (represented by dark green in Figure 11.1), which are likely the only areas where the old-growth forest is located, have been reduced to a small fraction of the total land area and are likely vulnerable to further reduction because these areas are widely dispersed and accessible to human activities. The vegetation shows considerable inter-annual variations (Perez and Comiso, 2014).

In addition to the negative impacts on the environment, the cost of damage from fire events has been prohibitively expensive. The frequency of fires and area burned over the natural forest, grassland, and plantations from 1992 to 2009 as well as the associated cost of damage are presented in Table 11.1 (DENR, 2014, 2016, 2019). It is apparent that there are years when the fire was unusually destructive, especially

TABLE 11.1
Fire Occurrences, Area Burned, and Total Cost of Damage by Year from 1992 to 2009

Year	Fire Occurrences	Natural Forest	Grassland	Plantation	Total	Total Cost (Pesos)	(Hectares/ Incident)
			Area (ha) Burned				
1992	1106	5064.34	14,403.93	31,842.11	51,310.38	185,453,558.30	46.39
1993	595	311.81	415.00	14,603.07	15,329.88	165,866,132.73	25.76
1994	218	647.77	2509.00	4564.00	7720.77	20,685,649.80	35.42
1995	280	1369.82	2055.10	7284.77	10,709.69	50,668,410.53	38.25
1996	194	5.00	890.65	4567.77	5463.42	111,263,443.05	28.16
1997	147	409.57	371.88	2779.98	3,561.43	56,066,969.52	24.23
1998	941	17,044.10	18,893.71	18,893.71	54,831.52	216,547,415.75	58.27
1999	36	257.22	3500.00	2344.35	6101.57	32,638,962.34	169.49
2000	38	828.80	1122.75	1185.83	3137.38	5,451,855.00	82.56
2001	26	128.70	700.00	752.93	1581.63	18,973,718.68	60.83
2002[a]	180	1970.98	2237.69	15.75	4224.42	45,803,806.50	23.47
2002[b]	227	1983.12	2619.22	4146.65	8748.99	59,077,568.09	38.54
2003	208	57.58	507.05	5235.75	5800.38	30,114,806.69	27.89
2004[c]	-	396.69	-	979.47	1376.16	20,518,383.18	-
2005	-	1203.47	-	1511.34	2714.81	17,361,684.50	-
2006	-	2355.00	-	649.67	3004.67	28,971,808.04	-
2007	-	360.76	-	541.78	902.54	3,871,514.90	-
2008	-	50.00	-	26.58	76.58	75,472.63	-
2009	-	574.00	-	626.32	1200.34	2,760,648.00	-

Sources of Data: Special Action and Information Division (SAID), Special Concerns Office (Adapted from Comiso et al., 2014) with additional data for 2004 to 2009 from DENR.

[a] Statistics from 1992 to 2002 are sourced from data submitted by regional offices compiled and consolidated.

[b] Statistics from 2002 to 2003 are sourced from submitted reports by regional offices in compliance with OUFO Memo – Dated April 27, 2004.

[c] Statistics from 2004 to 2009 are sourced from the Compendium of ENR Statistics Volume 1 Forest Management, 2014.

during El Niño years, such as in 1992, 1998, and 2002, when the grounds were unusually warm and dry. But even during non-El Niño years, the data show that thousands of hectares of land are being burned, and the cost can be very high as in 1993. The fires associated with the *kaingin* system are difficult to separate from the available data, but it has been estimated by the Department of Environment and Natural Resources (DENR) that this system was responsible for the burning of about 752 ha in 2002 that cost the destruction of property amounting to P8,131,940. Similar but less costly events occurred with the burning of 875 ha in 2000 and 105 ha in 2008 that destroyed about P970,000 and P2,150,000 worth of properties, respectively (DENR, 2019).

Monitoring Techniques of Fire and Its Interannual Variability

Monitoring fires in the Philippines has historically been done from the ground through direct observations but sometimes done from a distance through observation of smoke formation. When done from the ground, quantitative assessment of the extent and the rate of spread of the fire is very difficult, and there is always a good chance of missing some of the fires even with the use of aircraft and drones. A more robust technique for fire detection is through satellite remote sensing data, starting with Landsat and SPOT data, which became available in the 1970s but were not popularly adopted early on because of minimal global coverage, especially in regions like the Philippines. The launch of sensors like MODIS and VIIRS that provide comprehensive global coverage enabled the detection of fires worldwide almost daily. But although the detection of fire locations/thermal hotspots can be done effectively at $0.5\,m^2$, detection of burned areas is limited to only about 250–375 m, and when the ability to detect small fires is needed, Landsat and even higher resolution data have to be utilized concurrently. When satellite data are used, algorithms such as those developed by Giglio et al. (2003), Schroeder et al. (2014), and Schroeder et al. (2016) are needed. Also, some quality control and data screening need to be implemented to obtain reliable results. Furthermore, algorithms for daytime conditions should provide consistent results as those designed for nighttime conditions.

It should be noted that the frequency of fire changes from one season to another, as clearly depicted in Figure 11.2. Color-coded maps of the fire events in the entire country, as derived from Terra MODIS, Aqua MODIS, and VIIRS, separately, for two months in 2018 are presented in Figure 11.2. The number of fires detected during the warm month of April 2018 is shown to be much larger (i.e., about five times more) than that during the cold rainy month of October 2018. In April, the fire locations are also more evenly distributed, especially as detected by VIIRS data, and go far beyond residential and industrial regions. In October, the fires are farther away from historically forested areas like the Sierra Madre and the Cordilleras. It is difficult to determine unambiguously which fire events are due to slash and burn, which ones are due to crop residue burning, and which ones are accidental or due to arson, despite the availability of multichannel data. But some useful information associated with motivation can be partly inferred through the use of time series of such data and knowledge about the general location of crops, grassland, and forestry (see Figure 11.1). It also helps to know when crops are being harvested and when grasslands are usually set on fire.

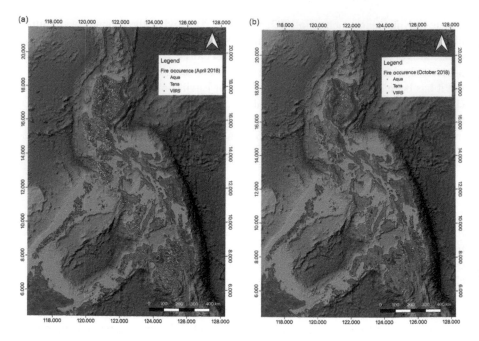

FIGURE 11.2 Location of fires in the Philippines during the months of (a) April 2018 and (b) October 2018 as derived from Aqua MODIS, Terra MODIS and VIIRS data. The area of each data point is exaggerated for visibility.

Figure 11.3 shows plots of fire frequency in the forest, crop, and grassland areas as derived separately from Terra MODIS, Aqua MODIS, and VIIRS over the periods when data are available for each sensor. There are significantly higher values in the frequency of fire as detected by VIIRS than by MODIS, as reported previously (Vadrevu et al., 2017; Vadrevu and Lasko, 2018). VIIRS is expected to provide a more comprehensive and accurate representation of the actual frequency because of better coverage (wider swath) and higher resolution (350 m vs. 1 km) that enables identification of relatively small fires that are not observed by MODIS. The time-series data from the three sensors, however, show generally the same patterns of variability. This means that the sampling statistics provided by MODIS are likely robust enough to assess the interannual variability and trend that the shorter VIIRS record could not provide. The longer data record also enables the ability to identify the effect of serious drought periods, such as the enhancements associated with El Niño in 2002 and 2010. It is encouraging that the MODIS data from Terra and Aqua show good consistency during the period of overlap from 2002 to 2018. Compared to the VIIRS data, there is also a general consistency except for the first 2 years of data (i.e., 2012 and 2013), where VIIRS shows a decline while MODIS shows increases. The reason for such inconsistency is currently unknown, but VIIRS has a broader swath width, and there may be relatively many more fires detected by VIIRS than by MODIS in 2012.

Interestingly, there is a large decline in the frequency of fires during all observational periods. It is also apparent that the most recent data available (i.e., 2017

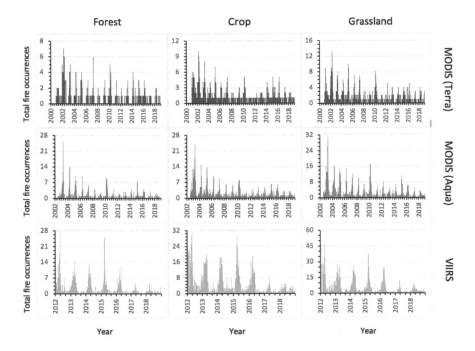

FIGURE 11.3 Total fire occurrences detected by Terra MODIS from 2000 to 2018; Aqua MODIS from 2002 to 2018; and VIIRS from 2012 to 2018 in Philippine forest, agricultural/crop areas, and grassland areas.

and 2018) show considerably lower frequencies than those in previous years. The observed decline is encouraging since it is likely an indication that the swidden/*kaingin* system is being abandoned for the more advanced agricultural systems where the productivity is maintained through the use of fertilizers, pesticides, and irrigation. Such a phenomenon has been reported to be happening already in some parts of the Philippines (Dressler and Pulhin, 2010) and attribute it to agrarian reform, commercialization of agriculture, and political pressure to adapt intensive modern agricultural practices. The wealthy and moderately wealthy can adapt readily, but the less affluent and indigenous farmers tend to stay on to swidden partly because of their inability to come up with the investment money needed. The ongoing biomass burning events are also reflected in Figure 11.3. A similar phenomenon about changes in swidden practices has also been reported in the entire Southeast Asian region (Mertz et al., 2009; Vadrevu et al., 2018).

BIOMASS BURNING EMISSION INVENTORY

In the absence of good air quality monitoring system, Emission Inventory (EI) could provide the desired estimate of pollutant emissions from various sources, including biomass burning and other fires. EI is a comprehensive database of geographically referenced datasets of atmospheric emission sources, and used as part of a broader air quality management (AQM) program. The goals are to enhance the ability to prepare strategies and regulations, evaluate trends, use air quality models, review the impact of new sources of pollution, ensure compliance with regulatory actions, and revise current air quality regulations and strategies. In estimating emissions, the EI uses emission factors, activity data, and, if available, any factor (numerical value) that can account for controlling the emissions. The activity data, reflecting the frequency and amount of biomass burned, and emission factors are available from the EMEP/EEA CORINAIR (based on European conditions) or US EPA AP-42 (derived from US conditions). However, such emission factors of the biomass burned to reflect Europe's and the USA's common practices and might not be applicable in developing countries where burning practices and technologies (if there are) are not as advanced, and pollutant sources are unique. Thus, pollutants from developing countries would need to have an emission factor that would better represent their biomass burning and fire emissions.

Gibe and Cayetano (2017) suggested a method for estimating emissions from agricultural biomass waste, rice straw in particular. This is based on the locally generated emission factor of rice straw, annual rice straw production, and the land area used for rice planting across the country. Emissions for agricultural waste burning were estimated with the formula shown in Equation (11.1):

$$E_{agricultural} = \left(\frac{RS}{RA} \right) \times EF \times SF \qquad (11.1)$$

where E is the calculated emissions from agricultural biomass burning and RS is the amount of rice straw produced per year, divided by RA, which is the total area in hectares ($0.01\,km^2$) used for growing rice. The emission factor is the in-house obtained emission factor for rice straw burning (Cayetano et al., 2014) for $PM_{2.5}$ per kilogram of rice straw burned and converted to kilogram $PM_{2.5}$ per year square kilometer. SF is the survey factor representing the farming area's percentage where the burning of rice straw as agricultural waste is used. This reduction factor is taken from the study of Launio et al. (2013). The result is a map of the estimated emissions of $PM_{2.5}$ from the burning of rice straw (Figure 11.4). The map indicates large spatial variability in $PM_{2.5}$ with the highest values in Isabela, Pangasinan, Tarlac, and Nueva Ecija.

We tested the method on a small-scale regional map, considering Cabanatuan City, Nueva Ecija, as the study site. The equation is applied to estimate $PM_{2.5}$ emissions for each cell, determined by its land cover type (in this case, agriculture). After the estimated emissions for each cell have been calculated, they were mapped using ArcMap (ArcGIS 10.1) software (ESRI, 2020). All cells with estimated $PM_{2.5}$ greater than zero are plotted for each land cover type.

CASE STUDIES

SWIDDEN FARMING IN BATANES

Batanes is a province composed of several islands situated in the northernmost part of the country. The biggest island is Itbayat, with an area of $83.13\,km^2$ and a household population of 2843 (2015 Census, PSA). The island's coastline is characterized

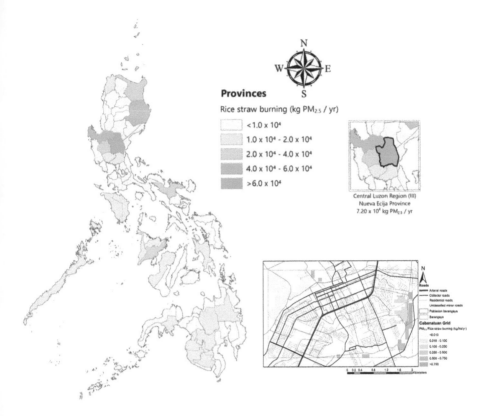

FIGURE 11.4 Estimated PM$_{2.5}$ emissions from rice straw burning by province, national scale. (Inset) Nueva Ecija Province in the central Luzon has the highest PM$_{2.5}$ emissions. The region contains provinces with the highest annual rice production in the country. (Adapted from Gibe and Cayetano, AOGS poster presentation 2017a, b.)

by cliffs and the absence of beaches or any sloping terrain toward the sea. This, together with generally rough seas most of the year, makes sea access to and from the island difficult. The island's relative isolation makes its inhabitants, the indigenous group of Ivatans, develop a strong culture of resiliency and self-sufficiency with swidden agriculture and pasture as its main livelihood (Bellwood and Dizon, 2013).

The Ivatans mainly cultivate root crops for personal consumption. Among the major crops are sweet potato (*Ipomoea batatas*), lesser yam (*Dioscorea esculenta*), purple yam (*Dioscorea alata*), and taro (*Colocasia esculenta*). Other crops such as corn are utilized for intercropping with root crops (De Guzman et al., 2014). The Ivatans practice crop rotation and multiple cropping systems based on farming areas' season and soil characteristics. There are two prominent seasons: summer, which lasts from March to May, and winter, which lasts from November to February. Crops are planted in either the grasslands or forests, with the land being allowed to regenerate for 3–5 years after being utilized for cropping. The subsequent planting will involve cutting shrubs or trees and clearing of weeds as part of the land preparation. This swidden farming practice ensures sufficient nutrient availability in the soil.

The *kaingin* system in Batanes is illustrated in a sequence of high-resolution images shown in Figure 11.5. What appears to be a cultivated area or grassland in 2008 (Figure 11.5a, enclosed blue box) was eventually covered with dense vegetation 8 years later (Figure 11.5b). The time in between these two periods includes a full cropping cycle and the regeneration period. Figure 11.5c includes a burnt area that appears to be being prepared for the next cropping cycle. A closer look at a plot in Itbayat that shows a recently cleared forest area with remnants of trees and ashes resulting from *kaingin* is presented in Figure 11.5d. This is an example of a small-scale event likely meant for one farmer.

Despite its adverse effects on the environment, *kaingin* is a continuing practice in Itbayat, primarily driven by the inhabitants' self-sufficiency. The government provided a Certificate of Ancestral Domain Title to the indigenous group, which allowed limited use of the island's natural resource and continued practice of swidden farming, which is recognized as part of their culture (RA 8371, 1997).

Fire Disturbances in NGP Sites

The NGP was established by the Philippine government in 2011 to combat the continuous decline of Philippine forests and harmonize various reforestation and biodiversity

FIGURE 11.5 Satellite images using very high-resolution (VHR) RGB data accessed through Google Earth Pro depict the swidden agriculture practice (inside blue box) in Itbayat. The images were captured on the following dates: (a) June 21, 2008; (b) November 7, 2016; and (c) April 29, 2018. (d) Burnt site in Itbayat located in 20.77°N, 121.84°E where swidden agriculture is in progress. (Photo (in d) is courtesy of Harry Merida, taken on April 9, 2018.)

initiatives in the last 25 years. The aim is to plant 1.5 billion trees in 1.5 million hect-ares of land within 6 years. The NGP is lauded to be generally successful and expanded to cover the remaining 1.7 million hectares of unproductive, denuded, and degraded forest lands for 12 more years (2016–2028). Despite its claimed success, the program is confronted with several challenges. One of them is the forest disturbances due to fire.

Barely a year since the program has started, there were already reports of fire inci-dents at different NGP sites. In 2012, around 10,000 young fruit and mahogany trees were razed by fire in Isabela, which led to the destruction of 10 ha of reforestation site (Catindig, 2012). The fire was reported to have spread from the adjacent farm where rice stalks and dried grass are being burned to plant. This number pales compared to the 350 ha of plantation forests in Mountain Province that were destroyed by 26 forest fire incidents just within the first quarter of 2014. Nearly 135,000 seedlings, saplings, and fully grown trees were torched by fires resulting from human-made and natural activities (See, 2014).

Another common cause of fire incidents in NGP sites is grass fire. In the uplands, the majority of the vegetation surrounding the reforestation area consists of highly flammable grasses. Once started, fire from these grasses can spread quickly to the plantation or reforestation area, especially during the dry season. Such was the case in 2015 when a grass fire expanded to engulf 120 ha of NGP plantation in Oslob, Cebu (Baquero, 2018). The intense warming caused by the 2015–2016 El Niño is also believed to be the reason for the rapid spread of grass fire that sets ablaze 107,000 seedlings, saplings, and trees in the Western Visayas region. The fires originated from various reasons, including incendiarism and throwing off a cigarette butt, which razed 676 ha of NGP plantation sites, and *kaingin* that damaged 8.5 ha of the plantation (DENR R6, 2019). In Pangasinan, 1614 ha out of 2860 ha of NGP sites were also damaged by several fire incidents in 2016, which is likely exacerbated by El Niño as well (DENR R1, 2019). Aside from the few reports such as those cited above, comprehensive information on fire occurrences at NGP sites is limited, making it difficult to monitor the extent of damages on a national scale. The plan is to use satellite data such as those from MODIS and VIIRS to provide a complete evaluation of fire incidents in NGP sites.

FIRE AND SMOKE POLLUTION: HEALTH IMPACTS

With two out of three Filipino farmers burning their agricultural biomass waste, the practice of open burning is still the most common method of agricultural biomass waste disposal by rice farmers in the Philippines (Launio et al. 2014), despite its environmental and health risks. The farmers are aware that burning agricultural bio-mass wastes contributes to air pollution, in the same way as they are also aware of the local (municipal or provincial) ordinances that discourage this practice. However, the national law under the Philippine Clean Air Act of 1999 (Section 20) exempts the burning of agricultural biomass waste.[1] This contradicts the local ordinances and

---

[1] RA 8749 Section 20: "B*an on Incineration.* – Incineration, hereby defined as burning of municipal, bio-medical and hazardous wastes, which process emits poisonous and toxic fumes, is hereby prohib-ited: *Provided, however,* that the prohibition shall not apply to traditional small-scale method of com-munity/neighborhood sanitation 'siga,' traditional, agricultural, cultural, health and food preparation and crematoria...."

puts the farmers and their families at risk of inhaling higher levels of air pollution, as burning agricultural wastes produces $PM_{2.5}$ that can travel long-range distances (Cayetano et al., 2011, Cayetano et al., 2014, Zhang, et al., 2012), and is directly linked to cardiopulmonary diseases (Chen et al., 2016), birth defects (Goto et al., 2016), and cancer (Cassidy et al., 2007). Idolor, et al. in their 2011 study determined the prevalence of chronic obstructive pulmonary disease (COPD) in rural communities in Nueva Ecija to be 20.8% for GOLD[2] Stage I and 16.7% for GOLD Stage II standard, with greater prevalence in men than women and increased prevalence between ages 40 and >70 years. This reported prevalence is associated with the Nueva Ecija households' use of firewood for cooking, working on a farm, and a history of tuberculosis, aside from smoking. Apparently, Nueva Ecija also belongs to the province that has the highest rice straw burning emissions.

The particulate matter's size and characteristics depend on the natural condition, types of material burnt (Rosales et.al, 2021), moisture content, and weather condition (Bagtasa et al., 2018). The wide varieties of gases released during fire include carbonaceous compounds such as CO, CO_2, and CH_4, and components of particulate matter such as organic sugars (levoglucosan to mannosan) and organic carbon, some of which are due to field burning of rice straw from the Philippines to southeast coastal China (Zhang et al., 2012).

SUMMARY AND CONCLUSION

The Philippines has a total land area of $300,000\,km^2$, about 90% of which was once covered by rain forest. As the population increased, the demand for more open space for agriculture led to swidden farming, referred to as *kaingin*, and a slow decline in forest cover. But over the brief period from the 1970s to the 1990s, the forest cover was drastically reduced by about 86% in part because of excessive logging to satisfy the strong global and national demand for timber and in part because of a natural or intentional forest fire. In this study, fire on the reduction of forest cover and environmental changes, including atmospheric pollution, is examined. Fire can be caused by natural events, such as lightning and volcanic eruptions, by arson, or by farmers for economic gains through *kaingin* or biomass burning. *Kaingin* has been a primary reason for forest fires, but historical records from in situ observations indicate that fires' frequency has been declining significantly. Records show that the total areas burned in 1992 were 51,310 ha with 1106 occurrences, while in 2002, the area burned was 5236 ha with 208 occurrences, and in 2008, the area burned reached a low point of only 76 ha. Satellite data from MODIS and VIIRS on fires show a continuation of the decline. The more comprehensive satellite data show a strong seasonality with the frequency about five times higher in the summer dry season than in the wet rainy season. The data also indicate that most of the fire originates from the agricultural areas suggesting the burning of crop residues after the harvest. Nevertheless, data show that *kaingin* goes on, especially in the upland and Batanes, and primarily by indigenous people who cannot adapt to modern agricultural practices.

---

[2] GOLD stands for Global Initiative for Chronic Obstructive Disease, a global standard for measuring the prevalence of COPD.

Fire is also shown to be one of the main causes of some of the NGP's setbacks. Although the program has been regarded as generally successful with more than a billion seedlings planted, a large fraction of the NGP sites have been destroyed by fire likely related to *kaingin*. The atmospheric pollution caused by a fire is examined, and although the burning of agricultural waste is known to cause adverse effects on health, about 2 out of every three farmers still do biomass burning. The source of biomass burning also varies considerably with the highest $PM_{2.5}$ emissions from rice straw occurring in Isabela, Pangasinan, Tarlac, and Nueva Ecija provinces. These emissions are also contributing to human-made greenhouse effects, and hence the changing climate, while some are toxic gases that affect the respiratory and other systems of the body. Thus, there is a strong need to determine the extent of air pollution from agricultural biomass waste burning. The results may help in addressing possible co-benefits and to implement mitigation options to curb biomass burning pollution.

ACKNOWLEDGMENTS

The authors would like to acknowledge the assistance of Mr. Harry Merida and Mr. Hezron Gibe for their contributions in the figures used in the chapter. The authors would like to thank the National Mapping and Resource Information Authority (NAMRIA) for the data used in Figure 11.1; and the GIST Research Institute (GRI) grant funded by the Gwangju Institute of Science and Technology in 2016, 2019, and 2020 for Figure 11.4. The authors would also like to thank the VIIRS fire product developers for freely sharing the datasets.

REFERENCES

Bagtasa, G., Cayetano, M. G. and Yuan, C.-S. 2018. Seasonal variation and chemical characterization of $PM_{2.5}$ in northwestern Philippines. *Atmospheric Chemistry and Physics.* 18, 4965–4980. doi:10.5194/acp-18-4965-2018.

Baquero, R.A.B. 2018. DENR: Keep forests, ecotourism sites safe from fire, Philippine Information Agency, 12 April 2018. https://pia.gov.ph/news/articles/1006974, (Date Accessed: 7 May 2019).

Bellwood, P., and Dizon, E. Z. 2013. 4000 years of migration and cultural exchange: The archaeology of the Batanes Islands, Northern Philippines. In Terra Australis 40: 254 pp, ANU Press.

Biswas, S., Lasko, K.D. and Vadrevu, K.P. 2015. Fire disturbance in tropical forests of Myanmar—Analysis using MODIS satellite datasets. *IEEE Journal of Selected Topics in Applied Earth Observations and Remote Sensing.* 8(5), 2273–2281.

Cassidy, B.E., Alabanza-Akers, M.A., Akers, T.A., Hall, D.B., Ryan, P.B., Bayer, C.W. and Naeher, L.P. 2007. Particulate matter and carbon monoxide multiple regression models using environmental characteristics in a high diesel-use area of Baguio City, Philippines. *Science of the total Environment.* 381(1–3), 47–58.

Catibog-Sinha, C. and Heaney L. R. 2006. *Philippine Biodiversity: Principles and Practice.* Haribon Foundation for the Conservation of Natural Resources, Inc., Quezon City, Philippines.

Catindig, R. 2012. Fire razes 10-hectare NGP site in Isabela, Philippine Star, 3 May 2012, https://www.philstar.com/nation/2012/05/03/802622/fire-razes-10-hectare-ngp-site-isabela, (Date Accessed: 7 May 2019).

Cayetano, M.G., Kim, Y.J., Jung, J.S., Batmunkh, T., Lee, K.Y., Kim, S.Y., Kim, K.C., Kim, D.G., Lee, S.J., Kim, J.S. and Chang, L.S. 2011. Observed chemical characteristics of long-range transported particles at a marine background site in Korea. *Journal of the Air & Waste Management Association*. 61(11), 1192–1203.

Cayetano, M.G., Hopke, P.K., Lee, K.H., Jung, J., Batmunkh, T., Lee, K. and Kim, Y.J. 2014. Investigations of the particle compositions of transported and local emissions in Korea. *Aerosol and Air Quality Research*. 14(3), 793–805.

Chen, R., Hu, B., Liu, Y., Xu, J., Yang, G., Xu, D. and Chen, C. 2016. Beyond $PM_{2.5}$: The role of ultrafine particles on adverse health effects of air pollution. *Biochimica et Biophysica Acta (BBA)-General Subjects*. 1860(12), 2844–2855.

Comiso, J.C., et al. 2014. *Changing Philippine Climate: Impacts on Agriculture and Natural Resources*. The University of the Philippines Press, Manila. p. 104.

De Guzman, L.E.P., Zamora, O.B., Talubo, J.P.P. and Hostallero, C.D.V. 2014. Sustainable agricultural production systems for food security in a changing climate in Batanes, Philippines. *Journal of Developments in Sustainable Agriculture*. 9(2), 111–119.

Department of Environment and Natural Resources. 2014. Compendium of ENR statistics volume 1 forest management. https://portal.denr.gov.ph/LIB/, (Date Accessed: 7 May 2019).

Department of Environment and Natural Resources Region 1. 2016. ENR statistical profile CY 2016. http://r1.denr.gov.ph/images/2018-updates/2018PMD/September/denr1statprofile2016.pdf, (Date Accessed: 7 May 2019).

Department of Environment and Natural Resources Region 6. 2019. Forest and grass fires affected some reforestation projects. https://r6.denr.gov.ph/index.php/86-region-news-items/744-forest-and-grass-fires-affected-some-reforestation-projects, (Date Accessed: 7 May 2019).

Dressler, W. and Pulhin, J. 2010. The shifting ground of Swidden agriculture on Palawan Island, the Philippines. *Agriculture and Human Values*. 27, 445–459.

European Environmental Agency (EEA). 2007. EMEP/CORINAIR emission inventory guidebook.

Gibe, H. and Cayetano, M.G. 2017a. Air particulate matter (APM) emissions from agricultural waste (rice straw) burning in the Philippines: Emission factor testing and estimation of nationwide $PM_{2.5}$ emissions. *14th Annual Meeting, Asia Oceania Geosciences Society*, 6–11 August, 2017, Singapore (AS42–A008).

Gibe, H.P. and Cayetano, M.G. 2017b. Spatial estimation of air $PM_{2.5}$ emissions using activity data, local emission factors and land cover derived from satellite imagery. *Atmospheric Measurement Techniques*. 10(9), 3313.

Giglio, L., Descloitres, J., Justice, C.O. and Kaufman, Y.J. 2003. An enhanced contextual fire detection algorithm for MODIS. *Remote Sensing of Environment*. 87(2–3), 273–282.

Goto, D., Ueda, K., Ng, C.F.S., Takami, A., Ariga, T., Matsuhashi, K. and Nakajima, T. 2016. Estimation of excess mortality due to long-term exposure to $PM_{2.5}$ in Japan using a high-resolution model for present and future scenarios. *Atmospheric Environment*. 140, 320–332.

Hayasaka, H. and Sepriando, A. 2018. Severe air pollution due to peat fires during 2015 super El Niño in Central Kalimantan, Indonesia. In: *Land-Atmospheric Research Applications in South/Southeast Asia*. Vadrevu, K.P., Ohara, T., and Justice, C. (Eds). Springer, Cham. pp. 129–142.

Idolor, L.F., De Guia, T.S., Francisco, N.A., Roa, C.C., Ayuyao, F.G., Tady, C.Z., Tan, D.T., Banal-Yang, S.Y.L.V.I.A., Balanag Jr, V.M., Reyes, M.T.N. and Dantes, R.B. 2011. Burden of obstructive lung disease in a rural setting in the Philippines. *Respirology*, 16(7), 1111–1118.

Israr, I., Jaya, S.N.I, Saharjo, H.S., Kuncahyo, B. and Vadrevu, K.P. 2018. Spatio-temporal analysis of land and forest fires in Indonesia using MODIS active fire dataset. In: *Land-Atmospheric Research Applications in South/Southeast Asia*. Vadrevu, K.P., Ohara, T. and Justice, C. (Eds). Springer, Cham. pp. 105–128.

Justice, C., Gutman, G. and Vadrevu, K.P. 2015. NASA land cover and land use change (LCLUC): An interdisciplinary research program. 148(15), 4–9.

Lasko, K. and Vadrevu, K.P. 2018. Improved rice residue burning emissions estimates: Accounting for practice-specific emission factors in air pollution assessments of Vietnam. *Environmental Pollution.* 236(5), 795–806.

Launio, C., Asis, C., Manalili, R. and Javier, E. 2013. Economic analysis of rice straw management alternatives and understanding farmer's choices. *Cost-Benefit Studies of Natural Resource Management in Southeast Asia.* 1, 93–111. doi:10.1007/978-981-287-393-4_5.

Launio, C., Asis, C., Manalili, R., Javier, E. and Belizario, A.F. 2014. What factors influence choice of waste management practice? Evidence from rice straw management in the Philippines. *Waste Management & Research*, 32(2), 140–148.

Manuel, W.V. 2014. Land cover data in the Philippines. *Presentation at the 5th UNREDD regional lessons learned workshop on monitoring systems and reference levels for REDD+,* October 20–22, 2014, Hanoi, Vietnam.

Mertz, O., Padoch, C., Fox, J., Cramb, R.A., Leisz, S.J., Lam, N.T. and Vien, T.D., 2009. Swidden change in Southeast Asia: understanding causes and consequences. *Human Ecology,* 37(3), 259–264.

Moya, T.B. and Malayang, B.S. 2004. Climate variability and deforestation-reforestation dynamics in the Philippines. *Environment Development and Sustainability.* 6, 261–277.

Perez, G.J.P. and Comiso, J.C., 2014. Seasonal and interannual variabilities of Philippine vegetation as seen from space. *Philippine Journal of Science,* 143(2), 147–155.

Perez, G.J. and Comiso, J.C. 2015. Monitoring Philippine vegetation using satellite NDVI and EVI data. *Journal of the Philippine Geoscience and Remote Sensing Society.* 1(1), 27–40.

Prasad, V.K., Kant, Y. and Badarinath, K.V.S. 2001. CENTURY ecosystem model application for quantifying vegetation dynamics in shifting cultivation areas: A case study from Rampa Forests, Eastern Ghats (India). *Ecological Research.* 16(3), 497–507.

Prasad, V.K., Kant, Y., Gupta, P.K., Elvidge, C. and Badarinath, K.V.S. 2002. Biomass burning and related trace gas emissions from tropical dry deciduous forests of India: A study using DMSP-OLS data and ground-based measurements. *International Journal of Remote Sensing.* 23(14), 2837–2851.

Predo, J. 1985. What motivates farmers? Tree growing and land use decisions in grasslands of Claveria, Philippines. Research Report No 2003-RR7. EEPSEA and the International Development Research Centre. http://web.idrc.ca/en/ev-51044-201-1-DO-TOPIC.html.

Ramachandran, S. 2018. Aerosols and climate change: Present understanding, challenges and future outlook. In: *Land-Atmospheric Research Applications in South/Southeast Asia.* Vadrevu, K.P., Ohara, T. and Justice, C. (Eds). Springer, Cham. pp. 341–378.

Republic Act No. 8371, Official Gazette, Republic of the Philippines, 29 October 1997. https://www.officialgazette.gov.ph/1997/10/29/republic-act-no-8371/, (Date Accessed: 7 May 2019).

Rosales, C.M.F., Jung, J., Cayetano, M.G. (2021). Emissions and chemical components of PM2.5 from simulated cooking conditions using traditional cookstoves and Fuels under a dilution tunnel system. *Aerosol Air Qual. Res.* https://doi.org/10.4209/aaqr.200581.

Schroeder, W., Oliva, P., Giglio, L. and Csiszar, I.A. 2014. The New VIIRS 375 m active fire detection data product: Algorithm description and initial assessment. *Remote Sensing of Environment.* 143, 85–96. doi:10.1002/jgrd50873.

Schroeder, W., Oliva, P., Giglio, L., Quayle, B., Lorenz, E. and Morelli, F. 2016. Active fire detection using Landsat-8/OLI data. *Remote Sensing of Environment.* 185, 210–220. doi:10.1016/j.rse.2015.08.032.

See, D.A. 2014. National Greening Program sites damaged by forest fires, Baguio Midland Courier, 30 March 2014. http://baguiomidlandcourier.com.ph/environment.asp?mode= archives/2014/march/3-30-2014/env1.txt, (Date Accessed: 7 May 2019).

The 2015 Census of Population, Philippine Statistics Authority. http://www.psa.gov.ph, (Date Accessed: 7 May 2019).

Vadrevu, K.P. and Lasko, K. 2018. Intercomparison of MODIS AQUA and VIIRS I-Band fires and emissions in an agricultural landscape—Implications for air pollution research. *Remote Sensing*. 10(7), 978. doi:10.3390/rs10070978.

Vadrevu, K.P., Ohara, T. and Justice, C. 2017. Land cover, land use changes and air pollution in Asia: A synthesis. *Environmental Research Letters*. 12(12), 120201.

Vadrevu, K.P., Ohara, T. and Justice, C. (Eds). 2018. *Land-Atmospheric Research Applications in South and Southeast Asia*. Springer, Cham.

Vadrevu, K.P., Lasko, K., Giglio, L., Schroeder, W., Biswas, S. and Justice, C. 2019. Trends in vegetation fires in south and Southeast Asian countries. *Scientific Reports*. 9(1), 7422. doi:10.1038/s41598-019-43940-x.

Zhang, Y.-N., Zhang, Z.-S., Chan, C.-Y., et al. 2012. Levoglucosan and carbonaceous species in the background aerosol of coastal southeast China: Case study on transport of biomass burning smoke from the Philippines. *Environmental Science and Pollution Research*. 19, 244–255. doi:10.1007/s11356-011-0548-7.

Section III

Climate Drivers and Biomass Burning

12 Fire Danger Indices and Methods: An Appraisal

Krishna Prasad Vadrevu
NASA Marshall Space Flight Center, USA

CONTENTS

INTRODUCTION

Vegetation fires are a common phenomenon in most regions of the world, including South/Southeast Asian countries. Fire behavior, fire duration, intensity, pattern, and the amount of fuel consumption and bioenergy loss can all influence injury and mortality of plants and their subsequent recovery (Vadrevu and Lasko, 2015; Justice et al., 2015; Vadrevu et al., 2017). The post-fire vegetation responses can also depend on the characteristics of the plant species on the site, their susceptibility to fire, and the means by which they recover after the fire (Miller, 2000; Biswas et al., 2015a, b). Specific to soils, fires can alter soil physical and chemical properties by impacting the structure and the loss of soil organic matter and nutrients thus reducing porosity, increased pH and runoff (Prasad et al., 2001; 2002; 2004; Certini, 2014). Fires not only alter the landscape structure and function but also affect the atmosphere through the release of radiatively active greenhouse gas emissions and aerosols (Prasad et al., 2000; Kant et al., 2000; Gupta et al., 2001; Badarinath et al., 2008a, b; Vadrevu and Justice, 2011; Vadrevu et al., 2017; Vadrevu and Lasko, 2018). Vegetation fire emissions are composed of large amounts of trace gases and aerosols. Important gases emitted include carbon dioxide (CO_2), carbon monoxide (CO), methane (CH_4), a series of volatile organic carbon compounds (VOCs), and inorganic species (Andreae et al., 1988; Prasad et al., 2003; Vadrevu et al., 2013; 2014a, b; 2015; Lasko et al., 2017; 2018; Lasko and Vadrevu, 2018; Vadrevu et al., 2019). Besides, secondary species such as ozone may be formed by the chemical reactions within the smoke plumes. The particulate matter (PM) emitted in the form of aerosols can cause significant respiratory and cardiovascular diseases (Sulong et al., 2017). Further, many of these pollutants have a long residence time; thus, they can travel long distances leading to large-scale modification of atmospheric composition and radiation budget (Vadrevu et al., 2018). Also, vegetation fires can pose a significant risk to life and property. Thus, an understanding of the critical factors controlling fires is fundamental to its management and mitigation including climate and land use/cover change research (Justice et al., 1993; Justice and Vadrevu, 2015; Vadrevu et al., 2019)

Important terminology relating to fire management specific to fire hazard, danger, risk, and vulnerability was reviewed in detail by Chuvieco et al. (2014) and others (Merril and Alexander, 1987; NWCG, 2003) and thus not reviewed here. Most of the fire danger indices integrate a variety of weather, meteorological, and vegetation-related parameters into one or more qualitative or quantitative indices for addressing fire prevention (Chandler et al., 1983; Chuvieco et al., 2003; Ayanz et al., 2003; de Groot et al., 2015; Schunk et al., 2017). Identifying fire danger in different regions of the world is vital as the results can aid fire mitigation efforts.

Over the past few decades, various fire danger indices, methods, and rating systems have been developed for different landscapes around the world integrating meteorological, vegetation, or soil moisture-related information (Deeming et al., 1977; Fosberg, 1978; Van Wagner and Forest, 1987; Burgan et al., 1998; Carlson et al., 2002; Soares and Batista, 2007). Of the different parameters, the meteorological parameters are important as they provide valuable information on fire initiation and spread, including the relative wetness or dryness of vegetation, which further can impact fire regime characteristics (frequency, intensity, duration, and spread). The literature on the fire indices, methods, and rating systems is highly scattered. Thus, in this study, a variety of fire danger assessment methods were compiled at one place. In addition, an overview on the commonly used remote sensing variables useful for fire danger assessment is provided. Also, some potentials and limitations of these indices and rating systems have been highlighted.

FIRE DANGER INDICES

A brief description of the major fire danger indices has been provided in this section. Due to limitations in space, additional references are provided where elaborate algorithm descriptions were not possible. Details of some of the indices are given below:

ANGSTRÖM INDEX

The Angström index (Willis et al., 2001) was developed in Sweden. The index is one of the simplest drought indices used in the fire-risk assessment. It is based on the temperature and relative humidity, and both are measured at 13 hours. It is a non-cumulative index and given as

$$B = 0.05H \times (T - 27/10) \tag{12.1}$$

where B is the Angström index, H is the relative humidity in %, and T is the air temperature in degrees centigrade. The index >4.0 indicates fire occurrence potential as unlikely; 3.0–4.0, fire occurrence potential unfavorable; 2.5–2.9, fire conditions favorable; 2.0–2.4, fire conditions more favorable; and <2.0, fire occurrence very likely.

BAUMGARTNER INDEX

This index quantifies drought and was used for fire-risk assessment in Germany (Skvarenina et al., 2003). The main inputs are the potential evapotranspiration and precipitation of the last 5 days (Baumgartner et al., 1967; Raphael, 2011) and given as

$$I_{Bg} = \sum_{d-4}^{d} p - \sum_{d-4}^{d} \mathrm{etp} \tag{12.2}$$

where I_{Bg} is the Baumgartner index, p is the precipitation (mm), etp is the potential evapotranspiration (mm), and d is the day of the calculation. Due to the evaporation, the susceptibility to ignition in a forest increases with the dryness of the fuel source and the intensity of evaporation on the incident solar radiation, wind speed, atmospheric humidity, and the amount of precipitation. The potential evaporation is calculated based on Penman (1963). The index values are incorporated into the Baumgartner system, and the index values obtained are then arranged according to the respective months.

BRUSCHEK INDEX

This index was developed at the Potsdam Institute for Climate Impact Research (PIK) to be used for the state of Brandenburg (Raphael, 2011; Bruschek, 1994). This is an annual drought index that uses daily maximum temperature and the sum of daily precipitation during the vegetation growing period and is given as

$$I_{Br} = \sum_{i=v_b}^{v_e} sd_i \Bigg/ \sum_{i=v_b}^{v_e} p_i \tag{12.3}$$

where I_{Br} is the Bruschek index; $sd = 1$ if the daily maximum air temperature is $\geq 25.0°C$, and 0 in all other cases; p is the daily sum of precipitation (mm); v_b is the beginning of the vegetation period (April 1); v_e is the end of the vegetation period (September 30); and "i" is the days between v_b and v_e.

NESTEROV INDEX

Nesterov (Nesterov, 1949; Shetinsky, 1994) proposed a fire-risk rating index to be used in the former Soviet Union, and the index establishes a range of discrete fire-risk levels. The index inputs include daily temperature, precipitation, and dew point temperature for its calculation. The number of days since last precipitation above 3 mm is a determining factor of fire risk, and the index is set to zero every time the rainfall exceeds this threshold. It is calculated as

$$I_N = \sum_{i=1}^{W} (T_i - D_i) T_i \tag{12.4}$$

where I_N is the Nesterov index value, T is the temperature at 13:00 hours, W is the number of wet days since last rainfall > 3 mm, D is the dew point temperature (°C), and i is the current day.

MODIFIED M-68 INDEX

This index was previously used to produce daily fire danger maps for Germany. The M-68 is a modified version of the index developed by Käse (1969) and follows the same principle as the Russian Nesterov Index. Flemming (1994) later modified the index to include phenology and precipitation correction factors (Raphael, 2011). Also, the index values are fed into a forest growth model (Schaber et al., 1999), and the results were used to define five danger classes. Essential inputs of the index include maximum temperature and saturation deficit at noon. The daily index is calculated as

$$I_{M,d} = k_{prec} I_{M,d-1} + k_{phen} (T_{max} + 10) \Delta vp \tag{12.5}$$

where $I_{M,\,d}$ is the modified M-68 index value for day "d," k_{prec} is the precipitation correction factor, k_{phen} is the phenology correction factor, T_{max} is the daily maximum temperature (°C), and Δvp is the saturation deficit at noon. The saturation deficit Δvp is calculated as

$$\Delta vp = svp - vp \tag{12.6}$$

$$svp = 6.1078\ e^{\frac{17.62 T_{max}}{243.12 T_{max}}} \tag{12.7}$$

$$vp = svp \cdot \frac{h_r}{100} \tag{12.8}$$

T_{max} represents the maximum temperature (°C), vp is the vapor pressure at noon, and h_r is the relative humidity and svp is the saturation vapor pressure at noon. More details about the correction factors can be found in Raphael (2011).

Fosberg Fire Weather Index (FFWI)

This index combines a dry-bulb temperature, relative humidity, and wind speed data in a linear relationship to relate to fire behavior. The FFWI is based upon equilibrium moisture content and wind speed (Fosberg, 1978) and given as

$$FFWI = \frac{\eta \sqrt{1 + U^2}}{0.3002} \tag{12.9}$$

where U is the wind speed in miles per hour. The moisture damping coefficient η is given as

$$\eta = 1 - 2\left(\frac{m}{30}\right) + 1.5\left(\frac{m}{30}\right)^2 - 0.5\left(\frac{m}{30}\right)^3 \tag{12.10}$$

The equilibrium moisture content (m) is given as a function of temperature (degrees Fahrenheit) (T) and relative humidity in percent (h) (Goodrick, 2002):

$$m = \begin{cases} 0.03229 + 0.281073h - 0.000578hT & \text{for } h < 10\% \\ 2.22749 + 0.160107h - 0.01478T & \text{for } 10\% < h \leq 50\% \\ 21.0606 + 0.005565h^2 - 0.00035hT - 0.483199h & \text{for } h > 50\% \end{cases} \tag{12.11}$$

Additional details about this index can be found in Goodrick (2002).

Haines Index or Lower Atmospheric Severity Index (LASI)

The Haines Index characterizes the potential impact of dry, unstable air ~1–3 km above the surface on wildfire behavior and growth (Haines, 1988; Winkler et al.,

2007). The index is the sum of a stability (*A*) component and a humidity (*B*) component, the former representing the change of temperature with height and the latter dew point depression, respectively. The index can vary based on the location, as "low," "mid," and "high," to address the surface elevation variations. Also, for "low," "mid," and "high" variants, the temperature and humidity differences are assigned an ordinal value (1, 2, or 3) for equal weighting of the two components, and then, the values are added. The Haines Index values vary from 2 (very low potential of large or erratic plume-dominated behavior) to 6 (very high potential) (Winkler et al., 2007). The calculation details of the Haines Index are given in Table 12.1.

PALMER DROUGHT SEVERITY INDEX (PDSI)

This index is most widely used for quantifying the droughts and has been used for quantifying the fire danger too (Baisan and Swetnam, 1990; Heyerdahl et al., 2008). Positive values of PDSI indicate wetter-than-normal conditions, and negative values suggest to drought (Palmer, 1965). PDSI analyzes either a weekly or monthly water budget and assumes that evapotranspiration (ET) occurs close to the potential monthly ET (PE) until a certain amount of the available water is depleted, after

TABLE 12.1

Haines Index Calculation Based on Stability and Humidity Components

Elevation	Calculation	Categories
	Stability (*A*) Component	
Low	950-hPa temperature to 850-hPa temperature	$A=1$ if $<4°C$
		$A=2$ if $4°C–74°C$
		$A=3$ if $\geq84°C$
Mid	850-hPa temperature to 700-hPa temperature	$A=1$ if $<64°C$
		$A=2$ if $6°C–104°C$
		$A=3$ if $\geq114°C$
High	700-hPa temperature to 500-hPa temperature	$A=1$ if $<184°C$
		$A=2$ if $18°C–214°C$
		$A=3$ if $\geq224°C$
	Humidity (*B*) Component	
Low	850-hPa temperature to 850-hPa dew point	$B=1$ if $<64°C$
		$B=2$ if $6°C–94°C$
		$B=3$ if $\geq104°C$
Mid	850-hPa temperature to 850-hPa dew point	$B=1$ if $<64°C$
		$B=2$ if $6°C–124°C$
		$B=3$ if $\geq134°C$
High	700-hPa temperature to 700-hPa dew point	$B=1$ if $<154°C$
		$B=2$ if $15°C–204°C$
		$B=3$ if $\geq214°C$

which the actual ET is less than PE. The PDSI uses the following equations to compute the moisture transfer between soil layers (Ntale and Gan, 2003):

$$L_s = \min\{S_s, (\text{PE} - P)\} P \leq \text{PE} \tag{12.12}$$

$$L_u = \big(\big([(\text{PE} - P) - L_s]S_u\big)/\text{AWC}\big)L_u \leq \text{PE} \tag{12.13}$$

where P is the precipitation and L_s and L_u (S_s and S_u) are the moisture loss in the upper or surface layers and the upper or surface and underlying layer(s) respectively at the start of the month. The available water capacity (AWC) is the combined available field capacity of all soil layers. PE is calculated by Thornthwaite's (1948) method, which has an approximate daily absolute error of 35%, but PE could be in error by over 100% on some individual days. As the timescale being considered increases to about 2 weeks or longer, the error decreases to about 10% to 15%, as being acceptable for the climatological analysis of moisture requirements (Palmer, 1965; Ntale and Gan, 2003). More details about the potential recharge (PR), potential loss (PL) of soil moisture to ET during dry periods, and potential runoff (PRO) are given in Palmer (1965).

KEETCH–BYRAM DROUGHT INDEX (KBDI)

This index represents an empirical approximation for moisture depletion in the upper soil and surface litter levels (Keetch and Byram, 1968). It is a drought index that is considered as a conventional tool for estimating fire potential (Janis et al., 2002; Garcia-Prats et al., 2015). The KBDI is based on a simple daily water balance and accounts for cumulative soil water depletion from evapotranspiration and precipitation effects on deep duff and upper soil layers. It ranges from 0 to 203.2 when rainfall is expressed in mm (from 0 to 800 when it is expressed in inches), from low to high fire risk, or from no soil water depletion to very dry conditions. The amount of water lost in a forested or wildland area is calculated as (Garcia-Pratis et al., 2015)

$$dQ = \frac{[800 - Q] \times \left[0.968 \times e^{(0.0486 \times T)} - 8.30\right] \times dt}{1 + 10.88 \times e^{-0.0441 \times R}} \times 10^{-3} \tag{12.14}$$

The original equation does not use SI units; temperature is expressed in F and rainfall in inches. The equation can be rewritten as

$$dQ = \frac{[203.2 - Q] \times \left[0.968 \times e^{(0.0875 \times T + 1.5552)} - 8.30\right] \times dt}{1 + 10.88 \times e^{(-0.001736 \times R)}} \times 10^{-3} \tag{12.15}$$

where dQ is a drought factor or soil water depletion (in mm) during a period of time dt, with the 1-day time step recommended by the authors; Q is the accumulated soil water depletion (in mm); T is the daily maximum temperature (in °C); R is mean annual rainfall (in mm); and 203.2 is the field capacity of soil expressed in mm (203.2 mm = 800 hundredths inches). Due to the exponential nature of the index, mathematically reaching the 203.2 point would require an infinite time (Keetch and Byram, 1968).

The potential evapotranspiration (ETP) is estimated daily as the ratio of an exponential function of the daily maximum temperature (T) to an exponential function of the mean annual rainfall (R):

$$ETP = \frac{\left[0.968 \times e^{(0.0875 \times T + 1.5552)} - 8.30\right] \times dt}{1 + 10.88 \times e^{(-0.001736 \times R)}} \times 10^{-3} \qquad (12.16)$$

The numerator describes a general curve that calculates ETP as a function of daily maximum temperature, which is then adjusted to a specific region using the exponential function of the mean annual rainfall located in the denominator. The ETP is converted to actual evapotranspiration as a linear function of soil water depletion; i.e., ETP is reduced as the soil dries.

$$dQ = ([203.2 - Q]) \times ETP \qquad (12.17)$$

Once dQ is calculated, the KBDI for today (KBDI$_t$) is obtained by adding dQ to yesterday's KBDI value (KBDI$_{t-1}$) minus the net rainfall on a current day (P_n). If the result is negative, then the KBDI is equated to zero. Note that KBDI$_{t-1}$ is equal to Q.

$$KBDI_t = KBDI_{t-1} + dQ - P_n \qquad (12.18)$$

The net rainfall is computed by subtracting 5.08 mm (0.2 in.) from the value of daily rainfall. If there are consecutive wet days (no drying between showers), 5.08 mm is subtracted only once, on the day when cumulative rainfall exceeds 5.08. A wet period ends when two rainy days are separated by one day without measurable rainfall; thus, 5.08 must be subtracted again in the next rain period (Garcia-Prats et al., 2015).

Standardized Precipitation Index (SPI)

The SPI was designed by McKee et al. (1993) to quantify the precipitation deficit for multiple timescales. It is calculated as "the difference of precipitation from the mean for a specified time period divided by the standard deviation" (McKee et al., 1993). If the precipitation amounts are not normally distributed, they must be first converted to a normal distribution (Lloyd-Huges and Saunders, 2002). SPI is given as

$$SPI = \frac{x_i - \bar{x}}{s} \qquad (12.19)$$

where x_i is the monthly rainfall amount and $\bar{x}$ is the mean and s is the standard deviation of rainfall calculated from the whole time series of values. Since the monthly precipitation data may consist of zero values, it is expected that precipitation values do not follow the normal distribution. Thus, the data need to be transformed to arrive at a normal distribution (Geerts, 2002; Hayes et al., 1999). Therefore, a theoretical cumulative probability distribution function is adjusted on the precipitation data.

Edwards (1997) applied the two-parameter gamma distribution for the calculation of SPI, while Guttman (1999) applied the Pearson type-III distribution. Thus, the cumulative distribution of the precipitation data is transformed to a normal. In this way, the calculated SPI values are percentages of the standard deviation, while the extreme values have about the same low frequency of occurrence in all places for a long time period (Livada and Assimakopoulos, 2007).

LIVE FUEL MOISTURE CONTENT (FMC) AND EQUIVALENT WATER THICKNESS (EWT) INDICES

Live fuel moisture content (FMC) is one of the important variables in fire danger index estimation, as it affects fire ignition and propagation (Van Wagner and Forest, 1987) and is therefore widely used in fire danger rating systems (Chuvieco et al., 2003). The following equation is used to estimate both the live and dead fuel moisture contents (with a caution that most of the fire danger indices refer FMC to dead fuels only) (Chuvieco et al., 2003):

$$FMC = \left((W_u - W_d)/W_d \right) 100 \qquad (12.20)$$

where W_u is the wet weight and W_d is the dry weight of the same sample. Commonly, this variable is obtained through field sampling using gravimetric methods. Instead of FMC, most of the ecologists use the relative water content (RWC), which is defined as the water proportion over the saturated water content (Slavík, 1974) and indicator of water stress (Chuvieco et al., 2003). When remotely sensed data are used, the equivalent variable is equivalent water thickness (EWT), which is defined by the amount of leaf water per unit area (A):

$$EWT = (W_w - W_d)/A \qquad (12.21)$$

EWT can be directly estimated from a ratio of shortwave infrared (SWIR) and near-infrared (NIR) reflectance (Ceccato et al., 2001). Since EWT is difficult to measure, in particular, Chuvieco et al. (2003) combined FMC and EWT as

$$EWT = (W_w - W_d/A)(W_d/A) = FMC \; X \; S_w \qquad (12.22)$$

where S_w represents the specific weight for each species and is assumed to be constant through time, and FMC variations can be estimated using remote sensing data. FMC for grasslands was more efficiently estimated by remote sensing data than for other fuels since water variations in grasslands have a more direct effect on chlorophyll content and are more sensitive to seasonal variations than shrubs or trees (Chuvieco et al., 2003).

FIRE POTENTIAL INDEX (FPI)

The FPI integrates meteorological factors (i.e., temperature and relative humidity), land-use variables (fuel maps), and live vegetation status to quantify the fire danger

(Burgan et al., 1998). The FPI uses the fuel model and the moisture of live and dead vegetation in estimating the relative fire potential, including the use of remote sensing variables. The index is scaled from 1 to 100. In the remote sensing literature, there are several vegetation indices that can help characterize vegetation. The green leaves absorb most of the incident visible radiation, particularly in the red band, and reflect and transmit most incident infrared light (NIR). Thus, several vegetation indices use RED and NIR band combinations. The FPI uses remotely sensed data for deriving vegetation status. Some of the details of the FPI model are provided below.

The relative greenness index (RG) is used to determine the live fuel load at a pixel level as

$$RG = (ND_0 - ND_{mn})/(ND_{mx} - ND_{mn}) \times 100 \tag{12.23}$$

where ND_0 is the highest observed Normalized Difference Vegetation Index (NDVI) value for a 1-week composite period, ND_{mn} is the historical minimum NDVI value for a given pixel, and ND_{mx} is the historical maximum NDVI value for a given pixel. The relative greenness fraction is calculated as

$$RG_f = RG/100 \tag{12.24}$$

where RG_f is the relative greenness fraction. The relative greenness is used to determine the current live fuel load for the model assigned to the pixel as

$$LL_p = RG_f \times LL_{fm} \tag{12.25}$$

where LL_p is the live fuel load for the pixel and LL_{fm} is the live load for the fuel model. The proportion of live load that is cured plus the dead load for the fuel model is assigned to the pixel.

$$DL_p = (1 - RG_f) \times LL_{fm} + DL_{fm} \tag{12.26}$$

where DL_p is the dead fuel load for the pixel and DL_{fm} is the dead fuel load for the fuel model.

The remainder of the calculations is performed only if the DL_p is greater than zero. The calculated fraction of the total fuel model load that is live is given as

$$L_f = (LL_p/LL_{fm} + DL_{fm}) \tag{12.27}$$

The fraction of the total fuel model load that is dead is given as

$$D_f = (DL_p/LL_{fm} + DL_{fm}) \tag{12.28}$$

The fractional ten-hour time lag fuel map is normalized based on dead fuel moisture of extinction (MX_d) for the fuel model expressed as a percent for different vegetation types. The dead fuel moisture of extinction is defined as the dead fuel moisture content at which a fire will not spread (Rothermel, 1972), which varies from one fuel type to another. Ten-hour fuel moisture is normalized to the moisture of extinction for the vegetation of any given climate, to scale it to the same as fractional relative greenness (0–1).

$$TN_f = \left(FM_{10}/MX_d\right) \tag{12.29}$$

where TN_f is the fractional ten-hour fuel moisture, FM_{10} is the ten-hour moisture (%), and MX_d is the dead fuel extinction moisture.

$$FPI_u = 100 - \left(RG_f\,L_f + TN_f D_f\right) \times 100 \tag{12.30}$$

where FPI_u is the uncorrected FPI. In the above equation, the TN_f limits the maximum value for the FPI depending on the dead fuel moisture of extinction as

$$FPI_{max} = 100 - \left(2/MX_d\right) \times 100 \tag{12.31}$$

where FPI_{max} is the maximum uncorrected FPI value.
The final FPI is given as

$$FPI = FPI_u + \left(2/MX_d\right) \times \left(FPI_u/FPI_{max}\right) \times 100 \tag{12.32}$$

The above model has been applied for the entire USA to produce a 1-km-resolution fire danger fuel model map by Burgan et al (1998).

The US National Fire Danger Rating System (NFDRS2016)

NFDRS2016 is a multiple-index system developed to provide information about current and predicted fire danger conditions. The National Fire Danger Rating System (NFDRS) was first introduced in 1964 and since then updated in different years, i.e., in 1972, 1978, 1988, and 2016, to integrate newer science and improved processing. Here, a brief description of the NFDRS2016 has been provided. Important inputs to the NFDRS include precipitation, temperature, relative humidity, wind speed, fuels, and slope (Deeming et al., 1977). The NFDRS2016 integrates as a variety of information and models to provide rich outputs to represent fire danger across US landscapes; the outputs primarily serve as a tool to aid fire management decision-making. The primary inputs for the NFDRS2016 include (1) hourly weather and solar radiation data; (2) fire occurrence data; and (3) weather station metadata. The NFDRS2016 uses the following models to generate meaningful outputs:

1. **The Nelson dead fuel moisture model**: This is a physical-based model that contains equations for heat and moisture transfer (Nelson, 2000). The inputs for the model include air temperature, relative humidity, solar radiation, and amount of rainfall since the last observation, whereas the outputs include

moisture content and temperature of the fuel at times corresponding to the weather data inputs (Carlson et al., 2007).

2. **The Growing Season Index (GSI)**: This is an indicator of canopy greenness and is calculated as

$$GSI = T_{min} * i_{VPD} * i_{Photo} \tag{12.33}$$

where T_{min} is the minimum temperature indicator, i_{VPD} is the vapor pressure deficit indicator, and i_{Photo} is the photoperiod indicator. The daily GSI is calculated as the 21-day moving average of the daily indicator to minimize any daily extreme events (https://www.wfas.net/index.php/growing-season-index-experimental-products-96).

3. **Five Fuel Models**: The NFDRS2016 includes five different fuel models representing different response types: (1) Grass (V); (2) Grass/Shrub (W); (3) Brush (X); (d) Timber (Y); and (e) Slash (Z). The response types refer to how quickly the fuel responds to changes in the weather, and fuel models for each of the response types are derived from existing fire behavior fuel models (Carlson et al., 2007; Scott and Burgan, 2005) with some modifications, including the addition of 1000-hour, and drought fuel load. The resulting fire danger fuel models are not site-specific but represent an entire fire danger rating area (i.e., large geographic area).

Fuel model weighting is done based on the Ignition Component (IC), Spread Component (SC), Burning Index (BI), and Energy Release Component (ERC). The IC is a function of the one-hour time lag of fuel moisture content in fine fuels, and the temperature of the fine receptive fuels and IC varies from 1 to 100 as a percentage or probability. SC is mainly based on the forward rate of spread of a head fire. SC integrates variables such as the effect of wind, slope, and fuel bed and fuel particle properties that influence the spread. SC daily variations are caused by the changes in the wind and moisture contents of the live fuels and the dead fuel time lag classes of 1, 10, and 100 hours. ERC provides an approximation of dryness based on estimates of fuel moisture (Andrews and Bevins, 2003). ERC is calculated from a suite of meteorological and site variables, including relative humidity, temperature, precipitation duration, latitude, and day of the year (Cohen, 1985). Further, ERC is calculated daily and is thus more dynamic than the PDSI and SPI as it is sensitive to daily relative humidity and precipitation timing and duration. The calculation of ERC is also affected by fuel loadings in different size classes. ERC(G) has been shown to have a strong relationship with fire occurrence in Arizona: The probability of fire increases with ERC(G) and can be quantified with logistic regression (Andrews et al., 2003). Therefore, the ERC(G) is used by US federal land agencies both operationally (Predictive Services) and in simulation models that predict fire size and probability, including FSPro and FSim (Finney et al., 2011; Riley et al., 2013) (https://sites.google.com/firenet.gov/nfdrs/the-models). Variations of the ERC can occur due to changes in the moisture content of the various fuel classes, including the 1000-hour time lag class. ERC is derived from the prediction of flaming combustion based on the rate of heat release per unit area and the duration of the burning, expressed in BTUs per square foot. BI, a measure of fire intensity, combines the SC and ERC to relate to the contribution of fire behavior to the effort of containing a fire. The BI has no units, but in general, it is ten times the flame length of a fire. More details about the NFDRS and definitions of terms including the sample outputs are given in the US Forest Service Wildland Fire Assessment System website (https://www.wfas.net/). Recently, Jolly et al. (2019) developed a new robust index entitled Severe Fire Danger Index (SFDI) combining two different NFDRS indices, i.e., ERC and BI, related to spread potential and fire intensity, respectively, and showed that the SFDI captures geographic and seasonal variations adequately in the USA and can be used as an early-warning decision support system by firefighters and managers.

METEOROLOGICAL RISK INDEX

This index was developed to delineate fire danger zones for southeastern Greece (Gouma and Chronopoulou-Sereli, 1998). Fire danger is characterized based on the meteorological risk (MR), which is a function of air temperature and relative humidity; increasing air temperature reduces relative humidity and implies high drying rates of fuels (Chandler et al., 1983). Fuel susceptibility is calculated based on inflammability indices of species and the fire occurrence based on historical data. Fire danger zones are then delineated by integrating Meteorological Risk (MR), Fuel Susceptibility (FS), and Fire Occurrence (FO) map layers.

CANADIAN FOREST FIRE DANGER RATING SYSTEM (CFFDRS)

The CFFDRS is a comprehensive system that includes environmental factors governing the ignition, spread, and behavior of wildland fires (Van Wagner and Forest, 1987). Details of this system can be found at https://www.nwcg.gov/publications/pms437/cffdrs/overview. The Fire Weather Index (FWI) of the CFFDRS is primarily based on the inputs from meteorology, i.e., air temperature, wind speed, relative humidity, and 24-hour cumulative rainfall, each measured at noon local time (Van Wagner and Forest, 1987) and used to calculate the Fine Fuel Moisture Code (FFMC), i.e., for the fine surface litter; the Duff Moisture Code (DMC), i.e., loosely compacted organic material; and the Drought Code (DC), i.e., the deeper organic layers/large surface fuels respectively of pine forest. The moisture codes are then used within the FWI system to calculate three different fire behavior indices: the Initial Spread Index (ISI) representing the potential rate of fire spread; the Buildup Index (BUI) indicating the total amount of available combustible fuel; and the "FWI." Further, FWI combines the ISI and BUI to provide a measure of the potential frontal intensity of a fire. Important equations for different components of CFFDRS are given below:

a. **Fine Fuel Moisture Code (FFMC)**: This varies from 0 (low flammability) to 100 (high flammability) and is a relative measure for the rating of moisture levels of the uppermost layers of forest litter and is a relative measure of the ease of ignition. FFMC has a time lag of about two-thirds of a day and is the fastest-changing component of the FWI. The first day's value of FFMC depends on snow cover and mean daily temperature. FFMC is given as

$$FFMC = 59.5 \times \frac{250 - m}{147.2 + m} \tag{12.34}$$

where m is the fine fuel content after drying.

b. **Duff Moisture Code (DMC):** It is a rating of the moisture content of the loosely compacted organic layer of 7 cm depth below the fine fuel layer. It has a time lag of about 12 days and has no upper limit. DMC of the first day is set to 6. DMC is a measure of the potential ignition by lightning and fuel consumption. If it has rained in the last 24 hours,

$$P = P_r + 100 * K \tag{12.35}$$

else,

$$P = P_0 + 100 * K \tag{12.36}$$

where P is the new DMC, P_0 is the previous day's DMC, P_r is the DMC after rain, and K is the log drying rate in DMC ($\log_{10}$m/day).

c. **Drought Code (DC):** It is a rating of the moisture content of the deeper and compact organic layer of 18 cm depth (Raphael, 2011). DC represents the seasonal drought effects on fire growth and has a time lag of about 52 days. It represents the smoldering potential of deeper layers and larger logs. If it has rained in the last 24 hours, DC is given as

$$D = D_r + 0.5 * K \tag{12.37}$$

else,

$$D = D_0 + 0.5 * K \tag{12.38}$$

where D is the new DC, D_0 is the previous day's DC, D_r is the DC after rain, and V is the potential evapotranspiration (mm).

d. **Initial Spread Index (ISI):** It combines the FFMC and wind information to quantify the expected rate of fire spread without taking into consideration the varying fuel quantity and exposure.

$$ISI = 0.208 * f(F) * f(W) \tag{12.39}$$

where ISI is the Initial Spread Index (ISI) value, $f(F)$ is a function of the FFMC, and $f(W)$ is a function of wind.

e. **Buildup Index (BUI)**: It reflects the amount of fuel available for combustion as a combination of DMC and DC. If the DMC is close to zero, the BUI is also close to zero. Thus, it reflects closely the actual fire danger of a day without exaggerating the smoldering potential.

$$\text{If} \quad \text{DMC} \leq 0.04 * \text{DC}$$

$$\text{BUI} = 0.8 * \text{DMC} * \text{DC}/(\text{DMC} + 0.4 * \text{DC}) \tag{12.40}$$

$$\text{If} \quad \text{DMC} > 0.4 * \text{DC}$$

$$\text{BUI} = \text{DMC} - \left(1 - \frac{0.8 * \text{DC}}{\text{DMC} + 0.4 * \text{DC}}\right) * \left(0.92 + (0.0111 * \text{DMC})^{1.7}\right) \tag{12.41}$$

f. **Fire Weather Index (FWI)**: It combines the ISI and BUI to produce a numerical rating of the potential intensity of a single fire.

$$\text{If} \quad \text{BUI} \leq 80$$

$$f(\text{DMC}) = 0.626 * \text{BUI}^{0.809} + 2 \tag{12.42}$$

$$\text{If} \quad \text{BUI} > 80$$

$$f(\text{DMC}) = 1000 / \left(25 + 108.64 * e^{0.023 * \text{BUI}}\right) \tag{12.43}$$

$$B = 0.1 * \text{ISI} * f(\text{DMC}) \tag{12.44}$$

$$\text{If} \quad B > 1$$

$$\text{FWI} = 2.72(0.434 * \log B)^{0.647} \tag{12.45}$$

$$\text{If} \quad B \leq 1$$

$$\text{FWI} = B \tag{12.46}$$

where B is an intermediate form of FWI.

THE FINNISH FIRE INDEX (FFI)

The FFI system is used for evaluating forest fire hazards in Finland by the Finnish Meteorological Institute (FMI). The institute issues a daily map of fire danger on its website, and the fire weather forecasts are specific for each province of Finland and are provided for the current date and subsequent 4 days (http://en.ilmatieteenlaitos. fi). The FFI values represent the estimated moisture content of a surface layer expressed as (Gouma and Chronopoulou-Sereli, 1998)

$$DW = E_{pot} X \ DE + P_i \qquad (12.47)$$

where DW ($m^3 m^{-3}$) is the change in the volumetric moisture content of the total surface layer, E_{pot} (mm) is the potential evapotranspiration (ET) that can be calculated using the Penman–Monteith equation (Monteith, 1981), DE is the drying efficiency, and P_i (mm) is the amount of water remaining in the surface layer. The equation can also be expressed as (Vajda et al., 2014)

$$DW = E_{pot} \times 0.757 \Big/ 1 + e^{(2.74-16.67\times(W_{vol}-0.1))} + 5.612 \times \left(1 - e^{(-P/5.612)}\right) \quad (12.48)$$

where E_{pot} (mm) is the potential evaporation, W_{vol} ($m^3 m^{-3}$) is the volumetric moisture of the surface layer, and P(mm) is the precipitation. The output from a numerical weather prediction model is integrated into the FFI, which calculates the humus content as a part of surface moisture for layers with a thickness of 3 and 6 cm. The FFI has six classes from very wet to very dry moisture conditions (from 1 to and 6), where the lower numbers refer to lower fire danger, and *vice versa* (Jurečka et al., 2019).

The Australian Forest Fire Danger Index (FFDI)

This is the most common Forest Fire Danger Index (FFDI) developed by McArthur (1967) and is used to assess the daily risk of fire in New South Wales, Australia (Verdon et al., 2004). The FFDI provides a measure of climatological conditions associated with the severe fire danger and an assessment of the chances of a fire starting, its rate of spread, fire intensity, and difficulty of suppression (Noble et al., 1980). A fire danger rating is determined daily, based on the FFDI, as follows: 0–5 – low; 5–12 – moderate; 12–24 – high; 24–50 – very high; and 50–100 – extreme. The daily FFDI is calculated as (Equation 12.49)

$$FFDI = 2.0 \times \exp(-0.450 + 0.987 \times \ln(D)$$
$$- 0.0345 \times H + 0.0338 \times T + 0.0234 \times V \qquad (12.49)$$

where H is the minimum relative humidity in percent, T is the maximum air temperature over a 24-hour period in degrees Celsius, V is the average wind velocity (km/h) in the open, at the height of 10 m, and D is the drought factor for that day, obtained using the KBDI (Keetch and Byram, 1968). The drought factor is derived as

$$D = 0.191 \times (I + 104) \times (N+1)^{1.5} \Big/ \left(3.52 \times (N+1)^{1.5} + P - 1\right) \qquad (12.50)$$

where I is the daily KBDI, P is the daily precipitation in mm, and N is the number of days since rain.

THE SPANISH WILDFIRE DANGER INDEX

This index considers essential risk factors related to fire occurrence, such as human activities or lightning (de Vicente and Crespo, 2012). Fire danger is modeled as a function of ignition danger and spread danger. The ignition danger is a function of ignition agents (roads, railroads, powerlines, agriculture, lightning), the historical occurrence of fires, and fuel conditions (based on probability of ignition and flammability). The spread danger is a function of spread rate and fuel conditions. Specific to natural ignition agents such as storm lightning, the forecasted probability of electric storm retrieved from the Ensemble Prediction System of the European Centre for Medium-Range Weather Forecasts (ECMWF) is used as a short-term risk indicator. The fuel conditions specific to the dead fuels are assessed based on the moisture content, which is inversely related to the probability of ignition and derived using the BEHAVE model. The spread danger based on the spread rate is also calculated using the BEHAVE model. Also, multi-criteria methods were used to incorporate expert opinions in the process of weighting the indicators and aggregating the components into the final index, which is used to map the probability of daily fire occurrence on a 0.5-km grid (de Vicente and Crespo, 2012).

INTEGRATED FIRE INDEX (IFI)

The IFI is developed for the Mediterranean region. It is operationally used in Sardinia for daily fire danger forecasting during the fire season (Sirca et al., 2007). The index is based on weather and physical fuel inputs, and the daily index is calculated as a sum of four subcomponents, i.e., Drought Code (DC), meteorological code (MC), solar radiation (R), and Fuel Code (FC), as follows:

$$\text{IFI} = \text{DC} + \text{MC} + \text{R} + \text{FC} \tag{12.51}$$

and

$$\text{DC} = \frac{e^{\left(0.261\frac{R_g T}{\lambda}\right)}}{1 + \sqrt{P_a} + \sqrt[3]{P_{c100}}} \tag{12.52}$$

In the above equation, R_g is the global daily radiation (in W/m²); λ is the latent heat of evaporation (J/g); T is the mean daily air temperature (°C); P_a represents the daily rainfall (mm); and P_{c100} is the rainfall of the last 4 days. DC values range from 0.1 (the lowest danger) to 5 (the highest danger conditions).

MC accounts for the meteorological conditions favoring the maximum fire spread during the day and is given as (Sirca et al., 2007)

$$\text{MC} = 0.14\left[\exp\left(0.0625 * T_x\right) + \exp\left(0.1 * \text{WS}\right) + \exp\left(-0.062 * \text{RH}_n\right)\right] \tag{12.53}$$

where T_x represents the maximum daily air temperature (°C), RH_n is the minimum daily air temperature, and WS is the maximum daily wind speed (km/h).

MC ranges from 0.4 to 5 representing minimum to maximum fire danger conditions, respectively.

The maximum daily solar radiation (RS_x) and the coefficient R are given as follows: If $RS_x < 400$ W/m^2, then the coefficient R is equal to 0.24; if 400 W/m$^2 \leq RS_x \leq 800$ W/m^2, then the coefficient R is equal to 0.32; and if $RS_x > 800$ W/m^2, then $R = 1$.

The Fuel Code (FC) is related to the structural and physical characteristics of fuel and is computed as a function of leaf area index (unitless), the leaf area density ($LAD = $ m^2 of foliage m^{-3}), and DW (the fuel moisture content):

$$FC = LAI * LAD * DW \qquad (12.54)$$

Further, DW represents a fraction of the fresh fuel weight ($0 = 100\%$ of water; $1 = 0\%$ of water). The calculation of FC requires the definition of specific fuel categories (*i.e.*, fuel models) based on the vegetation types, and the data can be derived from the field measurements or literature cited values. The daily IFI value can range from 1.05 (grasslands) to a maximum of 13.86 (tall shrublands).

THE BRAZILIAN MONTE ALEGRE FORMULA (FMA)

The FMA is a fire danger index used in the state of Paraná, other Brazilian regions, and some other South American countries (Soares and Batista, 2007). It is based on meteorological data and indicates the possibility of a fire starting on daily ignition conditions. It is given as

$$FMA = \sum_{i=1}^{n} \left(100/H_i\right) \qquad (12.55)$$

where FMA = Monte Alegre Formula; H = relative humidity (%); and n = number of days without rain. The index includes several precipitation qualifications; i.e., (1) if precipitation is ≤ 2.4 mm, do nothing; (2) if precipitation is 2.5–4.9 mm, subtract 30% from the calculated FMA and add ($100/H_i$) of the day; (3) if precipitation is 5.0–9.9 mm, subtract 60% from the calculated FMA and add ($100/H_i$) of the day; (4) if precipitation is 10.0–12.9, subtract 80% from the calculated FMA and add ($100/H_i$) of the day; and (5) if precipitation is >12.9, stop calculation (FMA = 0), and restart the next day or when the rain ceases. FMA fire danger classes include 0–1.0 – null; 1.01–3.0 – small; 3.1–8.0 – medium; 8.1–20.0 – high; and >20.0 – very high.

The original equation of the Monte Alegre Formula was later modified to include wind speed which can influence the fire spread (Nunes et al., 2009), and it is given as:

$$FMA^+ = \sum_{i=1}^{n} (100 / H_i) e^{0.04v} \qquad (12.56)$$

where FMA$^+$ = Monte Alegre Formula Altered; H = relative humidity (%); n = number of days without rain; v = wind speed in m/s; and e = base of natural logarithms (2,718,282). The index is cumulative and influenced by the precipitation constraints.

SIMPLE FIRE DANGER INDEX (SFDI)

Sharples et al. (2009) developed an SFDI in Australia. This fire danger rating system combines meteorological information with estimates of the moisture content of the fuel to produce a fire danger index. This index is calculated as follows:

$$F = \max\ (U_0, U)/\text{FMI} \qquad (12.57)$$

where F is the SFDI, U denotes wind speed in km/h, and U_0 is some threshold wind speed introduced to ensure that fire danger rating is greater than zero, even for zero wind speed. FMI is the fuel moisture index calculated as

$$\text{FMI} = 10 - 0.25(T - H) \qquad (12.58)$$

where T is the temperature (°C) and H is the relative humidity (%). The SFDI has a danger scale with five levels: 0–0.7 – low; 0.7–1.5 – moderate; 1.5–2.7 – high; 2.7–6.1 – very high; and >7 – extreme fire danger (Hamadeh et al., 2017). This index also showed a good performance in prediction when tested in Austria (Arpaci et al., 2013; Khan et al., 2019).

GRASSLAND FIRE DANGER INDEX (GFDI)

This index was developed by McArthur (1967) for grassland fuels in Australia. It is given as

$$\text{GFDI} = \begin{cases} 3.35w \exp\left(-0.097m_c + 0.0403u\right) \ m_c \le 18.8\% \\ 0.299w \exp\left(-1.686m_c + 0.0403u\right)\left(\ 30 - m_c\right) \ 18.8\% < m_c < 30\% \end{cases} \quad (12.59)$$

where w is the fuel weight (T/H_a), m_c is the fuel moisture content as a percentage, and u is the wind speed at 10 m height in km/h. The moisture content is modeled as

$$m_c = \frac{97.7 + 40.6\,\text{RH}}{T + 6} - 0.00854\,\text{RH} + \frac{300}{C} - 30 \qquad (12.60)$$

where RH is the relative humidity (%), T is the ambient temperature (°C), and C is the curing index (0–100%), a measure of the amount of dead material in the grassland. The GFDI is used to determine the rate of spread of the fire (RoS), which is used to model the intensity of the fire. The RoS is given as

$$\text{RoS} = 0.13\,\text{GFDI} \qquad (12.61)$$

The other details on the fire intensity calculation are given in Khan et al. (2019).

OTHER INDICES

In addition to the meteorological weather fire indices and rating systems listed above, several others exist in the literature, such as the Zhdanko Index in Russia (Zhdanko, 1965), Italian Fire Danger Index (Palmieri and Cozzi, 1983), Portuguese Index (Goncalves and Lourenco, 1990), Sol Numerical Risk in France (Sol, 1992; Drouet and Sol, 1990), Carrega Index in France (Carrega, 1991), Spanish ICONA method (ICONA, 1993), Italian Fire Danger Index (Palmieri et al., 1992), and IREPI index, Italy (Bovio and Camia, 1997).

THE GLOBAL WILDFIRE INFORMATION SYSTEM (GWIS)

The GWIS is a joint initiative of the Group on Earth Observations (GEO) and Copernicus work programs. The project is led by Dr. San-Miguel-Ayanz, Joint Research Center, Italy. GWIS is a comprehensive system and builds on the ongoing fire research activities of various international programs such as the European Forest Fire Information System (EFFIS), the Global Terrestrial Observing System (GTOS), Global Observation of Forest Cover–Global Observation of Land Dynamics (GOFC-GOLD) Fire Implementation Team (GOFC Fire IT), and the associated Regional Networks (RNs) (https://gofcgold.org/). The GWIS integrates a wide variety of information sources at the regional and national levels to provide a comprehensive assessment of fire regimes and fire effects at the global level and to provide tools to support operational wildfire management from national to global scales (https://gwis.jrc.ec.europa.eu/).

Fire forecast maps are produced from 1 to 9 days of fire danger levels, and the inputs from the numerical weather prediction model are used to compute the FWI described earlier (Van Wagner and Pickett, 1985). A variety of outputs relating to fire danger indices, i.e., the Australian McArthur Forest Fire Danger Index (Mark-5), the KBDI, and the NFDRS, are produced through the GWIS with the inputs from the global meteorological forecast data received daily from the ECMWF, and fire danger is mapped in six classes (very low, low, medium, high, very high, and extreme) with a spatial resolution of about 8 km (ECMWF data). More details about the system and its outputs can be accessed from the GWIS website (https://gwis.jrc.ec.europa.eu/).

DISCUSSION

The most important elements of fire include fuel, oxygen, and heat, which constitute the fire triangle or the combustion triangle (Countryman, 1972). The vegetation type, fuel moisture, and weather are considered critical parameters influencing fire ignition and spread. Thus, several fire indices integrate these parameters to model the fires (Turner and Lawson, 1978). In this study, no attempt has been made to classify different indices, methods, or rating systems into various categories as they are numerous and hugely diverse with varying inputs. A review of the indices and rating systems presented in this study suggests complexity and variations based on (1) the number of meteorological inputs considered; (2) inclusion of vegetation/phenology/fuel/soil characteristics along with meteorological parameters; (3) consideration of previous days or historical data in index calculation; and (4) integrated multiple fire indices that form fire danger rating systems. For example, indices such as the Angström, Baumgartner, Bruschek, M-68, Fosberg, Haines, Standardized Precipitation, Finnish Fire, Nesterov, Monte Alegre Formula, and SFDI require less than five meteorological parameters for the calculation. They do not directly consider vegetation or soil parameters while assessing the fire danger. In contrast, indices such as the FPI, KBDI, Spanish Wildfire Danger Index, and the IFI, and the rating systems such as CFFDRS, US NFDRS2016, and GWIS combine both the meteorological and phenology or vegetation/soil/fuel parameters for quantifying the fire danger.

Specific to the time-period aspects, the Angström Index considers a day period, while other indices such as the Baumgartner, Keetch–Byram, Standardized Precipitation, M-68, and Nesterov, and the rating systems such as CFFDRS, US NFDRS2016, and GWIS use accumulative meteorological parameters along different periods to assess the fire danger. Compared to the indices, the rating systems such as CFFDRS and US NFDRS2016 are comprehensive and robust and integrate a variety of indices relating to vegetation growing season characteristics, fuel quality, fuel moisture, the rate of spread of fire, energy release, drought, and other weather characteristics for quantifying the fire danger. Also, one should be cautious on the fire danger ranking systems, since the discrete fire danger levels used in fire danger rating applications often are based on continuous data produced by the actual index, and the ranking levels can be biased. In addition, compared to some of the rating systems which are country-specific, the recent GWIS is global in scope and assimilates extensive information at the regional and national levels for not only fire danger assessment, but also useful for quantifying fire regimes, and effects at global-to-national scales for operational wildfire management.

In addition, a review of the fire indices and rating systems outlined above suggests that historically the meteorological parameters such as precipitation, temperature, relative humidity, wind speed, wind direction, and radiation are obtained from local weather stations, whereas vegetation parameters such as fuel type, fuel load, and moisture content are obtained from field measurements. However, the density of weather stations can be sparse; thus, it may not be easy to retrieve such information, especially in the remote forested areas. Also, spatial interpolation techniques may not be suitable in complex terrain. For example, precipitation generally increases with elevation, and the mountain ranges also create "rain shadows" on the leeward side. Thus, if there are not enough weather stations capturing these variations, the spatial interpolation techniques such as kriging and co-kriging can fail (Phillips et al., 1992). Further, collecting vegetation and fuel parameters can be cumbersome and constrained due to a lack of funds. Thus, in addition to the weather station data, the use of remote sensing technology for fire danger estimation has immense potential. Satellite remote sensing technology with its unique multispectral, multitemporal, synoptic, and repetitive coverage capabilities can provide valuable information on various inputs of fire danger indices. Important parameters that can be retrieved using remote sensing techniques for fire danger assessment are given in Table 12.2. In this study, we provide only a broad overview of the retrieval techniques of different variables useful for fire danger assessment and not a detailed review on how these were already used by different researchers considering space limitations. However, we recognize the importance of such a review, which can be useful for fire research and management. Some specific

TABLE 12.2

Potential of Remote Sensing Technologies for Retrieving Fire Danger Variables Including Usefulness for Calibration and Validation

S. No.	Fire Weather and Danger Index Inputs	Remote Sensing Variables and Retrieval Techniques	Example Satellites
1	Air temperature	Satellite-derived land surface temperature from thermal channels and empirical relation with air temperature data from meteorological stations	Landsat, MODIS, VIIRS
2	Relative humidity	Empirical relationship between the precipitable water vapor channels (e.g., MODIS or VIIRS) from split-window techniques and then relating to RH	Landsat, MODIS, VIIRS
3	Wind speed	High-powered ultraviolet laser (e.g., Atmospheric Laser Doppler Instrument) or infrared energy signatures from satellites in combination with numerical weather forecast parameters or passive microwave sensors	AEOLUS satellite from the European Space Agency, GOES, MODIS, POES
4	Wind direction	As above	As above
5	Precipitation	Microwave imager in combination with dual-frequency precipitation radar; passive microwave, infrared, and visible channel combination	Global Precipitation Measurement Satellite (GPM), TRMM, geostationary satellites
6	Solar irradiance	Total solar irradiance (TSI) using electrical substitution cavity radiometers on board satellites	Total Irradiance Monitor from ERB, ACRIM, ERBS NASA missions including ESA's VIRGO
7	Earth radiation budget	TSI radiometers on board satellites	As above
8	Evapotranspiration	Map the infrared heat radiated from Earth, thus enabling us to distinguish the cool surfaces from the warm surfaces; typically thermal channel radiometers	MODIS, VIIRS, Landsat, ECOSTRESS
9	Topographic parameters (slope)	Radar measurements; laser altimeter satellites	SRTM, TanDEM-X, ICESat

(Continued)

TABLE 12.2 (*Continued*)

Potential of Remote Sensing Technologies for Retrieving Fire Danger Variables Including Usefulness for Calibration and Validation

S. No.	Fire Weather and Danger Index Inputs	Remote Sensing Variables and Retrieval Techniques	Example Satellites
10	Cloudiness or cloud cover fraction	Infrared radiation measured by cloudy scene from satellites and LiDAR	Polar and geostationary satellites; AVHRR, MODIS, GOES, and CALIOP on board CALIPSO
	Fire Detection		
11	Active fires	Visible, infrared, mid-infrared, and thermal bands	MODIS, VIIRS, LANDSAT, Sentinels, and GOES
12	Burnt areas	Mostly visible and shortwave infrared bands	As above
13	Fire intensity	Middle-infrared and thermal bands	As above
14	Date of onset/ offset	Visible, infrared, mid-infrared, and thermal bands	As above
	Vegetation Parameters		
15	Vegetation type	Visible, IR bands	As above
16	Fuel load	Correlating ground-based measurements with vegetation indices combining visible and IR bands from satellites or radar backscatter signatures	As above
17	Fuel moisture	Normalized Difference Moisture Index (NDMI) or Normalized Difference Water Index (NDWI) (Gao, 1996) as (NIR-SWIR/NIR + SWIR) with an increase in index with moisture; use of Synthetic Aperture Radar (SAR) data	As above, and SAR data (e.g., Sentinel-1, RADARSAT, PALSAR)
18	Biomass	Normalized Difference Vegetation Index calculated as NIR-RED/NIR+RED as a proxy and relating to ground-based measurements	MODIS, VIIRS, LANDSAT, Sentinels, and GOES

(*Continued*)

TABLE 12.2 (*Continued*)

Potential of Remote Sensing Technologies for Retrieving Fire Danger Variables Including Usefulness for Calibration and Validation

S. No.	Fire Weather and Danger Index Inputs	Remote Sensing Variables and Retrieval Techniques	Example Satellites
	Vegetation Parameters		
19	Fuel phenological stages	As above	As above
20	Foliar greenness	Multiple vegetation indices, for example, green chlorophyll index (GCI) measured as (NIR/Green)-1, used to estimate canopy chlorophyll content (Gitelson et al., 2003)	As above
21	Fuel classes	Visible, IR bands	As above
22	Canopy water content	Similar to NDMI or NDWI; use of SAR data	As above, and SAR data (e.g., Sentinel-1, RADARSAT, PALSAR)
23	Plant water content	Similar to NDMI or NDWI	As above
24	Tree height	LiDAR	GEDI
25	Vertical fuel distribution	LiDAR	As above
26	Understory characteristics	LiDAR	As above
27	Canopy base height of fuels for estimating crown and surface fires	LiDAR	As above
28	Single-tree estimates	LiDAR	As above
29	Crown variables depth, size (diameter, area, volume), biomass, and bulk density	LiDAR remote sensing in combination with allometry	As above
30	Foliage density indices	LiDAR	As above

examples can be found in the literature such as on the use of remote sensing data for forest fire danger forecasting systems (Akther and Hassan, 2011). More such studies are needed considering the latest developments in the remote sensing field.

With the robust improvements in sensor technology, together with the scientific advancements relating to various algorithms, it is increasingly easier to retrieve these parameters. A variety of multispectral, hyperspectral, laser, radar, and LiDAR instruments on board the satellites can be used for characterizing various biophysical and vegetation characteristics (Table 12.2). Some of the constraints in using remote sensing technologies include spatial and spectral resolutions, variations in satellite time of the pass, and frequency coinciding with the fire events. Also, cloud cover can constrain fire retrievals when the signal is retrieved using active remote sensing sensors, besides data accessibility and availability. Despite these constraints, remote sensing technology has significant advantages as compared to ground-based methodologies for a spatially explicit characterization of the surface conditions. In addition, a detailed review of the fire index literature suggests that most of them ignore the human component. Fires in several countries such as in South/Southeast Asia, South America, and the African continent are mostly human initiated. Thus, human risk factors of fires such as population densities and distance from roads and settlements should be effectively integrated with the meteorological variables and other information such as topography, land cover, and vegetation parameters for quantifying the fire danger (Prasad et al., 2008; Vadrevu, 2008; Biswas et al., 2015a,b).

For the same, geographic information systems (GIS) can be effectively used to display, manage, organize, and analyze spatial data. Moreover, there is a strong need to calibrate and validate the fire weather indices and rating systems in different regions of the world with the observed fire data. A detailed review on the existing studies regarding the performance and use of fire danger indices, methods, models, and rating systems can also offer some important insights. Also, developing spatial data infrastructure to integrate a variety of causative factors of fires is essential in addition to technology transfer and capacity building on the use of fire danger information for fire management and mitigation in different countries. Global programs such as GWIS (https://gwis.jrc.ec.europa.eu/) and the GOFC-GOLD Fire Implementation Team (https://gofcgold.org/) are leading such activities.

ACKNOWLEDGMENTS

This research was funded by the NASA Land Cover/Land Use Change Program, Grant Number 281945.02.58.03.01, South/Southeast Asia Research Initiative (SARI) project to the author.

REFERENCES

Akther, M.S. and Hassan, Q.K., 2011. Remote sensing-based assessment of fire danger conditions over boreal forest. *IEEE Journal of Selected Topics in Applied Earth Observations and Remote Sensing*. 4(4), 992–999.
Andreae, M.O., Browell, E.V., Garstang, M., Gregory, G.L., Harriss, R.C., Hill, G.F., Jacob, D.J., Pereira, M.C., Sachse, G.W., Setzer, A.W. and Dias, P.S. 1988. Biomass-burning

emissions and associated haze layers over Amazonia. *Journal of Geophysical Research: Atmospheres.* 93(D2), 1509–1527.

Andrews, P.L. and Bevins, C.D. 2003. November. BehavePlus fire modeling system version 2: overview. In *Proceedings of the Second International Wildland Fire Ecology and Fire Management Congress,* Coimbra, Portugal (pp. 17–20).

Andrews, P.L., Loftsgaarden, D.O. and Bradshaw, L.S. 2003. Evaluation of fire danger rating indexes using logistic regression and percentile analysis. International Journal of Wildland Fire. 12(2), 213–226.

Arpaci, A., Eastaugh, C.S. and Vacik, H. 2013. Selecting the best performing fire weather indices for Austrian ecoregions. *Theoretical and Applied Climatology.* 114(3–4), 393–406.

Badarinath, K.V.S., Kharol, S.K., Krishna Prasad, V., Kaskaoutis, D.G. and Kambezidis, H.D. 2008a. Variation in aerosol properties over Hyderabad, India during intense cyclonic conditions. *International Journal of Remote Sensing.* 29(15), 4575–4597.

Badarinath, K.V.S., Kharol, S.K., Prasad, V.K., Sharma, A.R., Reddi, E.U.B., Kambezidis, H.D. and Kaskaoutis, D.G. 2008b. Influence of natural and anthropogenic activities on UV Index variations–a study over tropical urban region using ground based observations and satellite data. *Journal of Atmospheric Chemistry.* 59(3), 219–236.

Baisan, C.H. and Swetnam, T.W. 1990. Fire history on a desert mountain range: Rincon Mountain Wilderness, Arizona, USA. *Canadian Journal of Forest Research.* 20(10), 1559–1569.

Baumgartner, A., Raschke, E., Klemmer, L. and Waldmann, G. 1967. Waldbrände in Bayern 1950–1959. *Allgemeine Forstzeitschrift.* 22, 220–22.

Biswas, S., Vadrevu, K.P., Lwin, Z.M., Lasko, K. and Justice, C.O., 2015a. Factors controlling vegetation fires in protected and non-protected areas of Myanmar. *PLoS One.* 10(4), 1–17. doi: 10.1371/journal.pone.0124346.

Biswas, S., Lasko, K.D. and Vadrevu, K.P. 2015b. Fire disturbance in tropical forests of Myanmar—Analysis using MODIS satellite datasets. *IEEE Journal of Selected Topics in Applied Earth Observations and Remote Sensing.* 8(5), 2273–2281.

Bovio, G and Camia, A. 1997. Meteorological indices for large fires danger rating. In E. Chuvieco (Ed.) *A Review of Remote Sensing Methods for the Study of Large Wildland Fires.* Universidad de Alcalá, Alcalá de Henares, Spain, pp. 73–91.

Bruschek, G. 1994. Waldgebiete und Waldbrandgeschehen in Brandenburg im Trockensommer 1992. Tech. Rep. 2, PIK.

Burgan, R.E., Klaver, R.W. and Klaver, J.M. 1998. Fuel models and fire potential from satellite and surface observations. *International Journal of Wildland Fire.* 8, 159–170.

Carlson, J.D., Burgan, R.E., Engle, D.M. and Greenfield, J.R. 2002. The Oklahoma Fire Danger Model: an operational tool for mesoscale fire danger rating in Oklahoma. *International Journal of Wildland Fire.* 11(4), 183–191.

Carlson, J.D., Bradshaw, L.S., Nelson, R.M., Bensch, R.R. and Jabrzemski, R. 2007. Application of the Nelson model to four time lag fuel classes using Oklahoma field observations: model evaluation and comparison with National Fire Danger Rating System algorithms. *International Journal of Wildland Fire.* 16(2), 204–216.

Carrega, P. 1991. A meteorological index of forest fire hazard in Mediterranean France. *International Journal of Wildland Fire.* I(2), 79–86.

Ceccato, P., Flasse, S., Tarantola, S., Jacquemoud, S. and Grégoire, J.M. 2001. Detecting vegetation leaf water content using reflectance in the optical domain. *Remote Sensing of Environment.* 77(1), 22–33.

Certini, G. 2014. Fire as a soil-forming factor. *Ambio,* 43(2), 191–195.

Chandler, C., Cheney, P., Thomas, P., Trabaud, L., Williams, D. 1983. *Fire in Forestry– Forest Fire Behaviour and Effects.* John Wiley & Sons, New York, Chichester, Brisbane, Toronto, Singapore.

Chuvieco, E., Aguado, I., Cocero, D. and Riaño, D. 2003. Design of an empirical index to estimate fuel moisture content from NOAA-AVHRR images in forest fire danger studies. *International Journal of Remote Sensing.* 24(8), 1621–1637.

Chuvieco, E., Aguado, I., Jurdao, S., Pettinari, M.L., Yebra, M., Salas, J., Hantson, S., de la Riva, J., Ibarra, P., Rodrigues, M. and Echeverría, M. 2014. Integrating geospatial information into fire risk assessment. International *Journal of Wildland Fire.* 23(5), 606–619.

Cohen, J.D. 1985. The National Fire-Danger Rating System: basic equations. USDA Forest Service, General Technical Report PSW-82.

Countryman, C.M. 1972. *The Fire Environment Concept. USDA Forest Service.* Pacific Southwest Forest and Range Experiment Station, Berkeley, CA, 12 pp.

de Groot, W.J., Wotton, B.M. and Flannigan, M.D. 2015. Wildland fire danger rating and early warning systems. In D. Paton (Ed.) *Wildfire Hazards, Risks and Disasters.* Elsevier, pp. 207–228.

de Vicente, J. and Crespo, F. 2012. A new wildland fire danger index for a Mediterranean region and some validation aspects. *International Journal of Wildland Fire.* 21(8), 1030–1041.

Deeming, J.E., Burgan, R.E. and Cohen, J.D. 1977. *The National Fire-Danger Rating System—1978,* Vol. 39. Intermountain Forest and Range Experiment Station, Forest Service, US Department of Agriculture.

Drouet, J.C. and Sol, B., 1990. Mise au point d un Indice numérique de risque météorologique d incendie. Revue Génerale de Securité, 92.

Edwards, D.C., 1997. Characteristics of 20th century drought in the United States at multiple time scales (No. AFIT-97-051). AIR FORCE INST OF TECH WRIGHT-PATTERSON AFB OH.

Finney, M.A., McHugh, C.W., Grenfell, I.C., Riley, K.L. and Short, K.C., 2011. A simulation of probabilistic wildfire risk components for the continental United States. *Stochastic Environmental Research and Risk Assessment.* 25(7), 973–1000.

Flemming, G. 1994. *Wald Wasser Klima | Einfuhrung in die Forstmeteorologie.* Deutscher Landwirtschaftsverlag, Berlin.

Fosberg, M.A. 1978. Weather in wildland fire management: the fire weather index. In *Proceedings of the Conference on Sierra Nevada Meteorology,* South Lake Tahoe, NV. pp. 1–4.

Gao, B.C., 1996. NDWI—A normalized difference water index for remote sensing of vegetation liquid water from space. *Remote Sensing of Environment.* 58(3), 257–266.

Garcia-Prats, A., Tarcísio, F.J. and Antonio, M.J. 2015. Development of a Keetch and Byram—based drought index sensitive to forest management in Mediterranean conditions. *Agricultural and Forest Meteorology.* 205, 40–50.

Geerts, B. 2002. *The Standardized Precipitation Index.* Western Regional Climate Center, Reno, Nevada, pp. 1–2.

Gitelson, A., Gritz, Y and Merzlyak, M. 2003. Relationships Between Leaf Chlorophyll Content and Spectral Reflectance and Algorithms for Non-Destructive Chlorophyll Assessment in Higher Plant Leaves. *Journal of Plant Physiology.* 160, 271–282.

Goncalves, Z.J. and Lourenco, L. 1990. Meteorological index of forest fire risk in the Portuguese mainland territory. In *Proc* International Conference on Forest Fire Research, Coimbra, Portugal, 1990, Vol. 7, pp. 1–14.

Goodrick, S.L. 2002. Modification of the Fosberg fire weather index to include drought. *International Journal of Wildland Fire.* 11(4), 205–211.

Gouma, V. and Chronopoulou-Sereli, A. 1998. Wildland fire danger zoning–a methodology. *International Journal of Wildland Fire.* 8(1), 37–43.

Gupta, P.K., Prasad, V.K., Sharma, C., Sarkar, A.K., Kant, Y., Badarinath, K.V.S. and Mitra, A.P. 2001. CH_4 emissions from biomass burning of shifting cultivation areas of tropical deciduous forests–experimental results from ground-based measurements. *Chemosphere-Global Change Science.* 3(2), 133–143.

Guttman, N.B. 1999. Accepting the standardized precipitation index: a calculation algorithm 1. *JAWRA Journal of the American Water Resources Association.* 35(2), 311–322.

Haines, D.A. 1988. A lower atmosphere severity index for wildland fires. *National Weather Digest.* 13, 23–27.

Hamadeh, N., Karouni, A., Daya, B. and Chauvet, P. 2017. Using correlative data analysis to develop weather index that estimates the risk of forest fires in Lebanon & Mediterranean: assessment versus prevalent meteorological indices. *Case Studies in Fire Safety.* 7, 8–22.

Hayes, M.J., Svoboda, M.D., Wilhite, D.A. and Varnyarkho, O.V. 1999. Monitoring the 1996 drought using the Standardized Precipitation Index. *Bulletin of the American Meteorological Society.* 80(3), 429–438.

Heyerdahl, E.K., McKenzie, D., Daniels, L.D., Hessl, A.E., Littell, J.S. and Mantua, N.J. 2008. Climate drivers of regionally synchronous fires in the inland Northwest (1651–1900). *International Journal of Wildland Fire.* 17(1), 40–49.

ICONA. 1993. *Manual de Operaciones Contra Incendios Forestales.* Ministerio de Agricultura, Pesca y Alimentación, Madrid, Spain.

Janis, M.J., Johnson, M.B. and Forthun, G. 2002. Near-real time mapping of Keetch-Byram drought index in the south-eastern United States. *International Journal of Wildland Fire.* 11(4), 281–289.

Jolly, W.M., Freeborn, P.H., Page, W.G. and Butler, B.W. 2019. Severe fire danger index: a forecastable metric to inform firefighter and community wildfire risk management. *Fire.* 2(3), 47.

Jurečka, F., Možný, M., Balek, J., Žalud, Z. and Trnka, M. 2019. Comparison of Methods for the Assessment of Fire Danger in the Czech Republic. *Acta Universitatis Agriculturae et Silviculturae Mendelianae Brunensis.* 67(5), 1285–1295.

Justice, C.O., Malingreau, J.P. and Setzer, A.W. 1993. *Satellite remote sensing of fires-Potential and limitations. Fire in the environment- The ecological, atmospheric, and climatic importance of vegetation fires.* John Wiley & Sons, Ltd, Chichester, United Kingdom. pp. 77–88.

Justice, C., Gutman, G. and Vadrevu, K.P. 2015. NASA land cover and land use change (LCLUC): an interdisciplinary research program. *Journal of Environmental Management.* 148(15), 4–9.

Kant, Y., Ghosh, A.B., Sharma, M.C., Gupta, P.K., Prasad, V.K., Badarinath, K.V.S. and Mitra, A.P. 2000. Studies on aerosol optical depth in biomass burning areas using satellite and ground-based observations. *Infrared Physics & Technology.* 41(1), 21–28.

Käse, H. (1969). Ein Vorschlag für eine Methode zur Bestimmung und Vorhesage der Waldbrandgefährdung mit Hilfe komplexer Kennziffern. Tech. Rep. 94, Abhandlungen des Meteorologischen Dienstes der Deutschen Demokratischen Re-publik, Berlin.

Keetch, J.J. and Byram, G.M. 1968. A drought index for forest fire control. USDA Forest Service Research Paper No. SE-38. pp. 1–32.

Khan, N., Sutherland, D., Wadhwani, R. and Moinuddin, K. 2019. Physics-based simulation of heat load on structures for improving construction standards for bushfire prone areas. *Frontiers in Mechanical Engineering.* 5, 35.

Lasko, K. and Vadrevu, K.P. 2018. Improved rice residue burning emissions estimates: accounting for practice-specific emission factors in air pollution assessments of Vietnam. *Environmental Pollution.* 236(5), 795–806.

Lasko, K., Vadrevu, K.P., Tran, V.T., Ellicott, E., Nguyen, T.T., Bui, H.Q. and Justice, C. 2017. Satellites may underestimate rice residue and associated burning emissions in Vietnam. *Environmental Research Letters.* 12(8), 085006.

Lasko, K., Vadrevu, K.P. and Nguyen, T.T.N. 2018. Analysis of air pollution over Hanoi, Vietnam using multi-satellite and MERRA reanalysis datasets. *PLoS One.* 13(5), e0196629.

Livada, I. and Assimakopoulos, V.D. 2007. Spatial and temporal analysis of drought in Greece using the Standardized Precipitation Index (SPI). *Theoretical and Applied Climatology.* 89(3–4), 143–153.

Lloyd-Hughes, B. and Saunders, M.A. 2002. A drought climatology for Europe. *International Journal of Climatology: A Journal of the Royal Meteorological Society.* 22(13), 1571–1592.

McArthur, A.G. 1967. Fire behavior in Eucalypt forest. Comm. Aust. For. Timb. Bur. Leaflet 107, 25pp.

McKee, T.B., Doesken, N.J. and Kleist, J. 1993. The relationship of drought frequency and duration to time scales. Proceedings of the 8th Conference on Applied Climatology. 17(22), 179–183. Zürich, Switzerland.

Merrill, D.F. and Alexander, M.E. (Eds.) 1987. *Glossary of Forest Fire Management Terms.* Fourth edition. National Resource Council Canada, Canadian Comm. Forest Management, Ottawa, Ontario. Publication NRCC No. 26516.

Miller, M. 2000. Fire autoecology. In J.K. Brown and J.K. Smith (Eds.) *Wildland Fire in Ecosystems.* UTA: RMRS-GTR-42, USDA Forest Service, Rocky Mountain Research Station, Ogden.

Monteith, J. 1981. Evaporation and surface temperature. *Quarterly Journal of the Royal Meteorological Society.* 107(451), 1–27.

Nelson Jr, R.M. 2000. Prediction of diurnal change in 10-h fuel stick moisture content. *Canadian Journal of Forest Research.* 30(7), 1071–1087.

Nesterov, V. 1949. *Combustibility of the Forest and Methods for Its Determination.* USSR State Industry Press. (in Russian).

Noble, J.C., Smith, A.W. and Leslie, H.W. 1980. Fire in the Mallee shrublands of western New South Wales. *The Rangeland Journal.* 2(1), 104–114.

Ntale, H.K. and Gan, T.Y. 2003. Drought indices and their application to East Africa. *International Journal of Climatology: A Journal of the Royal Meteorological Society.* 23(11), 1335–1357.

Nunes, J.R.S., Soares, R.V. and Batista, A.C. 2009. Analysis of the modified Monte Alegre Formula (FMA+) for the Parana State. *Floresta.* 39(3), 473–484.

NWCG. 2003. Glossary of wildland fire terminology. (National Wildfire Coordinating Group) Available at http://www.nwcg.gov/pms/pubs/glossary/pms205.pdf [Verified 3 April 2010]

Palmer, W.C. 1965. *Meteorological Drought,* Vol. 30. US Department of Commerce, Weather Bureau, Washington, DC.

Palmieri, S. and Cozzi, R. 1983. Il ruolo della meteorologica nella prevenzione e controllo degliincendi boschivi. Rivista Meteorologia Aeronautica XLIII, 4. In Italian.

Palmieri, S., Inghilesi, R., Siani, A.M. and Martellacci, C. 1992. Un indice meteorologico dirischio per incendi boschivi. *Bollettino Geofisico.* 15, 49–62.

Penman, H.L., 1963. Vegetation and hydrology. Technical Communication no. 53, Commonwealth Bureau of Soils, Harpenden, England.

Phillips, D.L., Dolph, J, Marks D. 1992. A comparison of geostatistical procedures for spatial analysis of precipitation in mountainous terrain. *Agricultural and Forest Meteorology.* 58, 119–141.

Prasad, V.K., Gupta, P.K., Sharma, C., Sarkar, A.K., Kant, Y., Badarinath, K.V.S., Rajagopal, T. and Mitra, A.P. 2000. NO_x emissions from biomass burning of shifting cultivation areas from tropical deciduous forests of India–estimates from ground-based measurements. *Atmospheric Environment.* 34(20), 3271–3280.

Prasad, V.K., Kant, Y. and Badarinath, K.V.S. 2001. CENTURY ecosystem model application for quantifying vegetation dynamics in shifting cultivation areas: a case study from Rampa Forests, Eastern Ghats (India). *Ecological Research.* 16(3), 497–507.

Prasad, V.K., Kant, Y., Gupta, P.K., Elvidge, C. and Badarinath, K.V.S. 2002. Biomass burning and related trace gas emissions from tropical dry deciduous forests of India: a study using DMSP-OLS data and ground-based measurements. *International Journal of Remote Sensing.* 23(14), 2837–2851.

Prasad, V.K., Lata, M. and Badarinath, K.V.S. 2003. Trace gas emissions from biomass burning from northeast region in India—estimates from satellite remote sensing data and GIS. *Environmentalist.* 23(3), 229–236.

Prasad, V.K., Badarinath, K.V.S., Yonemura, S. and Tsuruta, H. 2004. Regional inventory of soil surface nitrogen balances in Indian agriculture (2000–2001). *Journal of Environmental Management.* 73(3), 209–218.

Prasad, V.K., Badarinath, K.V.S. and Eaturu, A. 2008. Biophysical and anthropogenic controls of forest fires in the Deccan Plateau, India. *Journal of Environmental Management,* 86(1), 1–13.

Raphael, D.R.A. 2011. Evaluation of meteorological forest fire risk indices and projection of fire risk for German federal states (Doctoral dissertation, University of Applied Sciences, Germany).

Riley, K.L., Abatzoglou, J.T., Grenfell, I.C., Klene, A.E. and Heinsch, F.A. 2013. The relationship of large fire occurrence with drought and fire danger indices in the western USA, 1984–2008: the role of temporal scale. *International Journal of Wildland Fire.* 22(7), 894–909.

Rothermel, R.C. 1972. *A Mathematical Model for Predicting Fire Spread in Wildland Fuels,* Vol. 115. Intermountain Forest and Range Experiment Station, Forest Service, United States Department of Agriculture.

San-Miguel-Ayanz, J., Carlson, J.D., Alexander, M., Tolhurst, K., Morgan, G., Sneeuwjagt, R. and Dudley, M. 2003. Current methods to assess fire danger potential. In E. Chuvieco (Ed.), *Wildland Fire Danger Estimation and Mapping: The Role of Remote Sensing Data.* World Scientific Publishing, Singapore, pp. 21–61.

Schaber, J., Badeck, F.-W. and Lasch, P. 1999. Ein Modell der Sukzessionsdynamik europaischer Walder - Forest Ecosystems in a changing Environment (4C). In D. Pelz, O. Rau, and J. Saborowski (Eds.) *Freiburg: Deutscher Verband forstlicher Versuchsanstalten - Sektion forstliche Biometrie und Informatik,* vol. 11, pp. 212–217.

Schunk, C., Wastl, C., Leuchner, M. and Menzel, A. 2017. Fine fuel moisture for site-and species-specific fire danger assessment in comparison to fire danger indices. *Agricultural and Forest Meteorology.* 234, 1–47.

Scott, J.H. and Burgan, R.E., 2005. Standard fire behavior fuel models: a comprehensive set for use with Rothermel's surface fire spread model. USDA Forest Service, Rocky Mountain Research Station. General Technical Report RMRS-GTR-153. (Fort Collins, CO).

Sharples, J.J., McRae, R.H.D., Weber, R.O. and Gill, A.M. 2009. A simple index for assessing fire danger rating. *Environmental Modelling & Software.* 24(6), 764–774.

Shetinsky, E.A. 1994. Protection of forests and forest pyrology. Ecology, Moscow (in Russian), p. 209.

Sirca, C., Spano, D., Duce, P., Delogu, G., Cicalò, G.O. and Viegas, D.X. 2007. Performance of a newly developed integrated fire rating index in Sardinia, Italy. In *Proceedings of the 4th International Wildland Fire Conference.* Seville, Spain, pp. 13–17.

Skvarenina, J., Mindas, J., Holecy, J. and Tucek, J. 2003. Analysis of the natural and meteorological conditions during two largest forest fire events in the Slovak Paradise National Park. In *Proceedings of the International Scientific Workshop on Forest Fires in the Wildland–Urban Interface and Rural Areas in Europe: An Integral Planning and Management Challenge,* Athens, Greece (pp. 15–16).

Slavík, B. 1974. Methods of studying plant water relations. Academia (Czechoslovak Academy of Sciences).

Soares, R.V.; Batista, A.C. 2007. *Incêndios florestais: controle, efeitos e uso do fogo*. FUPEF, Curitiba. 264p.

Sol, B. 1992. Incendies de for& dans le Sud-Est de la France: le point sur l'estimation de la sCcheresse du sol. Note de travail SMIRISE no. 9, 29 p.

Sulong, N.A., Latif, M.T., Khan, M.F., Amil, N., Ashfold, M.J., Wahab, M.I.A., Chan, K.M. and Sahani, M. 2017. Source apportionment and health risk assessment among specific age groups during haze and non-haze episodes in Kuala Lumpur, Malaysia. *Science of the Total Environment*. 601, 556–570.

Thornthwaite, C.W. 1948. An approach toward a rational classification of climate. *Geographical Review*. 38(1), 55–94. Trockensommer 1992. Tech. Rep. 2, PIK.

Turner, J.A. and Lawson, B.D. 1978. Weather in the Canadian forest fire danger rating system. a user guide to national standards and practices. https://www.carbon.cfs.nrcan.gc.ca/publications?id=1843

Vadrevu, K.P. 2008. Analysis of fire events and controlling factors in eastern India using spatial scan and multivariate statistics. *Geografiska Annaler: Series A, Physical Geography*. 90(4), 315–328.

Vadrevu, K.P. and Justice, C.O. 2011. Vegetation fires in the Asian region: satellite observational needs and priorities. *Global Environment Research*. 15(1), 65–76.

Vadrevu, K.P., and Lasko, K.P. 2015. Fire regimes and potential bioenergy loss from agricultural lands in the Indo-Gangetic Plains. *Journal of Environmental Management*. 148, 10–20.

Vadrevu, K.P. and Lasko, K. 2018. Intercomparison of MODIS AQUA and VIIRS I-Band fires and emissions in an agricultural landscape—Implications for air pollution research. *Remote Sensing*. 10(7), 978. doi:10.3390/rs10070978.

Vadrevu, K.P., Giglio, L. and Justice, C. 2013. Satellite based analysis of fire–carbon monoxide relationships from forest and agricultural residue burning (2003–2011). *Atmospheric Environment*. 64, 179–191.

Vadrevu, K.P., Ohara T, Justice C. 2014a. Air pollution in Asia. *Environmental Pollution*. 12, 233–235.

Vadrevu, K.P., Lasko, K., Giglio, L. and Justice, C. 2014b. Analysis of Southeast Asian pollution episode during June 2013 using satellite remote sensing datasets. *Environmental Pollution*. 12, 245–256.

Vadrevu, K.P., Lasko, K., Giglio, L. and Justice, C. 2015. Vegetation fires, absorbing aerosols and smoke plume characteristics in diverse biomass burning regions of Asia. *Environmental Research Letters*. 10(10), 105003.

Vadrevu, K., Ohara, T. and Justice, C. 2017. Land cover, land use changes and air pollution in Asia: a synthesis. *Environmental Research Letters*. 12(12), 120201.

Vadrevu, K.P., Ohara, T. and Justice, C. (Eds.) 2018. *Land-Atmospheric Research Applications in South and Southeast Asia*. Springer, Cham.

Vadrevu, K.P., Lasko, K., Giglio, L., Schroeder, W., Biswas, S. and Justice, C. 2019. Trends in vegetation fires in south and southeast Asian countries. *Scientific Reports*. 9(1), 1–13.

Vajda, A., Venäläinen, A., Suomi, I., Junila, P. and Mäkelä, H.M. 2014. Assessment of forest fire danger in a boreal forest environment: description and evaluation of the operational system applied in F inland. *Meteorological Applications*. 21(4), 879–887.

Van Wagner, C.E. and Forest, P. 1987. Development and structure of the Canadian forest fire weather index system. In Can. For. Serv., Forestry Tech. Rep.

Van Wagner, C.E. and Pickett, T.L. 1985. Equations and FORTRAN program for the Canadian Forest Fire Weather Index System. Can. For. Serv., Ottawa, Ontario. For. Tech. Rep. 33. Burrows, ND 1984. Predicting blow-up fires in the jarrah forest. For Dep. WA, Perth, Western Australia. Tech. Pap. 12.

Verdon, D.C., Kiem, A.S. and Franks, S.W. 2004. Multi-decadal variability of forest fire risk—Eastern Australia. *International Journal of Wildland Fire*. 13(2), 165–171.

Willis, C., van Wilgen, B., Tolhurst, K., Everson, C., D'Abreton, P., Pero, L. and Fleming, G. 2001. *The Development of a National Fire Danger Rating System for South Africa.* Department of Water Affairs and Forestry, Pretoria.

Winkler, J.A., Potter, B.E., Wilhelm, D.F., Shadbolt, R.P., Piromsopa, K. and Bian, X. 2007. Climatological and statistical characteristics of the Haines Index for North America. *International Journal of Wildland Fire.* 16(2), 139–152.

Zhdanko, V.A. 1965. Scientific basis of development of regional scales and their importance for forest fire management. In *Contemporary Problems of Forest Protection from Fire and Firefighting.* Lesnaya Promyshlennost'Publ., Moscow, pp. 53–86.

13 Air Pollution Conditions near Peat Fire-Prone Areas during El Niño in Central Kalimantan, Indonesia

Hiroshi Hayasaka
Hokkaido University, Japan

Alpon Sepriando
Indonesian Agency for Meteorology, Indonesia

Aswin Usup
Palangka Raya University, Indonesia

Naito Daisuke
Kyoto University, Japan

CONTENTS

INTRODUCTION

Land use/cover changes (LCLUC) are among the most important pervasive changes impacting energy and emissions in different regions of the world (Justice et al., 2015). Of the different LCLUC, biomass burning is most common in South/Southeast Asian countries (Biswas et al., 2015a; Kant et al., 2000; Badarinath et al., 2008a, b; Vadrevu et al., 2018). Much of the interannual variability in global atmospheric carbon dioxide concentrations has been attributed to the variability of emissions from biomass burning (Page et al., 2002; Langenfelds et al., 2002; van der Werf et al., 2006; Field et al., 2009; Ito and Inatomo, 2018). The reasons for biomass burning vary. For example, in India, Myanmar, Laos, and Cambodia, fire is used as a land-clearing tool during the slash-and-burn agriculture (Prasad et al., 2000; 2002; Biswas et al., 2015b; Inoue, 2018), whereas clearing of forests for conversion to plantations is most common in Indonesia (Hayasaka and Sepriando, 2018; Israr et al., 2018). Since most of the countries in South/Southeast Asia are agrarian, several farmers use fire as a means, for example, for clearing the crop residues for planting the next crop (Lasko et al., 2017; Lasko and Vadrevu, 2018; Lasko et al., 2018a, b; Vadrevu and Lasko, 2018; Vadrevu et al., 2019). Specific to Indonesia, fires from peatlands and deforestation are most common and unique (Usup et al., 2004; Vadrevu et al., 2014a, b). Biomass burning from these land-clearing practices releases a significant amount of greenhouse gas emissions and aerosols, which can impact not only the local climate but also regional air quality due to the long-range transport (Prasad et al., 2001; 2003; 2008; Gupta et al., 2001). For example, fires in 1997–1998 in Indonesia consumed over 11 million ha of forest, resulting in economic losses and threats to biodiversity, water supplies, and other ecosystem services (Siegert et al., 2001; Taylor 2010; Hooijer et al., 2010). For quantifying the burnt areas and emissions, remotely sensed data could be quite useful (Justice et al., 2015).

Present flammable conditions in forests and peatlands of Indonesia are created by rapid deforestation. In Indonesia, annual primary forest loss (84,000 km²) was estimated to be higher than in Brazil (46,000 km²) by 2012. The world's highest deforestation rates are caused by "aggressive human activities" (Margono et al., 2014). Another type of deforestation is large-scale development, such as the large Mega Rice Project (MRP). MRP was initiated in 1996 near Palangkaraya, Central Kalimantan, to turn 1 million ha (9191 km²) of sparsely populated peat swamps into rice paddies. This type of large-scale land development involved extensive deforestation (Houterman and Ritzema, 2009).

The Indonesian haze disaster in 1997/1998 (Page et al., 2002; van der Werf et al., 2006) had a significant impact on regional air quality (Field et al., 2009). Average emissions from the region are smaller than those from Africa but are comparable to emissions from the Amazon basin (van der Werf et al., 2006; Field et al., 2009). The approximately 4000-km-long irrigation system in the MRP is not functioning properly, and many canals are now working as drainage systems. As a result, drained peatlands after deforestation are highly vulnerable to fires during the dry season in recent years.

Peat fires can worsen air pollution sources than aboveground biomass fires due to smoldering combustion of peat underground due to low temperature and poor oxygen conditions. Indonesia has about $225,000 \, km^2$ of peatlands (83% of the peatlands in Southeast Asia (SEA)), of which 57,600 and $30,100 \, km^2$ are in Kalimantan and Central Kalimantan, respectively (Hooijer et al., 2006). MRP peat, covering an area of roughly $4500 \, km^2$, is more than $3 \, m$ deep and is the primary source of fire-related air pollution. Because of the large amount of peat in the MRP area, it is one of the severe fire hotspots and density in Indonesia (Yulianti et al., 2012; Yulianti, 2013).

In SEA, El Niño is typically associated with drought, which induced some of the worst wildfires on record in Indonesia in 2015 (OCHA, 2015). The total burnt area in Indonesia during the previous super El Niño in 1997–1998 was estimated at $40,000 \, km^2$ (Siegert and Hoffmann, 1998). Thus, in Kalimantan, Indonesia, vegetation and peat fires were very active under El Niño and quasi-El Niño conditions, especially in recent years (2002, 2004, 2006, 2009, 2014, and 2015). These recent active fires may suggest vulnerable conditions in forests and peatlands in Kalimantan.

In particular, haze in Indonesia results in serious environmental and health problems. For example, in 1994 and 1997, when extensive forest fires were reported in Indonesia, pronounced peaks of total and tropospheric ozone were observed at Watukosek, Indonesia (Fujiwara et al., 1999). The aerosols released by the 1997 fires resulted in severe air pollution in SEA. Daily mean PM levels reached 2000×10^{-6} g/m^3 and higher during July through November, exceeding the ambient air quality standard by at least a factor of 7 in Indonesia. Also downwind to the fires, Malaysia and Singapore experienced severe air pollution exceeding the specific air quality standards (Heil, 2007). The Indonesian fires in 1997 released pyrogenic aerosols of around $55 \times 1012 \, g$ PM_{10} into the atmosphere, equivalent to about one-third of the global annual anthropogenic emissions of primary particles (Heil et al., 2007). Indonesia also formally admitted to contributing a considerable amount of CO_2 emissions from fires and deforestation (Hayasaka et al., 2014).

In 2013, Singapore's worst air pollution event in the past 16 years occurred on June 19, where the Pollutant Standards Index (PSI) reached 371 (International Business Times, Thursday, June 20, 2013, http://www.ibtimes.com/). Haze came from about $250 \, km$ away, where vegetation fires occurred in the east of Dumai, Riau Province, Sumatra, Indonesia. To reduce transboundary haze pollution from vegetation fires, information on the pollution sources and their impacts needs a thorough investigation (Vadrevu et al., 2006; 2012; 2018; Vadrevu and Lasko, 2018). There are several reports in Indonesia for urban air pollution sources, but only a few for vegetation fires (Pradani and Lestari, 2010).

To assess the impact of peat fires on air pollution, we analyzed PM_{10}, SO_2, CO, O_3, and NO_2 data obtained from the Air Pollution Center in Palangkaraya, in conjunction with Moderate Resolution Imaging Spectroradiometer (MODIS) satellite data fire hotspots and local weather data. We also carried out field surveys in severe peat fire areas to characterize haze behavior. This study discusses the air pollution sources in 2002 and photochemical smog occurrence in Palangkaraya based on field survey results (Hayasaka and Sepriando, 2018). We also summarize the fire and weather data and satellite imagery to discuss air pollution conditions near peat fire-prone areas of MRP in Central Kalimantan, and the peat fire-caused pollution.

DATA AND METHODOLOGY

Study Area

Figure 13.1 shows the peatland distribution and study area called "MRP+" in Kalimantan, and air pollution measuring sites in Palangkaraya (capital of Central Kalimantan). Peatlands in Kalimantan cover about $57,600\,km^2$, which is equivalent to peatlands of Sumatra. Central Kalimantan alone has $30,100\,km^2$ of peatlands. Deforested peatlands,

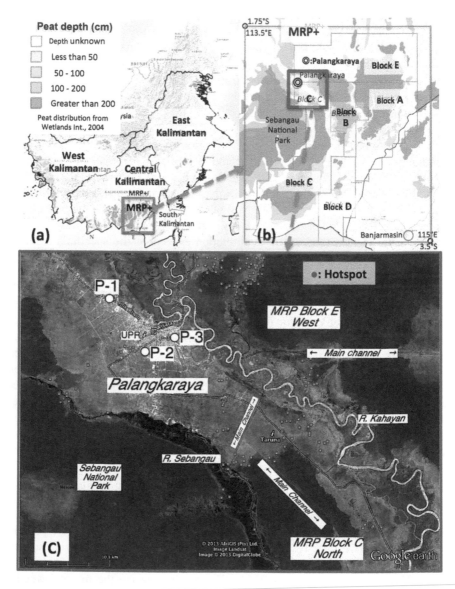

FIGURE 13.1　(a–c) Maps of peatland distribution in Kalimantan, MRP+, and Palangkaraya, Indonesia. (Original peatland map from Global Forest Watch and Google Earth.)

abandoned after the MRP, became major air pollution sources because of the large accumulation of peat and their combustion after human disturbances.

MRP+ covers the MRP area and its vicinity, including the Sebangau National Park (1.75–3.5°S, 113.5–115°E). MRP is divided into five blocks, block A to block E, as shown in Figure 13.1. Their actual boundaries are defined mainly by rivers. In this chapter, approximate boundaries defined by latitude and longitude lines are used for convenience. MRP+ was chosen as the study area simply because it is dominated by peatlands in Central Kalimantan and is one of the highest fire hotspot density areas in Indonesia (Hayasaka and Sepriando, 2015; 2018).

Palangkaraya is located in Central Kalimantan, as shown by a circle (2.207°S, 113.917°E) in Figure 13.1. The nearest coastline distance is about 100 km, and the average altitude is only around 10 m. The MRP was built on tropical swamp forest areas on the eastern and southern sides of Palangkaraya. Before the disturbance, the tropical swamp forest could hold enough water to stay wet even in the dry season. However, the newly constructed 4000-km-long MRP canal built for irrigation facilitated illegal logging and loss of water through drainage from most of the peatlands in the MRP area. These disturbances are the main reasons for severe fire occurrences in the MRP area.

AIR POLLUTION, PM$_{10}$ DATA, AND WEATHER DATA

Air pollution in Palangkaraya had been monitored by the Air Quality Management System and regional Air Quality Center in Palangkaraya from 2001 until around 2010. The locations of the three measuring stations are Tjilik Riwut (P-1), Tilung (P-2), and Murjani (P-3), and they are shown in Figure 13.1. Each station measured PM$_{10}$, SO$_2$, CO, O$_3$, and NO$_2$, by using automated continuous analyzers. The Air Quality Center processes all air pollution data automatically every 30 minutes and displays their values along with the Air Pollution Standard Index (APS) every day (update time at 15:00) located in the 100-m rotaries in Palangkaraya. Unfortunately, most of their gas analyzers were broken in 2011, and data were unavailable now. In this chapter, results during 2002 are mainly used (Hayasaka et al., 2014).

The weather data, including PM$_{10}$ and visibility, were measured hourly by the Palangkaraya weather station of the Indonesian Agency for Meteorology, Climatology, and Geophysics (BMKG) at Palangkaraya International Airport (Tjilik Riwut, 2.224°S, 113.946°E). This study used mainly PM$_{10}$ and visibility data from BMKG to evaluate air pollution in 2015.

HOTSPOT (FIRE) DATA AND SATELLITE IMAGERY

MODIS hotspot data from 2002 to 2015 were extracted from the Fire Information for Resource Management System (NASA EARTHDATA). Advanced Very High-Resolution Radiometer (AVHRR) hotspot data from 1997 to 2001 were obtained from the Japan International Cooperation Agency (JICA). The JICA Fire Project, Forest Fire Prevention and Management Project (FFPMP), was carried out from 1996 to 2001.

MODIS hotspot data were used to classify fire and non-fire years according to the number of monthly hotspots as done in our previous studies (Hayasaka et al., 2014;

Vadrevu et al., 2013; 2014a,b; 2015). Also, daily fire hotspots were used to find the primary air pollution sources of Palangkaraya and to determine their relationship with precipitation conditions in MRP+ and Kalimantan. Satellite imageries from MODIS on Aqua and worldview (EOSDIS Worldview) are used to infer haze (dense smoke) conditions due to peat fires.

RESULTS AND DISCUSSION

BACKGROUND OF AIR POLLUTION CAUSED BY PEAT FIRE

Fire History and El Niño

Extreme fire events are linked to the El Niño–Southern Oscillation (ENSO) climate phenomenon that results in the extended long, dry period, especially over Indonesia (Dennis, 1999). To find the relationship between hotspots (fires) and El Niño, we examine the number of hotspots, and the sea surface temperature (SST) anomalies of the NINO3.4 region (5°S–5°N, 170°W–120°W) obtained from the Climate Prediction Center, National Oceanic and Atmospheric Administration (NOAA, 2020). Both the values are plotted using double Y-axes in Figure 13.2. In Figure 13.2, only positive SST anomalies' positive values are shown to depict the El Niño occurrence.

NOAA has its own definition of El Niño. The index is defined as a 3-month average of SST departures from normal for a critical region of the equatorial Pacific (Niño 3.4 region; newest data on April 15, 2019, were used). El Niño is classified into four categories, weak, moderate, strong, and very strong, depending on the values of SST anomalies, as shown in Figure 13.2. In Figure 13.2, two horizontal lines for +0.5°C and +2.0°C are drawn to capture the occurrence of a weak and super (very

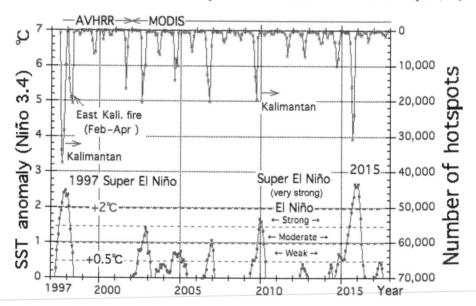

FIGURE 13.2 Hotspots in Kalimantan and El Niño 3.4 SST anomalies.

strong, >+2.0°C) El Niño. As two different hotspot data sources were used, we could not directly compare the number of hotspots or fire trends before 2002 with MODIS data. However, we were able to make useful inferences about fire activity and fire seasons before 2002.

In Figure 13.2, the peak months for each active fire year (1997, 2001, 2002, 2004, 2009, and 2015) are not clear. But their peak was mostly during September except for August in 2001. So, we could say that most fires in Kalimantan occur during the dry season or from June to October. Under super El Niño conditions in 2015, the SST anomaly and the number of hotspots (MODIS) were 2.12 and 30,710, respectively. These values of other active fire years were 0.6 and 19,429 in 2009, 0.7 and 14,266 in 2004, and 1.01 and 12,515 in 2002. From their relationship, we also infer a relatively strong correlation between the SST anomaly and the number of fire hotspots. When the SST anomaly exceeded 2°C, as it did in 1997 and 2015, the number of hotspots became larger, or the fire activity was aggressive. We also note the exceptional fires in East Kalimantan from February to April in 1998 under drought during super El Niño conditions from 1997.

Dry Season during El Niño Event

In Figure 13.3, accumulated daily precipitation for several years, as well as average values, was plotted. The daily average rainfall is calculated precipitation data from 1978 to 2014 and shown by the dotted line. Figure 13.3 clearly shows two precipitation periods (the dry and rainy seasons) in Palangkaraya. The daily precipitation rate clearly distinguishes the two seasons: 3.9 mm/day for the dry season and 10 mm/day for the rainy season.

Dry seasons during El Niño years in 1997, 2002, 2009, and 2015, shown in Figure 13.3, commonly had daily precipitation rates almost flat (<2.0 mm/day). We infer that an El Niño event results in about four dry months in Palangkaraya, Central Kalimantan. The dry season in 1997 was the longest (about 6 months), and

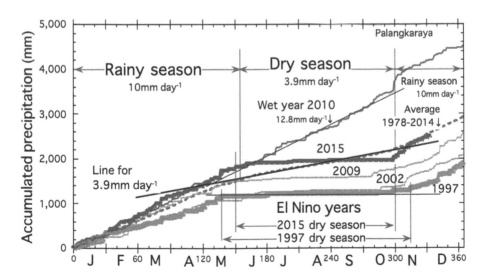

FIGURE 13.3   Dry and rainy season in Palangkaraya, Indonesia.

the daily precipitation rate was the smallest (=0.8 mm/day). During the 2015 super El Niño, the intense drought lasted about 127 days from June 20 to October 25 with lower precipitation (=0.49 mm/day, ratio to average = 0.49/3.9 = 0.125). In contrast, the La-Niña event in 2010 resulted in a very wet year, and the daily precipitation rate in the dry season was 12.8 mm/day (Figure 13.3). Under the very wet conditions created by the La Niña, fires were not so active (number of hotspots < 8000; Figure 13.2). The ignition probability of peat varies with the moisture content (Babrauskas, 2003). It increases rapidly from nearly zero to greater than 0.8 when the gravimetric moisture content of peat becomes less than 100% (Hayasaka et al., 2016). This implies a so-called threshold value for peat ignition, and peat becomes good fuel for the fire when it dries. This ignition property could explain the severe peat fire occurrences in the dry season. Usup et al. (2004) estimated that 40% of the study area had groundwater levels lower than 40 cm below the peat surface in July during the El Niño year 1997. In drought conditions during the El Niño event, the peatlands' groundwater level gradually drops to the critical threshold of 40 cm around mid-August (Wösten et al., 2008, Putra and Hayasaka, 2011) coinciding with the onset of fire activity in Central Kalimantan.

Air Pollution Caused by Peat Fire

Haze during El Niño Event

Satellite imagery in Figure 13.4 shows haze (smoke) caused by the peat fires in Southern Kalimantan in 2002 and 2015. The worst air pollution day of PM_{10} occurred several days after the active fire day (large hotspot day). Figure 13.4a shows the haze situation in 2002. The worst day with the highest PM_{10} air pollution occurred 2 days later of a large number of fire hotspots were detected. In 2015, the worst haze condition occurred after 7 days of a large number of hotspots detected later (Figure 13.4b). These time lags indicate that MODIS (satellite sensor) could not successfully detect the peat fires due to low temperature and underground fires in addition to dense haze from peat fires. The worst air pollution occurred on October 14, 2002, during a moderate El Niño event, and about 10 days before the onset of the rainy season (Hayasaka et al., 2014), one of the characteristics of a peat fire in this area (MRP+).

Air Pollutant Levels in 2002 during Moderate El Niño

Very severe peat fires occurred near Palangkaraya in 2002, 2 years after the Air Quality Management System started to measure air pollutants. Daily measured values of PM_{10}, SO_2, CO, O_3, and NO_2 during the fire season from August to November in 2002 are shown in Figure 13.5. Most air pollutants except NO_2 showed a similar trend. Maximum concentrations of PM_{10}, SO_2, CO, and O_3 were observed on October 14, 2002 (DN=287), and their values were 1905, 85.8, 38.3, and 1003×10^{-6} g/m³, respectively. Air Pollution Index (API) was also highest with 1805. These air pollutant peaks occurred just after the fire hotspots became active around October 12, 2002 (DN=285).

NO2 showed slightly different emission trends (Figure 13.5). Daily NO_2 varied greatly from 7.3 to 67.4×10^{-6} g/m³ during the fire season. To compare the NO_2 values in both fire and non-fire seasons, monthly averaged NO_2 values were calculated. The average NO_2 value during the fire season from August to October was 30.5×10^{-6} g/m^3. In contrast, the average NO_2 value during the non-fire season from January to June was 13.6×10^{-6} g/m^3. This difference of NO_2 between the fire and non-fire seasons may suggest two major air pollution sources: vegetation fires and another from vehicular combustion. In other words, NO_2 showed relatively

(a)

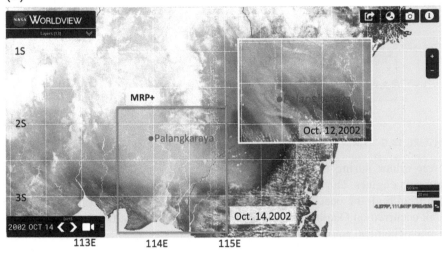

(b)

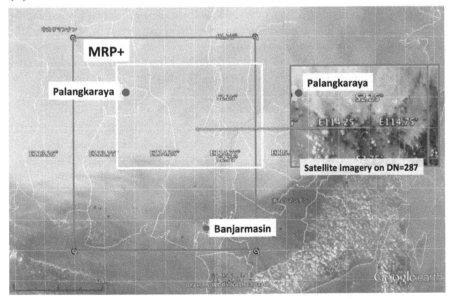

FIGURE 13.4   Satellite imagery (Worldview and Google Earth). (a) Haze (PM$_{10}$ worst day (DN = 287, October 14, 2002)). (b) Haze (PM$_{10}$ worst day (DN = 294, October 21, 2015.))

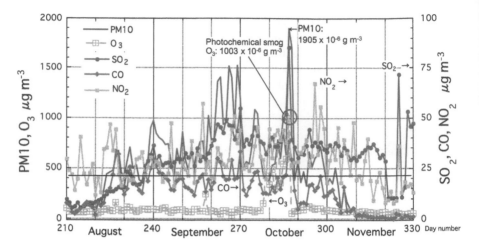

FIGURE 13.5 PM_{10}, SO_2, CO, O_3, and NO_2 during moderate El Niño conditions in 2002.

lower values during the non-fire season simply because the primary emission source is engines. During the fire season, NO_2 values were higher, thus contribution from both the peat fires and engines. Under the high-NO_2 conditions, the O_3 peak occurred on October 14 (DN = 287). This may suggest the serious formation of photochemical smog.

Large differences found in PM_{10}, SO_2, and CO during the fire season in Figure 13.5 also suggest a large contribution from peat fires. Particularly, increased SO_2 from around 5×10^{-6} g/m³ in early August to 85.8×10^{-6} g/m³ in mid-October (see Figure 13.5) was mostly from peat fires, as there is no other major source of SO_2 (such as industries) in the vicinity. Since most APIs were derived from PM_{10} in 2002, a horizontal line was drawn at $PM_{10} = 420 \times 10^{-6}$ g/m³ to show the hazardous level of API (>300). Thus from Figure 13.5, we infer that the hazardous condition lasted about 80 days from mid-August to late October in 2002.

Air Pollution in 2015 during Super El Niño

PM_{10} and minimum and average visibility measured at Palangkaraya are plotted in Figure 13.6. Minimum and average visibility gradually decreased with fire activity in MRP+. The daily maximum PM_{10} was 3245×10^{-6} g/m³, observed on October 21, 2015 (DN = 294). The hourly maximum PM_{10} was 3761×10^{-6} g/m³ and was also observed on the same day (DN = 294). The minimum and average visibility on this day (DN = 294) was nearly zero (hourly data varied from 10 to 150 m). A severe air pollution period ($PM_{10} > 420 \times 10^{-6}$ g/m³: hazardous level, API > 300) began in mid-September (about 4 months after the dry season started) and lasted until late October. Very severe air pollution (>1000×10^{-6} g/m³) occurred beginning in late September. This severe air pollution just before the rainy season also suggests that the primary source of air pollution was peat.

According to the recent news (The Wall Street Journal-India, November 2, 2013), the high PM_{10} concentration in New Delhi, India, is around 1940×10^{-6} g/m³ because of fuelwood biomass burning, fuel adulteration, vehicle emissions, and traffic congestion.

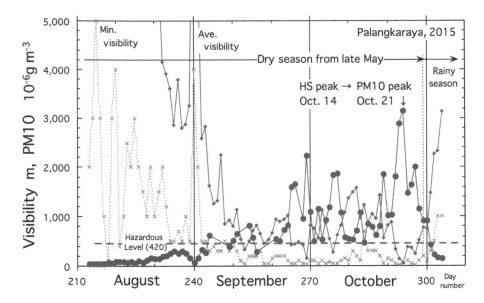

FIGURE 13.6 PM_{10} and minimum and average visibility in Palangkaraya, Indonesia.

The high PM_{10} concentration in Beijing, China, is around 1000×10^{-6} g/m³ (South China Morning Post, March 29 2015) because of coal combustion, vehicle emissions, traffic congestion, and dust storms in the spring. Compared to these regions, the air pollution in our study region (MRP+) may be the worst globally and is mainly due to peat fires.

The satellite imagery in Figure 13.4 clearly shows that a thick haze covers more than 60% of the northern part of the MRP+ area. Because of the thick haze and underground peat fires, the number of hotspots in Kalimantan and MRP+ was small, 724 on October 14 and 101 on October 14, respectively (Figure 13.4b). The worst air pollution occurring at the end of the dry season or just before the rainy season suggests that the air pollution source is peat because the underground water level becomes lowest at the end of the dry season, and this makes deep (<40 cm) peat highly combustible. Thus, attention should be paid to the dense haze at the end of the dry/fire season for pollution mitigation measures.

CONCLUSIONS

In this study, we summarized air pollution levels near peat fire-prone areas of MRP in Central Kalimantan during El Niño conditions. The background conditions of air pollution caused by peat fire were also clarified. Air pollution levels measured at Palangkaraya near MRP area were also discussed. During the moderate El Niño conditions in 2002, maximum concentrations of PM_{10}, SO_2, CO, and O_3 were observed on October 14, and their values were 1905, 85.8, 38.3, and 1003×10^{-6} g/m³, respectively. A maximum peak recorded for PM_{10} in 2002 was highest in the 10-year period from 2001 to 2010. High value of O_3 measured in 2002 suggested the formation of photochemical smog. During very strong El Niño in 2015, a period of severe air pollution ($PM_{10} > 420 \times 10^{-6}$ g/m³: hazardous level, API > 300) began in mid-September about 4 months after the dry season started and lasted until late October. The highest daily and hourly PM_{10} concentrations (3010 and 3760×10^{-6} g/m^3, respectively) were observed on October 20, 2015. These conditions occurred several days after the second largest hotspot peak day (1100 hotspots) in 2015. Both PM_{10} peaks in 2002 and 2015 that occurred at the end of dry season could be explained by the increased amount of ignitable dry peat related to the deepest underground water level. In overall, every El Niño event brought strong drought conditions to Kalimantan. The dry season lasts about 5 months from June to October compared with average precipitation (=3.9 mm/day). During moderate El Niño conditions in 2002, strong drought started from June 24 to October 13 (112 days in total) with lower precipitation (=1.1 mm/day, ratio to average = 1.1/3.9). Further, during the 2015 super El Niño, the dry season was longer and very dry; the strong drought lasted about 127 days, from June 20 to October 25, or late June to late October with lower precipitation (=0.49 mm/day, ratio to average = 0.49/3.9 = 0.125).

From the above results and discussion, we recommend strong conservation measures of peatlands in the MRP+ area in Central Kalimantan. In particular, we suggest limiting the construction of irrigation canals, which contribute to further development and subsequent drying of the land. Drastic land use/land management planning, including re-wetting and construction of a large-scale water reservoir in MRP area, should be carried out based on scientific knowledge and modern technology.

ACKNOWLEDGMENT

This study was partly supported by the Environment Research and Technology Development Fund 4–1506 of the Ministry of the Environment, Japan; and by the Center for International Forestry Research (CIFOR).

REFERENCES

Babrauskas, V. 2003. *Ignition Handbook*. Fire Science Publishers, Issaquah WA. Co-published by the Society of Fire Protection Engineers. ISBN: 10-0-9728111-3-3.

Badarinath, K.V.S., Kharol, S.K., Krishna Prasad, V., Kaskaoutis, D.G. and Kambezidis, H.D. 2008a. Variation in aerosol properties over Hyderabad, India during intense cyclonic conditions. *International Journal of Remote Sensing*. 29(15), 4575–4597.

Badarinath, K.V.S., Kharol, S.K., Prasad, V.K., Sharma, A.R., Reddi, E.U.B., Kambezidis, H.D. and Kaskaoutis, D.G. 2008b. Influence of natural and anthropogenic activities on UV Index variations–a study over tropical urban region using ground based observations and satellite data. *Journal of Atmospheric Chemistry*. 59(3), 219–236.

Biswas, S., Lasko, K.D. and Vadrevu, K.P. 2015a. Fire disturbance in tropical forests of Myanmar—Analysis using MODIS satellite datasets. *IEEE Journal of Selected Topics in Applied Earth Observations and Remote Sensing*. 8(5), 2273–2281.

Biswas, S., Vadrevu, K.P., Lwin, Z.M., Lasko, K. and Justice, C.O. 2015b. Factors controlling vegetation fires in protected and non-protected areas of Myanmar. *PLoS One*. 10(4), e0124346.

Dennis, R. 1999. *A Review of Fire Projects in Indonesia (1982–1998)*. Center for International Forestry Research (CIFOR), Bogor. Printed by SMT Grafika Desa Putera, Jakarta, ISBN 979-8764-30-7.

Field, R.D., Van Der Werf, G.R. and Shen, S.S. 2009. Human amplification of drought-induced biomass burning in Indonesia since 1960. *Nature Geoscience*. 2(3), 185–188.

Fujiwara, M., Kita, K., Kawakami, S., Ogawa, T., Komala, N., Saraspriya, S. and Suripto, A. 1999. Tropospheric ozone enhancements during the Indonesian forest fire events in 1994 and in 1997 as revealed by ground-based observations. *Geophysical Research Letters*, 26(16), 2417–2420.

Global forest watch http://fires.globalforestwatch.org/#v=home&x=115&y=0&l=5&lyrs=Active_Fires.

Gupta, P.K., Prasad, V.K., Sharma, C., Sarkar, A.K., Kant, Y., Badarinath, K.V.S. and Mitra, A.P. 2001. CH_4 emissions from biomass burning of shifting cultivation areas of tropical deciduous forests–Experimental results from ground-based measurements. *Chemosphere-Global Change Science*. 3(2), 133–143.

Hayasaka, H. and Sepriando, A. 2015. Severe Peat Fires and Air Pollution Near the Former Mega Rice Project Area in Central Kalimantan, Indonesia, Proceedings (A-421 Oral Presentation), *15th International Peat Congress 2016*, Kuching, Sarawak, Malaysia. pp. 323–326.

Hayasaka, H. and Sepriando, A. 2018. Severe air pollution due to peat fires during 2015 super El Niño in Central Kalimantan, Indonesia. In: Vadrevu, K.P., Ohara, T. and Justice, C. (Eds). *Land-Atmospheric Research Applications in South/Southeast Asia*. Springer, Cham. pp. 129–142.

Hayasaka, H., Noguchi, I., Putra, E.I., Yulianti, N. and Vadrevu, K. 2014. Peat-fire-related air pollution in Central Kalimantan, Indonesia. *Environmental Pollution*. 195, 257–266.

Hayasaka, H., Takahashi, H., Limin, S.H., Yulianti, N. and Usup, A. 2016. Peat fire occurrence. In M. Osaki and N. Tsuji (Eds). *Tropical Peatland Ecosystems*. Springer, Tokyo. pp. 377–395.

Heil, A. 2007. Indonesian Peat and Vegetation Fire Emissions: Emissions, Air Quality, and Human Health, *PhD thesis 2007*. International Max Planck Research School.

Heil, A., Langmann, B. and Aldrian, E. 2007. Indonesian peat and vegetation fire emissions: Study on factors influencing large-scale smoke haze pollution using a regional atmospheric chemistry model. *Mitigation and Adaptation Strategies for Global Change*. 12(1), 113–133.

Hooijer, A., Silvius, M., Wösten, H. and Page, S. 2006. PEAT-CO_2, Assessment of CO_2 Emissions from Drained Peatlands in SE Asia. Delft Hydraulics report Q3943.

Hooijer, A, Page S, Canadell, J.G., Silvius, M., Kwadijk, J., Wösten, H. and Jauhiainen J. 2010. Current and future CO_2 emissions from drained peatlands in Southeast Asia. *Biogeosciences*. 7(5), 1505–1514.

Houterman, J. and Ritzema, H. 2009. Technical Report No. 4 Affiliation, Master Plan for the Rehabilitation of the Ex-Mega Rice Project Area in Central Kalimantan. Euroconsult Mott MacDonald, Deltares, Delft. https://www.researchgate.net/publication/272510959_Land_and_water_management_in_the_Ex-Mega_Rice_Project_Area_in_Central_Kalimantan.

Inoue, Y. 2018. Ecosystem carbon stock, atmosphere and food security in slash and burn land use: A geospatial study in Mountainous region of Laos. In: *Land-Atmospheric Research Applications in South/Southeast Asia*. Vadrevu, K.P., Ohara, T. and Justice, C. (Eds). Springer, Cham. pp. 641–666.

Israr, I., Jaya, S.N.I, Saharjo, H.S., Kuncahyo, B. and Vadrevu, K.P. 2018. Spatio-temporal analysis of land and forest fires in Indonesia using MODIS active fire dataset. 2018. In: *Land-Atmospheric Research Applications in South/Southeast Asia*. Vadrevu, K.P., Ohara, T., and Justice, C. (Eds). Springer, Cham. pp. 105–128.

Ito, A., and Inatomo, M. 2018. Greenhouse gas budget of terrestrial ecosystems in Monsoon Asia: A process-based model study for the period 1901–2014. In: *Land-Atmospheric Research Applications in South/Southeast Asia*. Vadrevu, K.P., Ohara, T. and Justice, C. (Eds). Springer, Cham. pp. 223–232.

Justice, C., Gutman, G. and Vadrevu, K.P. 2015. NASA land cover and land use change (LCLUC): An interdisciplinary research program. *Journal of Environmental Management*. 148(15), 4–9.

Kant, Y., Ghosh, A.B., Sharma, M.C., Gupta, P.K., Prasad, V.K., Badarinath, K.V.S. and Mitra, A.P. 2000. Studies on aerosol optical depth in biomass burning areas using satellite and ground-based observations. *Infrared Physics & Technology*. 41(1), 21–28.

Langenfelds, R.L., Francey, R.J., Pak, B.C., Steele, L.P., Lloyd, J., Trudinger, C.M., Allison, C.E. 2002. Interannual growth rate variations of atmospheric CO_2 and its delta C_{13}, H_2, CH_4, and CO between 1992 and 1999 linked to biomass burning. *Global Biogeochemical Cycles*. 16. doi:10.1029/2001gb001466.

Lasko, K. and Vadrevu, K.P. 2018. Improved rice residue burning emissions estimates: Accounting for practice-specific emission factors in air pollution assessments of Vietnam. *Environmental pollution*. 236(5), 795–806.

Lasko, K., Vadrevu, K.P., Tran, V.T., Ellicott, E., Nguyen, T.T., Bui, H.Q. and Justice, C. 2017. Satellites may underestimate rice residue and associated burning emissions in Vietnam. *Environmental Research Letters*. 12(8), 085006.

Lasko, K., Vadrevu, K.P. and Nguyen, T.T.N. 2018a. Analysis of air pollution over Hanoi, Vietnam using multi-satellite and MERRA reanalysis datasets. *PLoS One*. 13(5), 0196629.

Lasko, K., Vadrevu, K.P., Tran, V.T. and Justice, C. 2018b. Mapping double and single crop paddy rice with Sentinel-1A at varying spatial scales and polarizations in Hanoi, Vietnam. *IEEE Journal of Selected Topics in Applied Earth Observations and Remote Sensing*. 11(2), 498–512.

Margono, B.A., Potapov, P.V., Turubanova, S., Stolle, F. and Hansen, M.C. 2014. Primary forest cover loss in Indonesia over 2000–2012. *Nature Climate Change*. 4(8), 730–735.

NASA EARTHDATA, Fire Information for Resources Management System, https://firms.modaps.eosdis.nasa.gov/download/.

NOAA. 2020. Monthly Atmospheric and SST Indices, Climate Prediction Center, National Oceanic and Atmospheric Administration, http://www.cpc.ncep.noaa.gov/data/indices/.

OCHA (United Nations Office for the Coordination of Humanitarian Affairs). 2015. El Niño in Asia, http://www.unocha.org/el-nino-asia-pacific.

Page, S.E., Siegert, F., Rieley, J.O., Boehm, H.D.V., Jaya, A. and Limin, S. 2002. The amount of carbon released from peat and forest fires in Indonesia during 1997. *Nature*. 420(6911), 61–65.

Pradani, M. and Lestari, P. 2010. Correlation between Hotspots and Ambient Air Quality as Impact of Forest. AP4, http://www.ftsl.itb.ac.id/kk/air_waste/wp-content/uploads/2010/11/PE-AP4-MAHARANI-PRADANI-15305022.pdf.

Prasad, V.K., Gupta, P.K., Sharma, C., Sarkar, A.K., Kant, Y., Badarinath, K.V.S., Rajagopal, T. and Mitra, A.P. 2000. NO_x emissions from biomass burning of shifting cultivation areas from tropical deciduous forests of India–estimates from ground-based measurements. *Atmospheric Environment*. 34(20), 3271–3280.

Prasad, V.K., Kant, Y. and Badarinath, K.V.S., 2001. CENTURY ecosystem model application for quantifying vegetation dynamics in shifting cultivation areas: A case study from Rampa Forests, Eastern Ghats (India). *Ecological Research*. 16(3), 497–507.

Prasad, V.K., Kant, Y., Gupta, P.K., Elvidge, C. and Badarinath, K.V.S. 2002. Biomass burning and related trace gas emissions from tropical dry deciduous forests of India: A study using DMSP-OLS data and ground-based measurements. *International Journal of Remote Sensing*. 23(14), 2837–2851.

Prasad, V.K., Lata, M. and Badarinath, K.V.S. 2003. Trace gas emissions from biomass burning from northeast region in India—estimates from satellite remote sensing data and GIS. *Environmentalist*. 23(3), 229–236.

Prasad, V.K., Badarinath, K.V.S. and Eaturu, A. 2008. Biophysical and anthropogenic controls of forest fires in the Deccan Plateau, India. *Journal of Environmental Management*. 86(1), 1–13.

Putra, E.I. and Hayasaka, H. 2011. The effect of the precipitation pattern of the dry season on peat fire occurrence in the Mega Rice Project area, Central Kalimantan, Indonesia. *TROPICS*. 19 (4):145–156.

Siegert, F. and Hoffmann, A.A. 1998. Evaluation of the 1998 Forest Fires in East-Kalimantan (Indonesia) Using Multitemporal ERS-2 SAR Images and NOAA-AVHRR Data. *Proceedings of The International Conference on Data Management and Modeling Using Remote Sensing and GIS for Tropical Forest Land Inventory*, Jakarta, Indonesia, 26–29 October 1998. Also available on http://smd.mega.net.id/iffm/n&r0.htm.

Siegert, F., Ruecker, G., Hinrichs, A. and Hoffmann, A.A. 2001. Increased damage from fires in logged forests during droughts caused by El Nino. *Nature*. 414(6862), 437–440.

Taylor, D. 2010. Biomass burning, humans and climate change in Southeast Asia. *Biodiversity and Conservation*. 19(4), 1025–1042.

Usup, A., Hashimoto, Y., Takahashi, H. and Hayasaka, H. 2004. Combustion and thermal characteristics of peat fire in tropical peatland in Central Kalimantan, Indonesia. Tropics, 14(1), 1–19.

Vadrevu, K.P. and Lasko, K. 2018. Intercomparison of MODIS AQUA and VIIRS I-Band fires and emissions in an agricultural landscape—Implications for air pollution research. *Remote Sensing*. 10(7), 978. doi:10.3390/rs10070978.

Vadrevu, K.P., Eaturu, A. and Badarinath, K.V.S. 2006. Spatial distribution of forest fires and controlling factors in Andhra Pradesh, India using spot satellite datasets. *Environmental Monitoring and Assessment*. 123(1–3), 75–96.

Vadrevu, K.P., Csiszar, I., Ellicott, E., Giglio, L., Badarinath, K.V.S., Vermote, E. and Justice, C. 2012. Hotspot analysis of vegetation fires and intensity in the Indian region. *IEEE Journal of Selected Topics in Applied Earth Observations and Remote Sensing*. 6(1), 224–238.

Vadrevu, K.P., Giglio, L. and Justice, C. 2013. Satellite based analysis of fire–carbon monoxide relationships from forest and agricultural residue burning (2003–2011). *Atmospheric Environment*. 64, 179–191.

Vadrevu, K.P., Ohara, T. and Justice, C. 2014a. Air pollution in Asia. *Environmental Pollution*. 12, 233–235.

Vadrevu, K.P., Lasko, K., Giglio, L. and Justice, C. 2014b. Analysis of Southeast Asian pollution episode during June 2013 using satellite remote sensing datasets. *Environmental Pollution*. 12, 245–256.

Vadrevu, K.P., Lasko, K., Giglio, L. and Justice, C. 2015. Vegetation fires, absorbing aerosols and smoke plume characteristics in diverse biomass burning regions of Asia. *Environmental Research Letters*. 10(10), 105003.

Vadrevu, K.P., Lasko, K., Giglio, L., Schroeder, W., Biswas, S. and Justice, C. 2019. Trends in vegetation fires in South and Southeast Asian countries. *Scientific reports*. 9(1), 7422. doi:10.1038/s41598-019-43940-x.

Vadrevu, K.P., Ohara, T. and Justice, C. (Eds). 2018. *Land-Atmospheric Research Applications in South and Southeast Asia*. Springer, Cham.

van der Werf, G.R., Randerson, J.T., Giglio, L., Collatz, G.J., Kasibhatla, P.S. and Arellano Jr. A.F. 2006. Interannual variability in global biomass burning emissions from 1997 to 2004. *Atmospheric Chemistry and Physics*. 6, 3423–3441.

Wösten, H., Clymans, E., Page, S., Rieley, J. and Limin, S.H. 2008. Peat-water interrelationship in a tropical peatland ecosystem in Southeast Asia. *Catena*. 73, 212–224.

Yulianti, N., Hayasaka, H. and Usup, A. 2012. Recent forest and peat fire trends in Indonesia, the latest decade by MODIS hotspot data. *Global Environmental Research. AIRIES*. 16(1), 105–116.

Yulianti, N. 2013. The Influence of Precipitation Patterns on Recent Peatland Fires in Indonesia. *PhD thesis*. Graduate School of Engineering, Hokkaido University.

14 Biomass Burning and Haze in Indonesia, Long-Term Climate Perspective, and Impact on Regional Air Quality

Sheila Dewi Ayu Kusumaningtyas
Indonesia Agency for Meteorology Climatology
and Geophysics (BMKG), Indonesia

Edvin Aldrian
Agency for Assessment and Implementation
of Technology (BPPT), Indonesia

Sunaryo Sunaryo and Roni Kurniawan
Indonesia Agency for Meteorology Climatology
and Geophysics (BMKG), Indonesia

CONTENTS

INTRODUCTION

Forest and land fires in Southeast Asia (SEA) are increasing and intensifying due to changing land-cover/land-use change and climate drivers. Forest and land fires in the tropics consume ~80% of the total biomass burned globally (Crutzen and Andreae, 1990). Furthermore, biomass burning that produces thick haze has caused considerable concern among scientists, governments, and the public. Biomass burning is prominent in several South/Southeast Asian countries (Prasad et al., 2000; 2002; 2008; Badarinath et al., 2008a, b; Biswas et al., 2015a, b; Vadrevu and Justice, 2011). The important sources of biomass burning in South/Southeast Asia include slash-and-burn agriculture, agriculture residue burning, and forest clearing for plantations, and most of the fires are human-dominated (Justice et al., 2015; Vadrevu et al., 2006; Vadrevu and Justice, 2011). Recent trends in the forest fires suggest that they are increasing in Southeast Asian countries with a significant release in air pollutants (Kant et al., 2000; Gupta et al., 2001; Prasad et al., 2003; 2004; Vadrevu, 2008; Vadrevu and Lasko, 2015; 2018; Lasko et al., 2018a, b; Vadrevu et al., 2017; 2018; 2019). Of the different Southeast Asian countries, fires in Indonesia dominate with frequent and severe biomass burning events, in particular in Sumatera and Kalimantan with a significant release of smoke aerosols (Israr et al., 2018; Hayasaka et al., 2014; Vadrevu et al., 2015; Hayasaka and Sepriando, 2018; Saharjo and Yungan, 2018).

Indonesia has the largest tropical peat swamp globally, estimated at 26 million hectares (Mha) or ~10.8% of Indonesia's land area. The peatland area in SEA covers about 26 Mha and is located mostly near the 175 coasts (Wösten et al., 2008). According to Heil (2007), Indonesian peatlands represent around 5%, 56%, and 80% of the world's total peatland area, the tropics, and Equatorial Asia (EQAS). Specifically, Sumatera and Kalimantan islands host ~7.2 and 4.7 Mha of peatlands, respectively, as per the Ministry of Environment and Forestry of the Republic of Indonesia.

Tropical peat swamp forests in their undisturbed state are resistant to drought and are an important reservoir of biodiversity, carbon, and water. However, natural peat swamp forests in Indonesia started to disappear due to increasing land demand. Since the 1980s, peat swamp forests have been converted to other purposes such as agriculture (e.g., rice), agroforestry (e.g., palm oil, rubber), and settlements. From 1990 to 2000, deforestation and peatland degradation increased rapidly. Massive peatland conversion in Central Kalimantan during 1995 was initiated by the Indonesian government, which is called "Mega Rice Project" (MRP) (Hayasaka and Sepriando, 2018). The project converted ~1.5 Mha of deep undisturbed peat swamp forests into agricultural areas, mostly rice fields (Heil, 2007). Fuller et al. (2004) showed that Kalimantan Island had experienced an annualized 2.0% deforestation rate between 1997 and 2002. In Sumatera from the 1990s to the 2000s, the annual average deforestation rate of peatlands increased from 4.6% to 5.2% with a reduction in the land cover of more than 60% of peat swamp forest; almost 85% of all deforested peatland turned into plantation areas and secondary forests (Miettinen et al., 2011). In the last decade, more and more peatlands are converted to palm oil plantations.

In Indonesia, fire is commonly used for clearing land and for other agricultural purposes since it is cheap, effective, and efficient (Miettinen et al., 2017; Anderson and Bowen, 2000; Heil and Goldammer, 2001; Gaveau et al., 2014; Vadrevu et al., 2014a, b; Lestari et al., 2014). During the extreme dry periods, the intensive land clearing from fires with peatland degradation is beyond control.

The major concern of scientists, governments, and the public is the smoke generated from forests and peatlands (Vadrevu et al., 2018). Smoke, especially from peatland fire, contains greenhouse gas (GHG) emissions and aerosols, which could affect climate, degrade air quality, impair visibility, and endanger human health. Smoke, which is commonly called "haze" in the Southeast Asian region, could also be transported thousands of kilometers to other areas and countries, thus causing transboundary air pollution, which further raises political tensions and impacts human health (Justice et al., 2015; Vadrevu and Lasko, 2015). Satellite data can be used extensively to capture the biomass burnt areas and the smoke-related pollution (Prasad et al., 2008; Vadrevu et al., 2012; 2013; Lasko et al., 2017). In this study, we present the historical fire and climate interactions and impact of forest and land fires on air quality, including some social and economic aspects. Emissions specific to the particulate matter and characterization of aerosols from biomass burning of forests and peatlands are also discussed.

HISTORY AND BACKGROUND OF FIRE IN INDONESIA

Meteorological forcing from the Australian continent during the Austral winter creates dry atmospheric conditions in the region. In this study, we present the connection between fire hotspots in Sumatera and Kalimantan and the El Niño–Southern Oscillation (ENSO) related to the influence of climate on the smoke haze distribution.

During El Niño (La Niña), colder (warmer) sea surface temperature (SST) brings less (more) evaporation, and thus drier (wetter) atmosphere that hinders (inhibit) fire incidences. This phenomenon is clearly seen from Figure 14.1a and b, where the SST over southeast Kalimantan as the primary source of air masses comes from there for

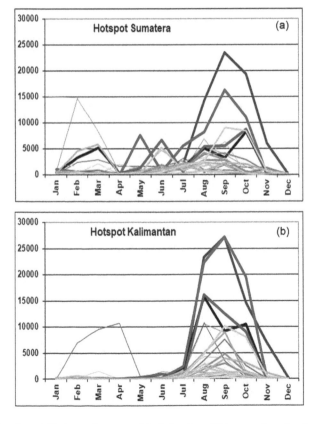

FIGURE 14.1   (a and b) Annual or monthly behavior of biomass burning in Sumatera and Kalimantan, data from 1997 to 2018.

Kalimantan in the dry season (August, September, and October). SST's importance indicates the amount of water vapor supply, as the cooler the SST, the less amount of water vapor. In Sumatera, most high-fire hotspot months are also in these months.

In Sumatera and Kalimantan, there is a large peak of fire hotspots during August and September, especially during El Niño years with the most extensive fire hotspots during 1997, 2006, 2004, and 2015; the drought prevails, and the dry atmosphere occupies most of the archipelago. The large peak during early 1998 over Kalimantan is due to prolonged El Niño episode 1997/1998 that lasted until May 1998. The major differences between fire hotspots over Sumatera and Kalimantan occur during the first half of the year when there are some peak fire hotspots over Sumatera around February and March. Interestingly, in Riau Province, fire hotspots during February and March 2005 contributed to 86% and 77% of total hotspot over Sumatera for those months, respectively. A close investigation suggests that the increase of fire hotspots in certain years is due to a strong cold surge episode that mainly takes place in February and March. People start to burn after one week of the dry period due to cold surge episode. The cold surge comes from dry and cold Siberia crossing the South China Sea (meridional wind). It brings a dry and cold atmosphere that cools the SST, brings dry air to Riau Province directly, and increases fire hotspots. In fact, during January 2005, the SST over the South China Sea stayed below 25°C for more than three weeks as the longest and lowest record for the last decade.

Meanwhile, the meridional wind for that period was also one extreme with an average for three consecutive weeks, the lowest 7 m/s southward. As a result, there were 13,613 fire hotspots in January and February 2005 alone. In contrast, in 1998, there was no fire in Riau Province, the SST was the highest, and the meridional wind was the weakest in the southward direction. Although many believe that forest fire is human-induced, there is a relatively strong link between meteorological conditions over the Pacific and the number of fire hotspots over that year. The SST anomalies represent the ENSO activities since the anomaly index is the NINO$_3$ index as well. From this empirical relationship, it is evident that forest fire events that occur in Sumatera and Kalimantan are induced or forced by the event over the Pacific. During the strong El Niño cases or El Niño years, such as in 1997, 2006, and 2015, there were strong fire hotspot episodes when the Pacific temperature increases. A strong connection between El Niño and fires can be seen between 1997 and 2004. However, that number drops quite a bit when we include 2005, 2006, or 2007 in our analyses. The most probable reason for such a drop is non-Pacific phenomenon such as the cold winter surge in early 2005.

There is evidence that cold surge plays a more significant role in the first half over Sumatera. Indication of the monthly behavior of fire hotspots is given in Figure 14.1. In the figure, most fires occur during August, September, and October in Sumatera and Kalimantan. Interestingly, the number of fire hotspots shows interannual variability related to the climatic factor; however, the hotspot numbers during 2005 and 2006 seem to deviate from the relationship. The number of fire hotspots during 2006 is very close (92.8%) to that of 1997 during an extreme El Niño year. From our fire hotspot history (Figures 14.2 and 14.3), there is a decrease in hotspot number after 2015 or after the largest and last El Niño year. In all provinces, the decrease in the number of fire hotspots is also reflected in the monthly hotspot figures.

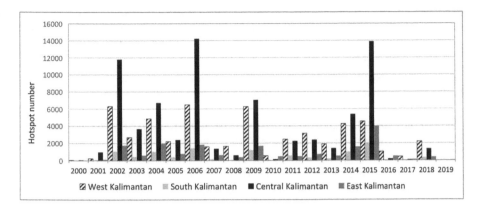

FIGURE 14.2 Hotspot distribution over Kalimantan Island. Hotspot count was derived from MODIS satellite Terra and Aqua with the level of confidence above 80%.

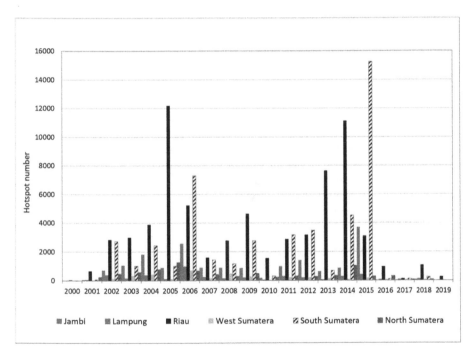

FIGURE 14.3 As in Figure 14.2, but for Sumatera Island.

The three largest fire hotspot numbers are, therefore, the years 2007, 2015, and 2006. Furthermore, from Figures 14.2 and 14.3, the three largest provinces contributing to the fires are Central Kalimantan, South Sumatera, and Riau provinces.

The peatlands of SEA lie within the intertropical convergence zone (ITCZ) that experiences a wet tropical climate with annual rainfall in excess of 2500 mm and two peaks and two troughs of rainy seasons. Seasonality of rainfall is not usually marked, but there is either a long "wet" season of 9–10 months (rainfall exceed evaporation) alternating with a shorter "dry" season of 2 or 3 months' duration (rainfall below evaporation) or two "monsoon" seasons (October–March and April–August) interspersed by two short "dry" periods (Rieley and Page, 2005).

Peatland areas during the last several decades have experienced extensive disturbance where land clearing, large-scale deforestation, and drainage drying have been occurring extensively (Page et al., 2009). This disturbance has led to massive fires in peatlands, which have been the main cause of haze episodes in SEA. According to Harrison et al. (2009), the main cause of peatland fires in SEA was illegal land-clearing activity with fire's cheap use. Often, the fires will become out of control and then spread to larger areas. Degraded peatlands are known to be susceptible to fire, especially during the dry season (Wösten et al., 2008). High organic matter in peatland, as either decomposed material or material continuing to decompose, is the factor that makes peatlands even more prone to fires (Usup et al., 2000). The exploitation of peatlands also reduces the water table in these areas, and lowering the water table increases the frequencies and extent of peat fires (Evers et al., 2017; Turetsky et al., 2015).

Peat is easily flammable in dry conditions, and when it burns, smoldering combustion will take place with fire underground (Budisulistiorini et al., 2017; Zaccone et al., 2014). Smoldering combustion is a process when an organic soil like peat burns steadily without flames and the burning slowly permeates into the soil (Rein et al., 2008). Smoldering combustion can last for several days or even weeks under low-temperature, low-oxygen concentration, and high-moisture content conditions (Turetsky et al., 2015). Figure 14.4 shows the annual distribution of fire hotspots in Jambi Province, central Sumatera. From 2005 onward, episodes of heavy fire occurred during 2006 and 2015. Fire incidents occurred in all places except in high mountain areas southwest of Jambi Province. This province shows the large shift of land-use change due to biomass burning. Jambi is also the province where the AERONET/Aerosol Robotic Network was installed to monitor the impact of air pollution due to biomass burning.

HAZE POLLUTION IN INDONESIA

Fire emissions affect many aspects of human life, as well as the environment. Smoke from forest and land fires is a significant source of GHGs and aerosols. Several trace gases and aerosols generated from peat fires burning are composed of CO_2, CO, CH_4, nonmethane hydrocarbons, oxygenated organic compounds, NH_3, HCN, NO_x, OCS, HCl, black carbon, and organic carbon (Stockwell et al., 2016; Heil and Goldammer, 2001).

Haze pollution is characterized by impairment in horizontal visibility and increased ambient air pollution levels due to high aerosol loading (Heil, 2007; Vadrevu et al., 2014a, b). According to the Association of Southeast Asian Nations Secretariat, haze consists of sufficient smoke, dust, moisture, and vapor suspended in the air to impair visibility. The terminology of haze also refers to a phenomenon that occurs when a sufficient concentration of aerosols in the atmosphere scatter visible light, and this results in a measurable reduction in visual range (Seinfeld and Pandis, 2006; Vadrevu and Lasko, 2018; Lasko and Vadrevu, 2018; Latif et al., 2018).

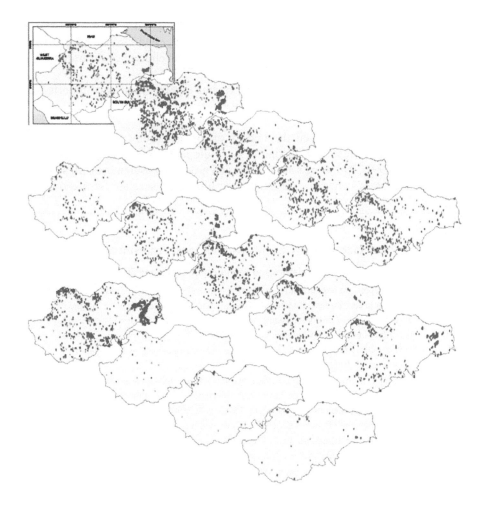

FIGURE 14.4   Example of fire episodes in Jambi Province in the middle of Sumatera from 2005 (upper left) down to 2017.

Aerosols are a complex mixture of organic and inorganic particles suspended in the atmosphere in the form of solid, liquid, or gaseous states (Gupta et al., 2006). Once airborne, aerosols undergo a physicochemical process, determining their concentration, composition, deposition, removal, and transport (Lazaridis, 2011; Seinfeld and Pandis, 2006). Aerosols could modify solar irradiation in both direct and indirect ways, thus affecting the climate system (Langmann et al., 2009; Boucher, 2015). Aerosols perturb the atmospheric radiation budget by scattering and absorbing solar radiation as part of their direct effect. Aerosols' indirect impact on climate is acting as cloud condensation nuclei (Hobbs and Radke, 1969; Seinfeld and Pandis, 2006) and, hence, modifying the microphysical and optical properties of clouds. Aerosol particles could interact with clouds, resulting in modifications of precipitation processes (Langmann et al., 2009; Lazaridis, 2011). Understanding aerosols' chemical and optical properties is important to determine their impact on regional air quality and health.

Air pollution degradation due to fires has been becoming more frequent in Indonesia since the last three decades. This pollution called "smoke haze" affects Indonesia and neighboring countries since the aerosols emitted could transport far way across boundaries.

Modified from Heil (2007), Table 14.1 presents several studies and reports from scientists, organizations, and media on smoke haze episodes in SEA from the 1960s to the most recent 2015.

Numerous studies addressed the impact of fire emissions on the atmospheric composition and air quality in Indonesia since the late 1990s after extreme smoke haze events in 1997/1998 using satellite data and ground measurements. Most studies emphasized aerosols' concentration and composition (Hayasaka and Sepriando, 2018;

TABLE 14.1
Reported Smoke Haze and Fire Events in Southeast Asia since 1960

Year	Reported Haze Episodes and Evidences of Corresponding Fire Events
1961	Written reports on haze are missing. There was most likely a haze episode over eastern parts of Sumatra and northern Borneo between July and November 1961 with impaired visibility. Written reports of fires are missing but are likely, because of abnormally dry conditions prevailing in 1961 (Chokkalingam et al., 2007).
1963	Fires in Indonesia. Visibility data show haze in Sumatra as well as in Singapore (Taylor et al., 1999).
1965	No written reports on haze. There was most likely a haze episode over eastern parts of Sumatra and Borneo between August and November 1963, as indicated by impaired visibility. Significant fires in Indonesia (Taylor et al., 1999).
1967	Haze in Singapore (Taylor et al., 1999). No written reports on fires.
1972	Severe haze over the eastern parts of Sumatra, Kalimantan, entire Malaysia, and Singapore between August and October 1972, leading to disruptions in air traffic (Brookfield et al., 1995, Taylor et al., 1999). Significant fires in southeastern Sumatra and the southern part of Borneo (Taylor et al., 1999).
1982	Severe haze stretching over Indonesia to Borneo and Peninsular Malaysia from September to December 1982. Daily mean particle concentrations (total suspended particles (TSP)) in Kuala Lumpur (Malaysia) ranged between 160 and 680 µg/m³ compared to a typical range of 50–80 µg/m³ (Aidid, 1985). Most of the fires occurred on Borneo, notably in East and West Kalimantan and Sarawak. Large fires also occurred in southeastern Sumatra (Taylor et al., 1999). In total, ~1 Mha burned in the year 1982 (Goldammer and Seibert, 1990, Zoumas et al., 2004).
1983	Severe haze over entire Malaysia in April 1983 (Soleiman et al., 2003) and September 1983 (Radojevic, 2003). Major fires occurred from March to May 1983. Most of the fires occurred on Borneo, notably in East Kalimantan, Sarawak and Sabah, and Brunei. In total, ~5 Mha burned in 1982/1983, of which around 4 Mha burned in 1983. Of the total burned area, an estimated 3.5 Mha burned in East Kalimantan only, 16% in peat swamp areas (Goldammer and Seibert, 1990).
1987	Extensive haze in September 1987 (Radojevic, 2003). Haze in Sumatra (Wang et al., 2004). Large fires in Indonesia exceeded the normal annual burning. As per official estimates, ~0.05 Mha burned (Makarim et al., 1998).

(Continued)

TABLE 14.1 (*Continued*)
Reported Smoke Haze and Fire Events in Southeast Asia since 1960

Year	Reported Haze Episodes and Evidences of Corresponding Fire Events
1991	Severe haze over entire Peninsular Malaysia and western Sarawak, notably in September and October 1991. For greater Kuala Lumpur region, the haze episode was considered to be unprecedented in severity and persistence. In the latter, daily mean particle concentrations (TSP) generally ranged between 200 and 570 $\mu g/m^3$ compared to a typical range of 50–80 $\mu g/m^3$ (Soleiman et al., 2003). Haze in Singapore during August–November 1991 with weekly mean particle concentration (TSP) range of 60–80 $\mu g/m^3$ compared to a typical range of 40–50 $\mu g/m^3$ (Chew et al., 1999). Large fires in Indonesia (Sumatra and Kalimantan) exceeded the normal annual burning. As per official estimates, ~0.12 Mha burned (Makarim et al., 1998).
1994	Severe haze over Indonesia, Malaysia, Brunei, and Singapore during August–October 1994. The maximum extent of the smoke haze layer was at least 3 million km^2 (Nichol, 1998). In Kuala Lumpur, TSP values were 2–4 times higher than the long-term mean. The highest daily value was 464 $\mu g/m^3$ (Soleiman et al., 2003). Severe haze observed during September and November 1994 in Kalimantan and Sumatra (Field et al., 2016). Severe haze also in Singapore during August and November 1994 with weekly mean particle concentrations (PM_{10}) between 50 and 90 $\mu g/m^3$ compared to a typical range of 30–50 $\mu g/m^3$ (Chew et al. 1999). The maximum PM_{10} concentration in Singapore was 255 $\mu g/m^3$ (Nichol, 1998). During the entire period, strong enhancements in tropospheric ozone were observed in Southeast Asia (Fujiwara et al., 1999). Large fires in the southern part of Sumatra and Kalimantan. Burned area estimates from literature are contradictory. According to the Nature Conservation Service of the Republic of Indonesia (PHPA), in total, 0.16 Mha was burned (Makarim et al., 1998). In contrast, Dennis (1999) and Goldammer and Hoffmann (2001) estimate 0.41–5.1 Mha as burned.
1997	Severe haze over Sumatra and Borneo during August and November 1997 with daily TSP levels generally between 150 and 4000 $\mu g/m^3$ compared to a normal background concentration of below 100 $\mu g/m^3$ TSP. Severe haze also affected Peninsular Malaysia and Singapore during August and October 1997, with daily TSP levels of 100 and 500 $\mu g/m^3$ (Heil and Goldammer, 2001). Major fires in southeastern Sumatra and the southern part of Borneo during July and November 1997. Officially, 0.3 Mha was declared as burned; however, scientific estimates vary between 4.6 and 13.6 Mha.
1998	Severe haze over entire Borneo and Peninsular Malaysia during February and April 1998. In Brunei, PM_{10} concentrations were ~50 and 650 $\mu g/m^3$, compared to a normal mean of 50 $\mu g/m^3$ (Yadav et al., 2003). Major fires in East Kalimantan from February to April 1998 burned a total area of around 5.2 Mha (Siegert and Hoffmann, 2000).
2000	Haze in Singapore from March to July 2000. Daily particle levels ($PM_{2.5}$) increased to 40–70 $\mu g/m^3$ compared to 10–40 $\mu g/m^3$ during non-haze episodes (Balasubramanian et al., 2003). Fires in Sumatra burned around 0.2–0.4 Mha during this period.
2001	Haze over Sumatra, Singapore, and parts of Peninsular Malaysia up to southern Thailand in July 2001. Visibility was down to 500 m in several areas due to haze. In some areas, schools were closed, and people were advised to wear protective masks when going outdoors. Fires in Sumatra in July burned ~0.08–0.1 Mha.

(Continued)

TABLE 14.1 (*Continued*)

Reported Smoke Haze and Fire Events in Southeast Asia since 1960

Year	Reported Haze Episodes and Evidences of Corresponding Fire Events
2002	Severe smoke haze in southern and western Kalimantan during August–November 2002. In Palangkaraya, the monthly mean PM_{10} concentration in September and October was 790 and 630 $\mu g/m^3$. Daily maxima of above 1300 $\mu g/m^3$ PM_{10} were frequently observed. In Pontianak, monthly mean PM_{10} concentration was ~300 $\mu g/m^3$ in July and August, with daily maxima reaching 850 $\mu g/m^3$. Also, in northern Sumatra, severe smoke haze conditions were observed. In Medan, monthly PM_{10} concentration increased to >450 $\mu g/m^3$. Typically, monthly mean PM_{10} concentrations are between 50 and 100 $\mu g/m^3$ (KLH, 2003). Moderate haze in Singapore from August to October 2002 with PM_{10} levels reaching 133 $\mu g/m^3$. Major fires in Kalimantan burning 1.2–1.5 Mha from August to November 2002.
2003	Severe haze during May–August over central Sumatra, namely the province Riau, in June 2003. The smoke haze also affected areas on the west coast of Peninsular Malaysia from Kuala Lumpur northward for a few days in June and August 2003. The smoke haze also reached southern Thailand. In southern Thailand, PM_{10} levels increased to twice the normal background concentration on a few days in June and August 2003 (i.e., between 90 and 110 $\mu g/m^3$ compared to a typical background concentration of between 30 and 50 $\mu g/m^3$) (Pentamwa, 2006). Light haze occurred in Singapore in June. Enhanced fire activity in Sumatra, notably in Riau, in June 2003, burning in total around 0.1–0.3 Mha.
2004	Slight-to-moderate haze in the months of June to September 2004 along the west coast of Peninsular Malaysia, mainly due to fires in Sumatra. The smoke plumes also reached southern Thailand for a few days in June, July, and August 2004. Particle concentrations rose to twice the typical background concentration of about 30–50 $\mu g/m^3$ PM_{10} (Pentamwa, 2006). Large fires in Sumatra during June–September 2004 burning in total around 0.2–0.7 Mha. Large fires also in Kalimantan during August–October 2004 burning in total around 0.4–0.5 Mha.
2005	Severe haze over Sumatra and Northern Peninsular Malaysia, no haze in Singapore during July to August 2005. The Malaysian government declared a state of emergency in August 2005 in two cities where the air pollution levels exceeded hazardous according to the Malaysian air quality classification. Schools in the capital Kuala Lumpur were closed, and people have been advised to stay indoors and wear masks if they go outside. The smoke plumes also reached southern Thailand for a few days in July and August 2005. Particle concentrations rose to twice the typical background concentration of about 30–50 $\mu g/m^3$ PM_{10} (Pentamwa 2006). Large fires in Sumatra during this July to August, burning in total around 0.2–0.5 Mha.
2006	Severe haze over Sumatra and Kalimantan from August to November 2006. Haze affected the neighboring countries of Singapore, Malaysia, and Brunei in October 2006. In Malaysia, PM_{10} levels were between 150 and 300 $\mu g/m^3$ PM_{10} on several days, while in Singapore, PM_{10} levels reached 161 $\mu g/m^3$. Large fires in Sumatra from August to October 2006 burning in total around 0.6–1.1 Mha. Abnormally large fires in Kalimantan during this period, namely in October 2006, burning around 0.8–1.6 Mha.
2009	Kalimantan and some provinces in Sumatera Island. According to the Indonesian non-governmental organization WALHI, land burning in total was 3626.4 ha (https://ekonomi. kompas.com/read/2009/09/08/15440098/walhi.3.626.ha.hutan.dan.lahan.terbakar).

(Continued)

TABLE 14.1 (*Continued*)

Reported Smoke Haze and Fire Events in Southeast Asia since 1960

Year	Reported Haze Episodes and Evidences of Corresponding Fire Events
2013	Forest and land fires in Riau Province in Sumatera during June caused high political tension due to transboundary smoke haze pollution in Singapore and Malaysia. According to Kusumaningtyas and Aldrian (2016), the Pollutant Standards Index (PSI) in Riau reached 175. Furthermore, the PSI in Singapore was 206, and $PM_{2.5}$ concentration far exceeded the ambient air quality standards, which was 244.89 µg/m^3.
2014	Riau Province in Sumatera experienced forest and land fires from February to March 2014 with more than 1200 fire hotspots. The Ministry of Environment of the Republic of Indonesia recorded that daily PSI rose to >500, indicating a hazardous level (Mulyana, 2014). Several national mass media reported the haze caused schools' closure, disrupted flights, and acute respiratory illness for more than 45,000 people (https://www.bbc.com/indonesia/berita_indonesia/2014/03/140313_kabut_asap_riau_kesehatan).
2015	Massive fires, coupled with El Niño in Indonesia from July to October, ranked as one of the worst after 1997, have devastated more than 2 Mha of land (Lohberger et al., 2017). South Sumatera became one of the worst provinces and suffered from the thick haze, followed by Central Kalimantan. According to the Agency for Meteorology, Climatology, and Geophysics of Indonesia (BMKG), PM_{10} peaked at 3145 µg/m^3. The haze extended to Singapore, Malaysia, and Thailand and caused millions of people exposed to hazardous low air quality for 2 months.

Source: Data from 1961 to 2006 are modified from Heil (2007).

Fujii et al., 2014; Hayasaka et al., 2014) both through a peat-fire sampling of satellite measurement. Nevertheless, a study on the impact of peat fire on air quality and some aspects of life (social and economic) using a combination of ground-based data, satellite measurement, meteorological data (rainfall and visibility), and secondary data is still scarce. Here, we present some evidence of air pollution decay due to forest and land fires over Sumatera and Kalimantan.

During 2013, thick haze pollution has blanketed Riau, Sumatera, and affected Singapore, Malaysia, and even the Philippines. In June, the Pollutant Standards Index (PSI) in Riau reached 175, indicating hazardous conditions (Kusumaningtyas and Aldrian, 2016). More than 3000 fire hotspots were found only in June when the lowest rainfall was observed. During the dry season (June–September), the prevailing southerly and southwesterly winds will blow from the south to north, transporting the fire emissions from Sumatera cross-boundary toward Singapore and Malaysia. The fires intensified during an extreme drought in 2015. Unlike any other fire-prone areas in Indonesia, Riau frequently suffers from fire from June to September and February to April.

Poor air quality in Singapore has been perceived as an impact of fire in Riau. The National Environmental Agency of Singapore (NEA) recorded a high concentration of PM_{10} with 244.89 µg/m^3 and beyond exceeding the ambient air quality standards in Singapore, including the World Health Organization's (2006) 25.89 µg/m^3. A significant correlation between the fire hotspots and PM_{10} concentrations was reported by several researchers (Gaveau et al., 2014; Anwar et al., 2010). Kusumaningtyas and Aldrian (2016) calculated a correlation $(r=0.483; p=0.023)$ between the number of fire hotspots detected and $PM_{2.5}$ in Riau with transboundary haze impacting Singapore. Artaxo et al. (1994) also concluded that an increase in the concentration of small particulate matter (diameter $< 2.00\,\mu m$) in the dry season is the result of biomass burning.

Figure 14.5 exhibits smoke trajectory using Hybrid Single-Particle Lagrangian Integrated Trajectory (HYSPLIT) model. This model is used to predict the course, dispersion, and concentration of pollutants from the point, line, and area sources. Figure 14.5 confirms the propagation and transportation of smoke haze from Riau toward Singapore (Kusumaningtyas and Aldrian, 2016).

Aouizerats et al. (2015) modeled a large fire-induced haze episode in 2006 stemming mostly from Indonesia using the Weather Research and Forecasting model coupled with chemistry (WRF-Chem) to simulate and analyze the aerosol particle emission and evolution due to biomass burning in Sumatera. They concluded that 21% of the total aerosol concentration was due to biomass burning in Sumatra during the four months of the simulation, and 48% when focusing on two weeks in October when smoke reaching Singapore was most intense.

Another perceived impact of air pollution is visibility degradation. According to the World Meteorological Organization (WMO), visibility, meteorological visibility (by day), and meteorological visibility at night are defined as the greatest distance at which a black object of suitable dimensions (located on the ground) can be seen and recognized when observed against the horizon sky during daylight or could be seen and recognized during the night if the general illumination were raised to the normal daylight level (WMO, 1992; 2010). The absorption and scattering of light reduce

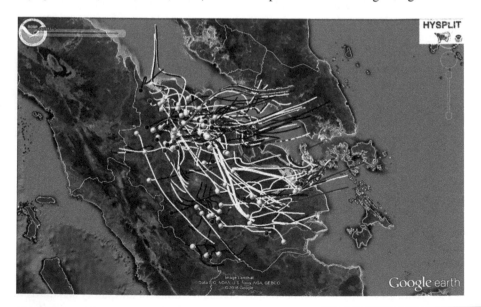

FIGURE 14.5 HYSPLIT forward trajectory ran 24 hours from June 1 to June 30, 2013, over Riau (Kusumaningtyas and Aldrian, 2016)

visibility by both gas and particles. The higher the aerosol loading, the stronger the impact on atmospheric transitivity, due to scattering and absorbing of incoming direct and diffuse solar radiation (Solomon et al., 2007; Jacobson, 2004; Seinfeld and Pandis, 2006). However, particles' light scattering is the most important phenomenon responsible for impairment of visibility (Seinfeld and Pandis, 2006). Declining visibility coinciding with the increase of $PM_{2.5}$ concentrations was recorded in Singapore due to fires in Riau, Sumatera, in June 2013. During hazy days, the lowest visibility observed was 1.01 km when the $PM_{2.5}$ concentration reached a peak at 266.62 μg/m^3 (Kusumaningtyas and Aldrian, 2016).

Characterization of aerosols' physical properties is important to describe their behavior about transport, sources, removal, and attribution to regional air quality. AERONET in situ measurements (Holben et al., 1998) provide an excellent and comprehensive database of aerosol properties. AERONET is a ground-based aerosol monitoring system that offers standardization for a ground-based regional-to-global-scale aerosol monitoring and characterization network and data archive. The AERONET is useful to study transport and radiation budget studies, radiative transfer modeling, improvement assessment of the impact of aerosols on climate forcing, and validation for aerosol information retrieval from satellites (Asmat et al., 2018). Furthermore, aerosol properties derived from AERONET could describe the source of the aerosols. To measure direct solar and diffuse sky radiation at several wavelengths, AERONET is using sun-photometer instruments to produce aerosol optical depth (AOD), which is the primary product.

AERONET in Indonesia was able to capture several haze events in Indonesia for several years. AERONET measurement in Palangkaraya, Central Kalimantan, showed a high AOD up to 6.0 attributed to peatland fires (Kusumaningtyas et al., 2016). Furthermore, during peak fire in 2014, they found that AOD above 3.0 can result in the reduction of visibility below 500 m (Figure 14.6)

Another research by Eck et al. (2019) characterized optical properties of aerosols emitted from the peat burning region of Central Kalimantan during September–October 2015 from the newly developed AERONET Version 3 algorithms (Giles et al., 2019) and reported AOD at 550 nm reaching ~11–13 during this event. According to them, these are the highest AODs ever estimated from the AERONET data. They also found that the single-scattering albedo (SSA) of the Indonesian biomass burning aerosols was too high in 2015, averaging ~0.975 with little wavelength dependence showing the smoke originated from peat fuel burning (smoldering combustion).

CONCLUSIONS

In this chapter, we presented the historical perspective of biomass burning in Indonesia. In Indonesia, biomass burning events are common in peatlands and rain forests, impacting the regional air quality. The annual behavior of the biomass burning events is closely related to climatology. Although most biomass burning is human-initiated, there is a high correlation with climate, as shown in our study, i.e., between the number of fire hotspots and the ENSO. The second phenomenon is the cold surge in the early year (January–March), which affects burning behavior in most of Sumatera and part of West Kalimantan.

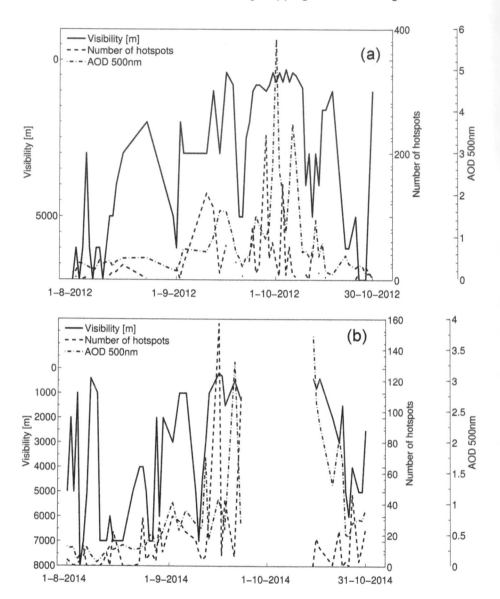

FIGURE 14.6 Relationship between monthly AOD, hotspot, and visibility. AOD at wavelength 500 nm and visibility versus AOD and hotspot in 2012 (a) and 2014 (b) fire episodes (notice gap of data during the peak of fire in 2014).

Biomass burning in Sumatera and Kalimantan has been degrading the air quality, including visibility, as evidenced by the AERONET data. Satellite remote sensing has proven to be a useful tool to monitor the air quality and the smoke/haze distribution. We infer that integrating atmospheric chemistry models and satellite data can aid in better monitoring of pollutants and aerosols from biomass burning in the region.

REFERENCES

Aidid, S.B. 1985. Multi-element studies of air particulates collected during the hazy periods in Kuala Lumpur, Malaysia. *Malaysian Journal of Physics.* 6(71), 71–80.

Anderson, I.P. and Bowen, M.R. 2000. Fire zones and the threat to the wetlands of Sumatra, Indonesia Forest Fire Prevention and Control Project www.fire.uni-freiburg.de/se_asia/projects/.

Anwar, A., Juneng, L., Othman, M.R. and Latif, M.T. 2010. Correlation between fire hotspots and air quality in Pekanbaru, Riau, Indonesia in 2006–2007. *Sains Malaysiana.* 39, 169–74.

Aouizerats, B., Van Der Werf, G.R., Balasubramanian, R. and Betha, R. 2015. Importance of transboundary transport of biomass burning emissions to regional air quality in Southeast Asia during a high fire event. *Atmospheric Chemistry & Physics.* 15(1), 363–373.

Artaxo, P., Gerab, F., Yamasoe, M.A. and Martins, J.V. 1994. Fine mode aerosol composition at three long-term atmospheric monitoring sites in the amazon basin. *Journal of Geophysical Research Atmospheres D.* 99, 22857–22868.

Asmat, A., Jalal, K.A. and Deros, S.N.M. 2018. Aerosol properties over Kuching, Sarawak from satellite and ground-based measurements. In: *Land-Atmospheric Research Applications in South and Southeast Asia.* Vadrevu, K.P., Ohara, T., and Justice, C. (Eds). Springer, Cham. pp. 447–469.

Badarinath, K.V.S., Kharol, S.K., Krishna Prasad, V., Kaskaoutis, D.G. and Kambezidis, H.D. 2008a. Variation in aerosol properties over Hyderabad, India during intense cyclonic conditions. *International Journal of Remote Sensing.* 29(15), 4575–4597.

Badarinath, K.V.S., Kharol, S.K., Prasad, V.K., Sharma, A.R., Reddi, E.U.B., Kambezidis, H.D. and Kaskaoutis, D.G. 2008b. Influence of natural and anthropogenic activities on UV Index variations–a study over tropical urban region using ground based observations and satellite data. *Journal of Atmospheric Chemistry.* 59(3), 219–236.

Balasubramanian, R., Qian, W-B, Decesari, S., Facchini, M.C. and Fuzzi, S. 2003. Comprehensive characterization of $PM_{2.5}$ aerosols in Singapore. *Journal of Geophysical Research.* 108(D169), 4523, doi:10.1029/2002JD002517.

Biswas, S., Vadrevu, K.P., Lwin, Z.M., Lasko, K. and Justice, C.O. 2015a. Factors controlling vegetation fires in protected and non-protected areas of Myanmar. *PLoS One.* 10(4), e0124346.

Biswas, S., Lasko, K.D. and Vadrevu, K.P. 2015b. Fire disturbance in tropical forests of Myanmar—Analysis using MODIS satellite datasets. *IEEE Journal of Selected Topics in Applied Earth Observations and Remote Sensing.* 8(5), 2273–2281.

Boucher, O. 2015. *Atmospheric Aerosols: Properties and Climate Impacts.* Springer, Dordrecht.

Brookfield, H., Potter, L. and Byron, Y. 1995. *In Place of the Forest. Environmental and Socio-economic Transformation in Borneo and the Eastern Malay Peninsula.* Vol. 893. United Nations University Press, Tokyo.

Budisulistiorini, S.H., Nenes, A., Carlton, A.G., Surratt, J.D., McNeill, V.F. and Pye, H.O.T. 2017. Simulating Aqueous-Phase Isoprene-Epoxydiol (IEPOX) Secondary Organic Aerosol Production During the 2013 Southern Oxidant and Aerosol Study (SOAS). *Environmental Science & Technology.* 51(9), 5026–5034.

Chew, F.T., Goh, D.Y.T., Ooi, B.C., Saharom, R., Hui, J.K.S. and Lee, B.W. 1999. Association of ambient air-pollution levels with acute asthma exacerbation among children in Singapore. *Allergy.* 54(4), 320–329.

Chokkalingam, U., Permana, R.P., Kurniawan, I., Mannes, J., Darmawan, A., Khususyiah, N. and Susanto, R.H. 2007. Community fire use, resource change, and livelihood impacts: The downward spiral in the wetlands of southern Sumatra. *Mitigation and Adaptation Strategies for Global Change.* 12(1), 75–100.

Crutzen, P.J. and Andreae, M.O. 1990. Biomass burning in the tropics: Impact on atmospheric chemistry and biogeochemical cycles. *Science*. 250, 1669–1678.

Dennis, R.A. 1999. A review of fire projects in Indonesia 1982–1998. Center for International Forestry Research, Bogor, Indonesia. *Science*. 250, 1669–1678.

Eck, T.F., Holben, B.N., Giles, D.M., Slutsker, I., Sinyuk, A., Schafer, J.S., Smirnov, A., Sorokin, M., Reid, J.S., Sayer, A.M. and Hsu, N.C. 2019. AERONET remotely sensed measurements and retrievals of biomass burning aerosol optical properties during the 2015 Indonesian burning season. *Journal of Geophysical Research: Atmospheres*. 124(8), 4722–4740.

Evers, S., Yule, C.M., Padfield, R., O'Reilly, P. and Varkkey, H., 2017. Keep wetlands wet: The myth of sustainable development of tropical peatlands–implications for policies and management. *Global Change Biology*. 23(2), 534–549.

Field, R.D., van der Werf, G.R., Fanin, T., Fetzer, E.J., Fuller, R., Jethva, H., Levy, R., Nathaniel, J., Livesey, J., Luo, M., Torres, O. and Worden, H.M. 2016. Indonesian fire activity and smoke pollution in 2015 show persistent non-linear sensitivity to El Niño-induced drought. *Proceedings of the National Academy of Sciences*. 113(33), 9204–9209.

Fujii, Y., Iriana, W., Oda, M., Puriwigati, A., Tohno, S., Lestari, P., Mizohata, A. and Huboyo, H.S. 2014. Characteristics of carbonaceous aerosols emitted from peatland fire in Riau, Sumatra, Indonesia. *Atmospheric Environment*. 87, 164–169.

Fujiwara, M., Kita, K., Kawakami, S., Ogawa, T., Komala, N., Saraspriya, S. and Suripto, A., 1999. Tropospheric ozone enhancements during the Indonesian forest fire events in 1994 and in 1997 as revealed by ground-based observations. *Geophysical Research Letters*. 26(16), 2417–2420.

Fuller, D.O., Jessup, T.C. and Salim, A. 2004. Loss of forest cover in Kalimantan, Indonesia, since the 1997–1998 El Nino. *Conservation Biology*. 18(1), 249–254.

Gaveau, D.L., Salim, M.A., Hergoualc'h, K., Locatelli, B., Sloan, S., Wooster, M., Marlier, M.E., Molidena, E., Yaen, H., DeFries, R. and Verchot, L. 2014. Major atmospheric emissions from peat fires in Southeast Asia during non-drought years: Evidence from the 2013 Sumatran fires. *Scientific Reports*. 4, 6112.

Giles, D.M., Sinyuk, A., Sorokin, M.G., Schafer, J.S., Smirnov, A., Slutsker, I., Eck, T.F., Holben, B.N., Lewis, J.R., Campbell, J.R. and Welton, E.J. 2019. Advancements in the Aerosol Robotic Network (AERONET) Version 3 database–automated near-real-time quality control algorithm with improved cloud screening for Sun photometer aerosol optical depth (AOD) measurements. *Atmospheric Measurement Techniques*. 12(1), 169–209.

Goldammer, J.G. and Hoffmann, A.A. 2001. Fire situation in Indonesia. In FAO (Food and Agriculture Organization) FRA Global Forest Fire Assessment 1990–2000. Forest Resources Assessment Programme, Working Paper 55. 132–144. FAO: Rome, Italy.

Goldammer, J.G and Seibert, B. 1990. The impact of droughts and forest fires on tropical low-land rain forest of Eastern Borneo. In *Fire in the Tropical Biota*. Ecosystem Processes and Global Challenges, Ecological Studies 84. Goldammer, J.G. (Ed). Springer-Verlag, Berlin-Heidelberg-New York. pp. 11–31.

Gupta, P.K., Prasad, V.K., Sharma, C., Sarkar, A.K., Kant, Y., Badarinath, K.V.S. and Mitra, A.P. 2001. CH_4 emissions from biomass burning of shifting cultivation areas of tropical deciduous forests–Experimental results from ground-based measurements. *Chemosphere-Global Change Science*. 3(2), 133–143.

Gupta, P., Christopher, S.A., Wang, J., Gehrig, R., Lee, Y.C. and Kumar, N. 2006. Satellite remote sensing of particulate matter and air quality assessment over global cities. *Atmospheric Environment*. 40(30), 5880–5892.

Harrison, M.E., Page, S.E., Limin, S.H. 2009, The global impact of Indonesian forest fires. *Biologist*. 56(3), 156–163.

Hayasaka, H. and Sepriando, A. 2018. Severe air pollution due to peat fires during 2015 super El Niño in Central Kalimantan, Indonesia. In: *Land-Atmospheric Research Applications in South/Southeast Asia.* Vadrevu, K.P., Ohara, T., and Justice, C., (Eds). Springer, Cham. pp. 129– 142.

Hayasaka, H., Noguchi, I., Putra, E. I., Yulianti, N. and Vadrevu, K. 2014. Peat-fire-related air pollution in Central Kalimantan, Indonesia. *Environmental Pollution.* 195, 257–266. doi:10.1016/j.envpol.2014.06.031.

Heil, A. 2007. Indonesian forest and peat fires: Emissions, air quality, and human health. *PhD diss.*, Univ of Hamburg.

Heil, A. and Goldammer, J.G. 2001. Smoke-haze pollution: A review of the 1997 episode in Southeast Asia. *Regional Environmental Change.* 2, 24–37.

Hobbs, P.V. and Radke, L.F. 1969. Cloud condensation nuclei from a simulated forest fire. *Science.* 163(3864), 279–280.

Holben, B.N., Eck, T.F., Slutsker, I.A., Tanre, D., Buis, J.P., Setzer, A., Vermote, E., Reagan, J.A., Kaufman, Y.J., Nakajima, T. and Lavenu, F., 1998. AERONET—A federated instrument network and data archive for aerosol characterization. *Remote Sensing of Environment.* 66(1), 1–16.

Israr, I., Jaya, S.N.I, Saharjo, H.S., Kuncahyo, B. and Vadrevu, K.P. 2018. Spatio-temporal analysis of land and forest fires in Indonesia using MODIS active fire dataset. In: *Land-Atmospheric Research Applications in South/Southeast Asia.* Vadrevu, K.P., Ohara, T., and Justice, C. (Eds). Springer, Cham. pp. 105–128.

Jacobson, M.Z. 2004. The short-term cooling but long-term global warming due to biomass burning. *Journal of Climate.* 17, 2909–2926.

Justice, C., Gutman, G. and Vadrevu, K.P. 2015. NASA land cover and land use change (LCLUC): An interdisciplinary research program. *Journal of Environmental Management.* 148(15), 4–9.

Kant, Y., Ghosh, A.B., Sharma, M.C., Gupta, P.K., Prasad, V.K., Badarinath, K.V.S. and Mitra, A.P. 2000. Studies on aerosol optical depth in biomass burning areas using satellite and ground-based observations. *Infrared Physics & Technology.* 41(1), 21–28.

KLH (Kementerian Negara Lingkungan Hidup). 2003. Status Lingkungan Hidup Indonesia 2002. Kementrian Lingkungan Hidup, Asdep Urusan Informasi, Jakarta, Indonesia. Available online at http://www.menlh.go.id/archive.php?action=info&id=25 (30 March 2004).

Kusumaningtyas, S.D.A. and Aldrian, E. 2016. Impact of the June 2013 Riau province Sumatera smoke haze event on regional air pollution. *Environmental Research Letters.* 11(7).

Kusumaningtyas, S.D.A., Aldrian, E., Rahman, M.A. and Sopaheluwakan, A. 2016. Aerosol properties in Central Kalimantan due to peatland fire. *Air Toxics Aerosol and Air Quality Research.* 16, 2757–2767. doi: 10.1088/1748-9326/11/7/075007

Langmann, B., Duncan, B., Textor, C., Trentmann, J. and van der Werf, G.R., 2009. Vegetation fire emissions and their impact on air pollution and climate. *Atmospheric Environment.* 43(1), 107–116.

Lasko, K. and Vadrevu, K.P. 2018. Improved rice residue burning emissions estimates: Accounting for practice-specific emission factors in air pollution assessments of Vietnam. *Environmental Pollution.* 236(5), 795–806.

Lasko, K., Vadrevu, K.P., Tran, V.T., Ellicott, E., Nguyen, T.T., Bui, H.Q. and Justice, C. 2017. Satellites may underestimate rice residue and associated burning emissions in Vietnam. *Environmental Research Letters.* 12(8), 085006.

Lasko, K., Vadrevu, K.P. and Nguyen, T.T.N. 2018a. Analysis of air pollution over Hanoi, Vietnam using multi-satellite and MERRA reanalysis datasets. *PLoS One.* 13(5), 0196629.

Lasko, K., Vadrevu, K.P., Tran, V.T. and Justice, C. 2018b. Mapping double and single crop paddy rice with Sentinel-1A at varying spatial scales and polarizations in Hanoi, Vietnam. *IEEE Journal of Selected Topics in Applied Earth Observations and Remote Sensing.* 11(2), 498–512.

Latif, M.T., Othman, M., Idirs, N., Juneng, L., Abdullah, A.M., Hamzah, W.P., Khan, M.F., Sulaiman, N.M.N., Jewaratnam, J., Aghamohammadi, N., Sahani, M., Xiang, C.J., Ahamad, F., Amil, N., Darus, M., Varkkey, H., Tangang, F. and Jaafar, A.B. 2018. Impact of regional haze towards air quality in Malaysia: A review. *Atmospheric Environment.* 177, 28–44.

Lazaridis, M. 2011. *First Principles of Meteorology and Air Pollution.* Springer, New York.

Lestari, R.K., Watanabe, M., Imada, Y., Shiogama, H., Field, R.D., Takemura, T. and Kimoto, M. 2014. Increasing potential of biomass burning over Sumatra, Indonesia induced by anthropogenic tropical warming. *Environmental Research Letters.* 9(10), 104010.

Lohberger, S., Stängel, M. and Atwood, E.C., et al. 2017. Spatial evaluation of Indonesia's 2015 fire affected area and estimated carbon emissions using Sentinel-1. *Global Change Biology.* 24(2), 644–654.

Makarim, N, Arba'I, Y.A., Deddy, A. and Brady, M. 1998. Assessment of 1997 land and forest fires in Indonesia: National coordination. *International Forest Fire News.* 18, 4–12.

Miettinen, J., Shi, C. and Liew, S.C. 2011. Deforestation rates in insular Southeast Asia between 2000 and 2010. *Global Change Biology.* 17, 2261–2270.

Miettinen, J., Hooijer, A., Vernimmen, R., Liew, S.C. and Page, S.E. 2017. From carbon sink to carbon source: Extensive peat oxidation in insular Southeast Asia since 1990. *Environmental Research Letters.* 12(2), 024014.

Mulyana, E. 2014. Bencana kabut asap akibat kebakaran hutan dan lahan serta pengaruhnya terhadap kualitas udara di Provinsi Riau Februari – Maret 2014 (Smoke haze disaster of land and forest fire in relation with air quality in Riau province during February - March 2014). *JSTI* 16(3). doi: 10.29122/jsti.v16i3.3417.

Nichol, J. 1998. Smoke haze event in Southeast Asia: A predictable recurrence. *Atmospheric Environment.* 32(14/15), 2715–2716.

Page, S., Hosciło, A., Wösten, H., Jauhiainen, J., Silvius, M., Rieley, J., Ritzema, H., Tansey, K., Graham, L., Vasander, H. and Limin, S., 2009. Restoration ecology of lowland tropical peatlands in Southeast Asia: Current knowledge and future research directions. *Ecosystems.* 12(6), 888–905.

Pentamwa, P. 2006. Air Quality in Southern Thailand during the 2005 Haze Episode. Yogyakarta, Indonesia. *Proceedings of the Better Air Quality (BAQ-2006) Conference,* 12 December 2006, Yogyakarta, Indonesia.

Prasad, V.K., Gupta, P.K., Sharma, C., Sarkar, A.K., Kant, Y., Badarinath, K.V.S., Rajagopal, T. and Mitra, A.P. 2000. NO_x emissions from biomass burning of shifting cultivation areas from tropical deciduous forests of India–estimates from ground-based measurements. *Atmospheric Environment.* 34(20), 3271–3280.

Prasad, V.K., Kant, Y., Gupta, P.K., Elvidge, C. and Badarinath, K.V.S. 2002. Biomass burning and related trace gas emissions from tropical dry deciduous forests of India: A study using DMSP-OLS data and ground-based measurements. *International Journal of Remote Sensing.* 23(14), 2837–2851.

Prasad, V.K., Lata, M. and Badarinath, K.V.S. 2003. Trace gas emissions from biomass burning from northeast region in India—estimates from satellite remote sensing data and GIS. *Environmentalist.* 23(3), 229–236.

Prasad, V.K., Badarinath, K.V.S., Yonemura, S. and Tsuruta, H. 2004. Regional inventory of soil surface nitrogen balances in Indian agriculture (2000–2001). *Journal of Environmental Management.* 73(3), 209–218.

Prasad, V.K., Badarinath, K.V.S. and Eaturu, A. 2008. Biophysical and anthropogenic controls of forest fires in the Deccan Plateau, India. *Journal of Environmental Management.* 86(1), 1–13.

Radojevic, M. 2003. Chemistry of forest fires and regional haze with emphasis on Southeast Asia. *Pure and Applied Geophysics.* 160, 157–187.

Rein, G., Cleaver, N., Ashton, C., Pironi, P. and Torero, J.L. 2008. The severity of smoldering peat fires and damage to the forest soil. *Catena*. 74(3), 304–309.

Rieley, J.O. and Page, S.E. 2005. Wise use of tropical peatlands: Focus on Southeast Asia. ALTERRA: Wageningen, the Netherlands) Available at http://www.restorpeat.alterra. wur.nl/download/WUG.pdf [Verified 28 May 2011].

Saharjo, B.H. and Yungan, A. 2018. Forest and land fires in Riau Province: A Case Study in Fire Prevention Policy Implementation with local concession holders. In: *Land-Atmospheric Research Applications in South/Southeast Asia*. Vadrevu, K.P., Ohara, T., and Justice, C. (Eds). Springer, Cham. pp. 143–170.

Seinfeld, J.H. and Pandis, S.N. 2006. *Atmospheric Chemistry and Physics, from Air Pollution to Climate Change 2nd*. Wiley, New York.

Siegert, F. and Hoffmann, A.A. 2000. The 1998 forest fires in East Kalimantan (Indonesia): A quantitative evaluation using high resolution, multi-temporal ERS-2 SAR images and NOAA-AVHRR hotspot data. *Remote Sensing Environment*. 72, 64–77.

Soleiman, A., Othman, M., Samah, A.A., Sulaiman, N.M. and Radojevic, M., 2003. The occurrence of haze in Malaysia: A case study in an urban industrial area. In G.V. Rao, Sethu Raman, M.P. Singh (Eds). *Air Quality*. Birkhäuser, Basel. pp. 221–238.

Solomon, S., Manning, M., Marquis, M. and Qin, D., 2007. Climate change 2007-the physical science basis: Working group I contribution to the fourth assessment report of the IPCC (Vol. 4). Cambridge university press.

Stockwell, C.E., Jayarathne, T., Cochrane, M.A., Ryan, K.C., Putra, E.I., Saharjo, B.H., Nurhayati, A.D., Albar, I., Blake, D.R., Simpson, I.J. and Stone, E.A. 2016. Field measurements of trace gases and aerosols emitted by peat fires in central Kalimantan, Indonesia, during the 2015 El Niño. *Atmospheric Chemistry and Physics*. 16(18), 11711–11732.

Taylor, D., Saksena, P., Sanderson, P.G. and Kucera, K. 1999. Environmental change and rain forests on the Sunda shelf of Southeast Asia: Drought, fire and the biological cooling of biodiversity hotspots. *Biodiversity and Conservation*. 8(9), 1159–1177.

Turetsky, M.R., Benscoter, B., Page, S., Rein, G., Van Der Werf, G.R. and Watts, A. 2015. Global vulnerability of peatlands to fire and carbon loss. *Nature Geoscience*. 8(1), 11.

Usup, A., Takahashi, H. and Limin, S.H. 2000. Aspect and mechanism of peat fire in tropical peat land: A case study in Central Kalimantan 1997. In *Proceedings of the International Symposium on Tropical Peatlands*. Bogor, Indonesia, Hokkaido University and Indonesian Institute of Science (pp. 79–88).

Vadrevu, K.P. 2008. Analysis of fire events and controlling factors in eastern India using spatial scan and multivariate statistics. *Geografiska Annaler: Series A, Physical Geography*. 90(4), 315–328.

Vadrevu, K.P. and Justice, C.O. 2011. Vegetation fires in the Asian region: Satellite observational needs and priorities. *Global Environmental Research*. 15(1), 65–76.

Vadrevu, K.P. and Lasko, K.P., 2015. Fire regimes and potential bioenergy loss from agricultural lands in the Indo-Gangetic Plains. *Journal of Environmental Management*. 148, 10–20.

Vadrevu, K.P. and Lasko, K. 2018. Intercomparison of MODIS AQUA and VIIRS I-Band fires and emissions in an agricultural landscape—Implications for air pollution research. *Remote Sensing*. 10(7), 978. doi:10.3390/rs10070978.

Vadrevu, K.P., Eaturu, A. and Badarinath, K.V.S. 2006. Spatial distribution of forest fires and controlling factors in Andhra Pradesh, India using spot satellite datasets. *Environmental Monitoring and Assessment*. 123(1–3), 75–96.

Vadrevu, K.P., Csiszar, I., Ellicott, E., Giglio, L., Badarinath, K.V.S., Vermote, E. and Justice, C. 2012. Hotspot analysis of vegetation fires and intensity in the Indian region. *IEEE Journal of Selected Topics in Applied Earth Observations and Remote Sensing*. 6(1), 224–238.

Vadrevu, K.P., Giglio, L. and Justice, C. 2013. Satellite based analysis of fire–carbon monoxide relationships from forest and agricultural residue burning (2003–2011). *Atmospheric Environment*. 64, 179–191.

Vadrevu, K.P., Ohara, T. and Justice, C. 2014a. Air pollution in Asia. *Environmental Pollution*. 12, 233–235.

Vadrevu, K.P., Lasko, K., Giglio, L. and Justice, C. 2014b. Analysis of Southeast Asian pollution episode during June 2013 using satellite remote sensing datasets. *Environmental Pollution*. 12, 245–256.

Vadrevu, K.P., Lasko, K., Giglio, L. and Justice, C. 2015. Vegetation fires, absorbing aerosols and smoke plume characteristics in diverse biomass burning regions of Asia. *Environmental Research Letters*. 10(10), 105003.

Vadrevu, K.P., Ohara, T. and Justice, C. 2017. Land cover, land use changes and air pollution in Asia: A synthesis. *Environmental Research Letters*. 12(12), 120201.

Vadrevu, K.P., Ohara, T. and Justice, C. (Eds). 2018. *Land-Atmospheric Research Applications in South and Southeast Asia*. Springer, Cham.

Vadrevu, K.P., Lasko, K., Giglio, L., Schroeder, W., Biswas, S. and Justice, C. 2019. Trends in vegetation fires in South and Southeast Asian countries. *Scientific Reports*. 9(1), 7422. doi:10.1038/s41598-019-43940-x.

Wang, Y., Field, R.D., and Roswintiarti, O. 2004. Trends in atmospheric haze induced by peat fires in Sumatra Island, Indonesia and El Niño phenomenon from 1973 to 2003. *Geophysical Research Letters*. 31, doi:10.1029/2003GL018853.

World Health Organization (WHO). 2006. Air quality guidelines for particulate matter, ozone, nitrogen dioxide and sulfur dioxide Global Update 2005: Summary risk assessment.

World Meteorological Organization. 1992. International Meteorological Vocabulary (WMO-No. 182). Geneva.

World Meteorological Organization (WMO). 2008. Guide to Meteorological Instruments and Methods of Observation (WMO-No. 8). Geneva.

World Meteorological Organization. 2010. Guide to the Global Observing System (WMO-No. 488). Geneva.

Wösten, J.H.M., Clymans, E., Page, S.E., Rieley, J.O. and Limin, S.H. 2008. Peat–water interrelationships in a tropical peatland ecosystem in Southeast Asia. *Catena*. 73(2), 212–224.

Yadav, A.K., Kumar, K., Kasim, M.H.A., Singh, M.P., Parida, S.K. and Sharan, M. 2003. Visibility and incidence of respiratory diseases during the 1998 haze episode in Brunei Darussalam. In *Air Quality*. Birkhäuser, Basel. pp. 265–277.

Zaccone, C., Rein, G., D'Orazio, V., Hadden, R.M., Belcher, C.M. and Miano, T.M., 2014. Smouldering fire signatures in peat and their implications for palaeoenvironmental reconstructions. *Geochemical et Cosmochimica Acta*. 137, 134–146.

Zoumas, A., Wooster, M. and Perry, G. 2004. Fire and El Niño in Borneo, SE Asia. *International Archives of Photogrammetry and Remote Sensing*. 35(7), 596–600.

15 Meteorological Drivers of Anomalous Wildfire Activity in the Western Ghats, India

Narendran Kodandapani
Center for Advanced Spatial and
Environmental Research (CASER), India

CONTENTS

INTRODUCTION

Fire activity at various spatial and temporal scales is influenced by several factors, including fuel, climate, climate variation, ignitions, human activity, and topography (Vadrevu et al. 2006; 2012; Prasad et al. 2008; Parisien et al. 2016). Recent studies have established the impacts of anthropogenic climate change on an increasingly diverse array of meteorological and hydrological phenomena (Mann and Gleick 2015). Since the 1970s, droughts in several tropical regions have been longer and more intense (Malhi and Wright 2004). Terrestrial evapotranspiration affects precipitation and surface temperatures, especially the duration and intensity of heat waves (Jung et al. 2010). Various evapotranspiration estimates indicate ecologically important aspects of climate, water balance, and plant productivity (Fisher et al. 2011). The difference in potential evapotranspiration (PET) due to temperature and actual transpiration, limited by available moisture, is an essential indicator of drought stress in ecosystems (Westerling, 2016). Water balance is affected by low rainfall and intense ecosystem evapotranspiration, which coincides with strong seasonal drought conditions.

Current climate change projections show an increased likelihood of enhanced fire risk in forests due to rising temperatures, increased drought, and longer fire seasons (Moritz et al. 2012; Jolly et al. 2015; Schoennagel et al. 2017). Climate change accelerates various fire regime components in terms of fire severity, intensity, frequency, extent, size, and prolonged fire season length (Bowman et al. 2017). Global teleconnections and anomalies in climate, such as the El Niño–Southern Oscillation (ENSO), have exacerbated wildfires in several forest areas (Siegert et al. 2001; Chen et al. 2017). An estimated 20 Mha of tropical forest in South America and Southeast Asia succumbed to drought-induced fires during the 1997–1998 ENSO event (Cochrane 2003). Synergies between climate change, land use/cover change, and socioeconomic factors have resulted in disastrous wildfires in different parts of the globe (Andela et al. 2017; Vadrevu and Justice 2011; Justice et al. 2015).

Several wildfires attributed to accelerating fire–weather conditions have assumed disastrous proportions due to the high economic, social, and environmental costs (Jolly et al. 2015). For example, in Western Europe, catastrophic wildfires were witnessed in Portugal, where about 120 people were killed in 2017 (Turco et al. 2019). In 2018, the Camp wildfire in California killed at least 83 people and left about 10,000 homeless (BBC 2018). During the 2018 Kurangani wildfire, 22 lives were lost in the Western Ghats due to a catastrophic fire in India (TOI 2018). These wildfires have occurred at the intersection of a complex set of factors, including human management of forests (Flannigan et al. 2009), climate change (Bowman et al. 2017), and change in land-use patterns (Cochrane et al. 1999). Local weather conditions largely drive landscape-scale fire behavior, and daily surface weather variables are closely related to the magnitude and extent of wildfire activity (Jolly et al. 2015). Consequently, fire prediction has received increasing attention among agencies responsible for the management of natural resources and forests around the globe.

Drought also plays a crucial role in driving fire activity in southwestern India (Renard et al. 2012). In the southern Indian state of Andhra Pradesh, models built with fire count data and explanatory variables such as population density, climate, topography, and demand for metabolic energy explained more than 60% of the variability in fire activity (Vadrevu et al. 2006). Recent studies in the Western Ghats have incorporated several explanatory variables to identify fire-prone areas through a two-step process, first at the regional scale covering the entire Western Ghats and subsequently at the local scale (Renard et al. 2012). They highlight the importance of interannual climate variables, especially the climatic data of the prior-year monsoon season, in driving fire patterns in the Western Ghats. However, finer-scale, subregional models, which included the type of vegetation, provided more robust estimates of the fire pattern due to their ability to capture forest fuel characteristics and degradation levels (Renard et al. 2012). Simultaneously, recent studies in the Western Ghats demonstrated increasing water deficits and drought stress at short timescales (e.g., 3–6 months) enhanced fire activity (Kodandapani et al. 2009; Kodandapani and Parks 2019). For example, short-term (3 months) drought exemplified by climatic variables, and local weather were important drivers of fire in tropical dry forests in the Western Ghats (Kodandapani and Parks 2019). Although most of these studies in the Western Ghats have assessed the effects of annual climate variation on wildfire activity, very little information exists on the impacts of daily meteorological conditions on landscape-scale fire behavior. In this study, we analyzed the effects of daily local weather conditions on fire occurrences in the Nilgiris landscape. Specifically, at a landscape scale, we aimed (1) to evaluate weather and wildfire's temporal variability and (2) to evaluate the relationships between daily precipitation, maximum temperature, minimum relative humidity, maximum wind speed, and fire occurrences.

STUDY AREA

Our landscape-scale analyses were conducted in the Western Ghats, a mountain range in southwest India that is one of the 34 global hotspots of biodiversity (Mittermeier et al. 2004). It is also the biodiversity hotspot with the highest human densities (Cincotta et al. 2000). The landscape-scale study site was nested within the central Western Ghats, the Nilgiris landscape (1545 km²; latitude = 11.7°N, longitude = 76.5°E) (Figure 15.1). The elevation of the Nilgiris landscape ranges from 0 to 1450 m. The estimated mean annual precipitation (MAP) ranges from 600 to 2000 mm/year (Hijmans et al. 2005). The Nilgiris landscape has a short dry season (rainfall < 50 mm/month) of 4 months (Pai et al. 2014). Nearly all fires in the landscape are accidentally or intentionally ignited by humans (Kodandapani et al. 2004; Kodandapani et al. 2008; Mehta et al. 2008; Mondal and Sukumar 2014; Vadrevu et al. 2018).

The Nilgiris landscape is comprised of tropical dry deciduous forests (65%), tropical moist deciduous forests (10%), tropical dry thorn forests (20%), and settlements (5%) (Kodandapani et al. 2004; Kodandapani et al. 2008). The dry season extends between January and March (Kodandapani et al. 2004). The Nilgiris landscape is comprised of three protected areas: the Bandipur Tiger Reserve, the Mudumalai Tiger Reserve, and the Wayanad Wildlife Sanctuary. Protected areas, in general, have various management objectives ranging from strict biodiversity conservation to permitting human activities in certain zones (Jones et al. 2018). In the Mudumalai Tiger Reserve, extraction of non-timber forest products (NTFPs) is banned, including fodder extractions and grazing except the eastern part of the reserve. Similarly, in the Bandipur Tiger Reserve, the extraction of NTFPs is forbidden. However, in the Wayanad Wildlife Sanctuary, the extraction of NTFPs is allowed through the issue of permits (Narendran et al. 2001). The Nilgiris landscape is especially critical from a biodiversity conservation standpoint, as it has the highest densities and largest population of two endangered species in the world, the Asian elephant (*Elephas maximus*) and Bengal tiger (*Panthera tigris*) (Mehta et al. 2008; Jhala et al. 2014).

TEMPORAL ANALYSIS

We evaluated daily variability in fire activity (defined here as the number of fire detections) and drought. Fire activity was measured using MODIS fire detection data spanning from 2004 to 2009 (Collection 6: MOD14A1, and MYD14A1; available at http://maps.geog.umd.edu/firms/) (Giglio et al. 2016). MODIS sensors record the location of thermal anomalies (i.e., fire detections) four times per day (resolution =1 km). Thermal sensors are designed to detect flaming and smoldering

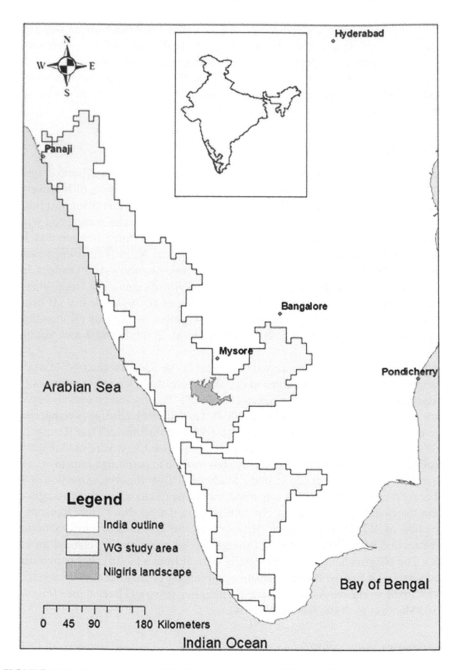

FIGURE 15.1 Location of the Nilgiris landscape in the Western Ghats, India.

fire hotspots from ~1000 m² in size. Elaborate algorithms have been developed to improve detection accuracy, especially smaller and cooler fires based on potential fire thresholds (Giglio et al. 2016). Recent studies have determined that the vast majority (85%) of MODIS fire detections from 2001 to 2015 in the Western Ghats occurred in January–March, with much lower fire activity in the remainder of the year (Kodandapani and Parks 2019). Hence, our analysis has been restricted to the first 90 days of the year (DOY).

Drought and surface weather variability was characterized using variables representing daily weather, obtained from the Mulehole (76° 27′E, 11° 44′N) weather station (Descloitres et al. 2008; Ruiz et al. 2010; Riotte et al. 2014; Chitra-Tarak et al. 2018). The Mulehole meteorological station provides a daily time series of humidity, temperature, radiation, and wind speed and has been in operation since 2003. Measures of water deficits and surface weather variables, with demonstrated links to fire activity (Littell et al. 2009; Abatzoglou and Kolden 2013; Williams et al. 2015), were computed from hourly temperature, relative humidity, global radiation, and wind speed data. Evapotranspiration was calculated using the Penman–Monteith approach (Allen et al. 1998).

RESULTS

About half of MODIS fire detections from 2004 to 2009 in the Nilgiris landscape occurred during 2 years, 2004 and 2009, with much lower fire activity during the other years. Nevertheless, the Nilgiris landscape exhibits substantial interannual variability in fire activity with a mean of 138 ± 61 counts (Table 15.1). For example, looking at the January–March MODIS fire detections, the 2 years with the highest fire activity exhibit about three times more fires than the year with the lowest fire activity. Interannual variability is also evident, with a majority of fire activity occurring in the tropical dry deciduous forests (115 ± 58), followed by the tropical thorn forests (10 ± 9), and the tropical moist deciduous forests (5 ± 7) in the Nilgiris landscape.

Total precipitation, PET, maximum temperatures, minimum relative humidity, and maximum wind speed also exhibited substantial interannual variability, corresponding to the fire activity. From 2004 to 2009, < 50 mm of total precipitation was received during the dry months (January–March) in 2004, 2007, and 2009. Interannual variability is evident with much higher total precipitation recorded during wet years, and mean total rainfall during the 6 years (January–March) was

---

TABLE 15.1

Landscape-Scale Seasonal (January–March) Pattern of MODIS Fire Detections in the Nilgiris Landscape

Year	Total	Tropical Moist Deciduous Forest	Tropical Dry Deciduous Forest	Tropical Dry Thorn Forest
2004	211	17	194	0
2005	94	0	69	25
2006	59	3	45	11
2007	111	1	96	14
2008	114	0	114	0
2009	186	6	170	10
Average	138	4.5	115	10

73 ± 48 mm. Similarly, mean cumulative (January–March) PET was 291 ± 57 mm, mean maximum temperature (January–March) was $33°C \pm 0.67°C$, mean minimum relative humidity (January–March) was $27\% \pm 4\%$, and mean maximum wind speed was 4.55 ± 0.24 m/s.

Similar trends were observed concerning weather and fire activity at daily timescales; for example, the months (January–March) have a mean daily total precipitation between 0.5 and 3.22 mm. The mean daily PET ranges from 1.99 to 3.78 mm; mean daily maximum temperatures range from 31.79°C to 33.46°C; mean daily minimum relative humidity ranges from 21.5% to 32.8%; and mean daily wind speed ranges from 4.24 to 4.91 m/s. Mean daily PET was highest in 2004, 3.78 ± 0.86 mm, and lowest in 2008, 1.99 ± 0.57 mm (Table 15.2). Similarly, daily fire activity ranged from 0.7 to 2.7 MODIS fire detections during the six years. The mean daily MODIS fire detections were also highest in 2004, whereas the lowest mean daily MODIS fire detections were in 2006, 0.68 ± 2. Both the mean daily PET and the mean daily MODIS fire detections were significantly ($P < 0.01$) higher in 2004 compared to all other years.

Daily MODIS fire detections in the Nilgiris landscape were high in 2004, 2007, and 2009 and found to be positively correlated with PET (Figure 15.2, Table 15.3). The daily variability in wildfire incidence in 2004 is strongly correlated with PET ($\rho = 0.69$, $P < 0.01$). Likewise, strong correlations were observed in 2007 ($\rho = 0.5$, $P < 0.01$) and 2009 ($\rho = 0.5$, $P < 0.01$), whereas correlations were weak to moderate for all other years (Table 15.3). Similarly, daily variability in wildfire incidence in 2004 was strongly correlated with daily maximum temperature ($\rho = 0.5$, $P < 0.01$). Likewise, daily variability in wildfire incidence in 2004 was inversely correlated with daily minimum relative humidity ($\rho = -0.62$, $P < 0.01$). Correlations between daily MODIS fire detections and daily maximum wind speed were weak to moderate.

Models relating to daily fire–weather and fire activity are shown in Table 15.4. Of the three drier years, daily weather metrics of 2004 demonstrated the strongest

TABLE 15.2
Summary of Daily MODIS Fire Detections and Fire-Weather Variables in the Nilgiris Landscape

Year	MODIS Fire Detections (Mean ± SD)	PPT (mm) (Mean ± SD)	PET (mm) (Mean ± SD)	Maximum Temperature (°C) (Mean ± SD)	Minimum Relative Humidity (%) (Mean ± SD)	Maximum Wind Speed (m/s) (Mean ± SD)
2004	2.5 ± 4.1	0.5 ± 2.9	3.8 ± 0.9	33.5 ± 2.3	23.8 ± 11	4.9 ± 1
2005	1.2 ± 3.4	1 ± 5.6	3.6 ± 0.8	33.3 ± 2.4	28.4 ± 12.1	4.3 ± 0.9
2006	0.7 ± 2.1	1.1 ± 4.2	3.2 ± 0.2	32.8 ± 2	29.1 ± 11.2	4.2 ± 1
2007	1.3 ± 2.5	0.1 ± 1.2	3.4 ± 0.7	32.9 ± 2.2	24.6 ± 8.3	4.6 ± 1
2008	1.3 ± 4.5	1.6 ± 5.9	2 ± 0.6	31.8 ± 2.4	32.9 ± 16.2	4.6 ± 1
2009	2.2 ± 4.7	0.5 ± 1.9	3.3 ± 0.7	33.7 ± 2.6	21.5 ± 2.5	4.5 ± 0.9

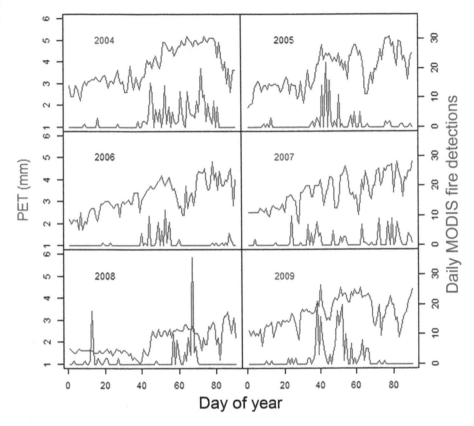

FIGURE 15.2   Trends in daily PET and number of MODIS fire detections between 2004 and 2009 in the Nilgiris landscape.

TABLE 15.3

Correlations between Fire-Weather Variables and MODIS Fire Detections in the Nilgiris Landscape

Year	MODIS Fire Detections (Mean ± SD)	PET (mm)	Maximum Temperature (°C)	Minimum Relative Humidity (%)	Maximum Wind Speed (m/s)
2004	2.5 ± 4.1	0.69[a]	0.5[a]	−0.62[a]	0.28[b]
2005	1.2 ± 3.4	0.36[a]	0.28[a]	−0.44[a]	0.19[b]
2006	0.7 ± 2.1	0.36[a]	0.37[a]	−0.43[a]	0.2[b]
2007	1.3 ± 2.5	0.5[a]	0.34[a]	−0.41[a]	0.17[b]
2008	1.3 ± 4.5	0.19[b]	0.38[a]	−0.57[a]	−0.18[ns]
2009	2.2 ± 4.7	0.5[a]	0.33[a]	0.25[b]	0.34[a]

ns, not significant; [a]P < 0.01; [b]P < 0.05

TABLE 15.4

Linear Models Relating Daily Fire–Weather Variables with the Number of MODIS Fire Detections during the Three Dry Years in the Nilgiris Landscape

Year	Model	R^2
2004	Fire = 2.74 × (PET) − 7.87	0.33
	Fire = 0.73 × (MAX TEMP) − 21.7	0.16
	Fire = −0.18 × (MIN RH) + 6.87	0.24
	Fire = 1.29 × (MAX WIND SPEED) − 3.84	0.08
2007	Fire = 1.59 × (PET) − 4.18	0.21
	Fire = 0.35 × (MAX TEMP) − 10.3	0.1
	Fire = −0.12 × (MIN RH) + 4.36	0.16
	Fire = 0.32 × (MAX WIND SPEED) − 0.18	0.01
2009	Fire = 3.3 × (PET) − 8.8	0.22
	Fire = 0.55 × (MAX TEMP) − 16.6	0.1
	Fire = 0.35 × (MIN RH) − 5.42	0.04
	Fire = 1.53 × (MAX WIND SPEED) − 4.77	0.08

relationships with daily fire activity. Of the weather metrics, daily PET in 2004 established the strongest association with daily MODIS fire detections ($R^2 = 0.33$; Table 15.4; Figure 15.3). Relationships between daily PET and daily fire activity were moderate during 2007 ($R^2 = 0.21$) and 2009 ($R^2 = 0.22$). Likewise, the daily maximum temperature in 2004 demonstrated a moderate relationship with daily fire activity ($R^2 = 0.16$) and weaker relationships during 2007 ($R^2 = 0.1$) and 2009 ($R^2 = 0.1$).

Similarly, minimum relative humidity in 2004 also showed moderate relationships with daily fire activity ($R^2 = 0.24$) and weaker relationships during 2007 ($R^2 = 0.16$) and 2009 ($R^2 = 0.04$). The daily maximum wind speed had a weak connection with daily fire activity for all years (Table 15.4). Thus, during the year 2004, three daily fire–weather variables (PET, maximum temperature, and minimum relative humidity) explained almost 73% of the variability in daily fire activity in the Nilgiris landscape. Overall, daily PET was the most consistent variable in terms of statistically significant relationships with fire activity. Relationships to fire activity were generally positive for daily PET, maximum temperature, and maximum wind speed and negative for daily minimum relative humidity.

DISCUSSION

Our results generally support the view that fire activity increases during hot, dry years (e.g., Liu et al. 2010; Abatzoglou and Kolden 2103; Riley et al. 2013; Aragão et al. 2018). However, the strength and ubiquity of the response between daily weather variation and fire activity in our study depend on the convergence of influential fire–weather conditions. For example, during the drought year of 2004, statistically significant relationships were observed for PET, maximum temperature, minimum relative humidity, and fire activity. In contrast, during the two other dry years (2007 and 2009), statistically

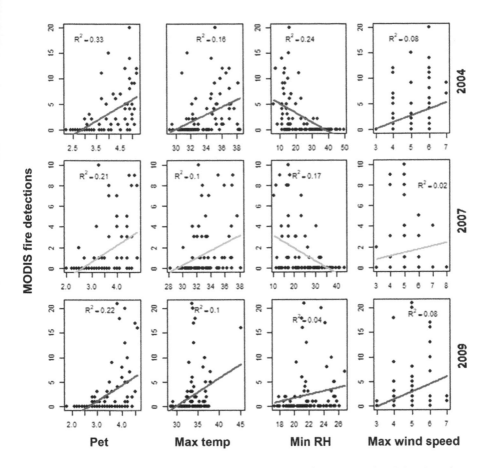

FIGURE 15.3 Plots show the model fits and the resulting R^2 for daily MODIS fire detections and corresponding fire–weather variable in Nilgiris landscape (top row, 2004; middle row, 2007; bottom row, 2009).

significant but weaker relationships were observed for PET, maximum temperature, minimum relative humidity, and fire activity. These results are potentially due to the dynamic changes in the duration and magnitude of PET, the recovery of PET through rainfall, and the timing of rain during the fire season, all of which are important factors that could amplify fire activity in these ecosystems (Figure 15.4). Weather is one of the most important variables and largest driver of regional fire activities; variables such as temperature, relative humidity, wind speed, and precipitation independently influence fire spread and intensity (Jolly et al. 2015). Our daily analyses of fire occurrences in the Nilgiris landscape indicate that meteorological events' co-occurrence is important for the amplification of fire conditions in these ecosystems.

Annual droughts in tropical dry deciduous forests in the Nilgiris are common, with differences in the strength and duration of drought conditions from one year to the next; fire activity is closely related to these drought conditions in tropical dry decidu-ous forests (Murphy and Lugo 1986; Kodandapani et al. 2004; Kodandapani et al. 2008; Kodandapani and Parks 2019). In contrast, fires in tropical moist deciduous

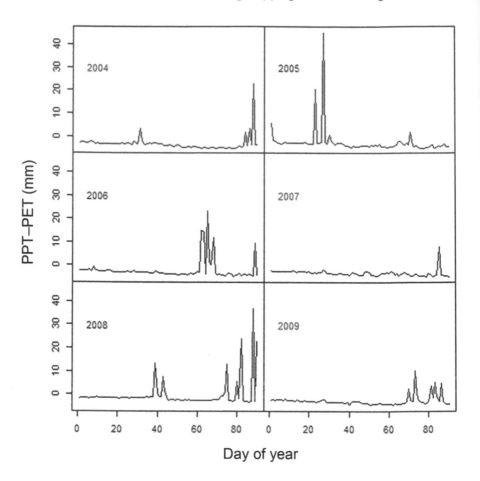

FIGURE 15.4   Changes in daily water balance metrics in the Nilgiris landscape between 2004 and 2009.

forests are less frequent; however, prolonged water deficit and drought conditions such as in 2004 are a precursor to increased wildfire activity (Kodandapani and Parks 2019). The rainfall occurrence, especially early or in the middle of the fire season as in 2005, 2006, and 2008, appears to have a moderating effect on PET, altering the latent heat flux (Figure 15.4). The 2004 accentuated fire activity in the Nilgiris landscape indicates one of the plausible effects of future projected climate change on wildfire, especially increasing duration and intensity of drought conditions in the tropics, increasing surface temperatures, and declines in seasonal rainfall (Schwalm et al. 2017; Malhi and Wright 2004). The enhanced fire activity in the dry forests and especially moist forests could have important implications for the conservation of moist forests in the Western Ghats, as these forests could be extremely vulnerable to frequent fires (Kodandapani and Parks 2019).

Globally, regionally, and locally in different parts of the world, disastrous wildfires have become more common, and there is a strong likelihood of such wildfire events in the future (Bowman et al. 2017). Accentuated wildfire behavior has been witnessed in several parts of the Nilgiris landscape during recent years when crowning of trees were observed (*personal observation*). This, combined with little awareness of fire–weather relationships, has left fire managers ill-prepared in managing extreme wildfire conditions. Monitoring fire–weather conditions at a short timescale and especially daily assessment of fire–weather such as the PET, rain-free days, maximum temperature, minimum relative humidity, and maximum wind speed in these forests would enable managers to mitigate the impacts of extreme and disastrous fires.

CONCLUSIONS

There was an increased fire activity in response to anomalous weather conditions in the Nilgiris landscape. Daily variation in fire–weather has a strong influence on fire activity, and daily PET, maximum temperature, and minimum relative humidity were significant drivers of fire activity. Using daily weather data and climate information as part of early-warning systems in the Western Ghats would help prepare agencies to mitigate the effects of disastrous fires on social, economic, and environmental costs in the short-term and long-term adaptation to the threat of frequent fires due to climate change. The risk of global warming and climate change is an especially urgent confounding factor concerning wildfire studies. Developing policies and management practices that increase communities' and ecosystems' adaptive capacity to heightened wildfire in the future are important.

ACKNOWLEDGMENTS

We gratefully acknowledge the support for this research by the Council of Scientific and Industrial Research, Government of India. We thank the forest departments of Karnataka, Kerala, and Tamil Nadu for permission to research the study areas. We thank the Indo-French Cell for Water Sciences, joint laboratory IISc/IRD, for kindly providing the weather data. We thank the University of Maryland for kindly providing the MODIS fire detection dataset.

REFERENCES

Abatzoglou, J.T. and Kolden, C.A. 2013. Relationships between climate and macroscale area burned in the western United States. *International Journal of Wildland Fire*. 22(7), 1003–1020.

Allen, R.G., Pereira, L.S., Raes, D., Smith, M. 1998. Crop evapotranspiration-Guidelines for computing crop water requirements. FAO Irrigation and drainage paper 56, FAO, Rome 300, D05109.

Andela, N., Morton, D.C., Giglio, L., Chen, Y., Van Der Werf, G.R., Kasibhatla, P.S., DeFries, R.S., Collatz, G.J., Hantson, S., Kloster, S. and Bachelet, D. 2017. A human-driven decline in global burned area. *Science*. 356(6345), 1356–1362.

Aragão, L.E., Anderson, L.O., Fonseca, M.G., Rosan, T.M., Vedovato, L.B., Wagner, F.H., Silva, C.V., Junior, C.H.S., Arai, E., Aguiar, A.P. and Barlow, J. 2018. 21st Century drought-related fires counteract the decline of Amazon deforestation carbon emissions. *Nature Communications*. 9(1), 1–12.

BBC. 2018. California wildfires: Camp fires nearly fully contained. https://www.bbc.com/news/world-us-canada-46315029 (last accessed 01/4/2020).

Bowman, D.M., Williamson, G.J., Abatzoglou, J.T., Kolden, C.A., Cochrane, M.A. and Smith, A.M. 2017. Human exposure and sensitivity to globally extreme wildfire events. *Nature Ecology & Evolution*. 1(3), 1–6.

Chen, Y., Morton, D.C., Andela, N., Van Der Werf, G.R., Giglio, L. and Randerson, J.T. 2017. A pan-tropical cascade of fire driven by El Niño/Southern Oscillation. *Nature Climate Change*. 7(12), 906–911.

Chitra-Tarak, R., Ruiz, L., Dattaraja, H.S., Kumar, M.M., Riotte, J., Suresh, H.S., McMahon, S.M. and Sukumar, R. 2018. The roots of the drought: Hydrology and water uptake strategies mediate forest-wide demographic response to precipitation. *Journal of Ecology*. 106(4), 1495–1507.

Cincotta, R.P., Wisnewski, J. and Engelman, R. 2000. Human population in the biodiversity hotspots. *Nature*. 404(6781), 990–992.

Cochrane, M.A., Alencar, A., Schulze, M.D., Souza, C.M., Nepstad, D.C., Lefebvre, P. and Davidson, E.A. 1999. Positive feedbacks in the fire dynamic of closed canopy tropical forests. *Science*. 284(5421), 1832–1835.

Cochrane, M.A. 2003. Fire science for rainforests. *Nature*. 421, 913–919.

Descloitres, M., Ruiz, L., Sekhar, M., Legchenko, A., Braun, J.J., Mohan Kumar, M.S. and Subramanian, S. 2008. Characterization of seasonal local recharge using electrical resistivity tomography and magnetic resonance sounding. *Hydrological Processes: An International Journal*. 22(3), 384–394.

Fisher, J.B., Whittaker, R.J. and Malhi, Y. 2011. ET come home: Potential evapotranspiration in geographical ecology. *Global Ecology and Biogeography*. 20(1), 1–18.

Flannigan, M.D., Krawchuk, M.A., de Groot, W.J., Wotton, B.M. and Gowman, L.M. 2009. Implications of changing climate for global wildland fire. *International Journal of Wildland Fire*. 18(5), 483–507.

Giglio, L., Schroeder, W. and Justice, C.O. 2016. The collection 6 MODIS active fire detection algorithm and fire products. *Remote Sensing of Environment*. 178, 31–41.

Hijmans, R.J., Cameron, S.E., Parra, J.L., Jones, P.G. and Jarvis, A. 2005. Very high resolution interpolated climate surfaces for global land areas. *International Journal of Climatology: A Journal of the Royal Meteorological Society*. 25(15), 1965–1978.

Jhala, Y.V., Qureshi, Q. and Gopal, R. 2014. *The Status of Tigers in India*. National Tiger Conservation Authority, The Wildlife Institute of India, Dehradun, 19pp.

Jolly, W.M., Cochrane, M.A., Freeborn, P.H., Holden, Z.A., Brown, T.J., Williamson, G.J. and Bowman, D.M. 2015. Climate-induced variations in global wildfire danger from 1979 to 2013. *Nature Communications*. 6(1), 1–11.

Jones, K.R., Venter, O., Fuller, R.A., Allan, J.R., Maxwell, S.L., Negret, P.J. and Watson, J.E., 2018. One-third of global protected land is under intense human pressure. *Science*. 360(6390), 788–791.

Jung, M., Reichstein, M., Ciais, P., Seneviratne, S.I., Sheffield, J., Goulden, M.L., Bonan, G., Cescatti, A., Chen, J., De Jeu, R. and Dolman, A.J. 2010. Recent decline in the global land evapotranspiration trend due to limited moisture supply. *Nature*. 467(7318), 951–954.

Justice, C., Gutman, G. and Vadrevu, K.P. 2015. NASA land cover and land use change (LCLUC): An interdisciplinary research program. *Journal of Environmental Management*. 148(15), 4–9.

Kodandapani, N., Cochrane, M.A. and Sukumar, R., 2004. Conservation threat of increasing fire frequencies in the Western Ghats, India. *Conservation Biology*. 18(6), 1553–1561.

Kodandapani, N., Cochrane, M.A. and Sukumar, R. 2008. A comparative analysis of spatial, temporal, and ecological characteristics of forest fires in seasonally dry tropical ecosystems in the Western Ghats, India. *Forest Ecology and Management*. 256(4), 607–617.

Kodandapani, N., Cochrane, M.A. and Sukumar, R. 2009. Forest fire regimes and their ecological effects in seasonally dry tropical ecosystems in the Western Ghats, India. In *Tropical Fire Ecology*. Springer, Berlin, Heidelberg. pp. 335–354.

Kodandapani, N. and Parks, S.A. 2019. Effects of drought on wildfires in forest landscapes of the Western Ghats, India. *International Journal of Wildland Fire*. 28(6), 431–444.

Littell, J.S., McKenzie, D., Peterson, D.L. and Westerling, A.L. 2009. Climate and wildfire area burned in western US ecoprovinces, 1916–2003. *Ecological Applications*. 19(4), 1003–1021.

Liu, Y., Stanturf, J. and Goodrick, S. 2010. Trends in global wildfire potential in a changing climate. *Forest Ecology and Management*. 259(4), 685–697.

Malhi, Y. and Wright, J. 2004. Spatial patterns and recent trends in the climate of tropical rainforest regions. *Philosophical Transactions of the Royal Society of London. Series B: Biological Sciences*. 359(1443), 311–329.

Mann, M.E. and Gleick, P.H. 2015. Climate change and California drought in the 21st century. *Proceedings of the National Academy of Sciences*. 112(13), 3858–3859.

Mehta, V.K., Sullivan, P.J., Walter, M.T., Krishnaswamy, J. and DeGloria, S.D. 2008. Ecosystem impacts of disturbance in a dry tropical forest in southern India. *Ecohydrology: Ecosystems, Land and Water Process Interactions, Ecohydrogeomorphology*. 1(2), 149–160.

Mittermeier, R.A., Gil, P.R., Hoffman, M., Pilgrim, J., Brooks, T., Mittermeier, C., Lamoreux, J., Da Fonseca, G.A.B. and Saligmann, P.A. 2004. Hotspots Revisited: Earth's Biologically Richest and Most Endangered Terrestrial Ecoregions Cemex. Mexico City.

Mondal, N. and Sukumar, R. 2014. Characterising weather patterns associated with fire in a seasonally dry tropical forest in southern India. *International Journal of Wildland Fire*. 23(2), 196–201.

Moritz, M.A., Parisien, M.A., Batllori, E., Krawchuk, M.A., Van Dorn, J., Ganz, D.J. and Hayhoe, K. 2012. Climate change and disruptions to global fire activity. *Ecosphere*, 3(6), 1–22.

Murphy, P.G. and Lugo, A.E. 1986. Ecology of tropical dry forest. *Annual Review of Ecology and Systematics*. 17(1), 67–88.

Narendran, K., Murthy, I.K., Suresh, H.S., Dattaraja, H.S., Ravindranath, N.H. and Sukumar, R. 2001. Nontimber forest product extraction, utilization and valuation: A case study from the Nilgiri Biosphere Reserve, southern India. *Economic Botany*. 55(4), 528–538.

Pai, D.S., Sridhar, L., Rajeevan, M., Sreejith, O.P., Satbhai, N.S. and Mukhopadhyay, B. 2014. Development of a new high spatial resolution (0.25×0.25) long period (1901–2010) daily gridded rainfall data set over India and its comparison with existing data sets over the region. *Mausam*. 65(1), 1–18.

Parisien, M.A., Miller, C., Parks, S.A., DeLancey, E.R., Robinne, F.N. and Flannigan, M.D. 2016. The spatially varying influence of humans on fire probability in North America. *Environmental Research Letters*. 11(7), 075005.

Prasad, V.K., Badarinath, K.V.S. and Eaturu, A. 2008. Biophysical and anthropogenic controls of forest fires in the Deccan Plateau, India. *Journal of Environmental Management*, 86(1), 1–13.

Renard, Q., Pélissier, R., Ramesh, B.R. and Kodandapani, N. 2012. Environmental susceptibility model for predicting forest fire occurrence in the Western Ghats of India. *International Journal of Wildland Fire*. 21(4), 368–379.

Riley, K.L., Abatzoglou, J.T., Grenfell, I.C., Klene, A.E. and Heinsch, F.A. 2013. The relationship of large fire occurrence with drought and fire danger indices in the western USA, 1984–2008: The role of temporal scale. *International Journal of Wildland Fire*. 22(7), 894–909.

Riotte, J., Maréchal, J.C., Audry, S., Kumar, C., Bedimo, J.B., Ruiz, L., Sekhar, M., Cisel, M., Tarak, R.C., Varma, M.R.R. and Lagane, C. 2014. Vegetation impact on stream chemical fluxes: Mule Hole watershed (South India). *Geochimica et Cosmochimica Acta*. 145, 116–138. doi:10.1016/j.gca.2014.09.015.

Ruiz, L., Varma, M.R., Kumar, M.M., Sekhar, M., Maréchal, J.C., Descloitres, M., Riotte, J., Kumar, S., Kumar, C. and Braun, J.J. 2010. Water balance modelling in a tropical watershed under deciduous forest (Mule Hole, India): Regolith matric storage buffers the groundwater recharge process. *Journal of Hydrology*. 380(3–4), 460–472. doi:10.1016/j.jhydrol.2009.11.020.

Schoennagel, T., Balch, J.K., Brenkert-Smith, H., Dennison, P.E., Harvey, B.J., Krawchuk, M.A., Mietkiewicz, N., Morgan, P., Moritz, M.A., Rasker, R. and Turner, M.G. 2017. Adapt to more wildfire in western North American forests as climate changes. *Proceedings of the National Academy of Sciences*. 114(18), 4582–4590.

Schwalm, C.R., Anderegg, W.R., Michalak, A.M., Fisher, J.B., Biondi, F., Koch, G., Litvak, M., Ogle, K., Shaw, J.D., Wolf, A. and Huntzinger, D.N. 2017. Global patterns of drought recovery. *Nature*. 548(7666), 202–205.

Siegert, F., Ruecker, G., Hinrichs, A. and Hoffmann, A.A. 2001. Increased damage from fires in logged forests during droughts caused by El Nino. *Nature*. 414(6862), 437–440.

TOI. 2018. Kurangani death toll 22; trekking club founder remains untraced. https://timesofindia.indiatimes.com/city/chennai/kurangani-death-toll-22-trekking-club-founder-remains-untraced/articleshow/63601961.cms (last accessed 1/4/2020).

Turco, M., Jerez, S., Augusto, S., Tarín-Carrasco, P., Ratola, N., Jiménez-Guerrero, P. and Trigo, R.M., 2019. Climate drivers of the 2017 devastating fires in Portugal. *Scientific Reports*. 9(1), 1–8.

Vadrevu, K.P., Eaturu, A. and Badarinath, K.V.S. 2006. Spatial distribution of forest fires and controlling factors in Andhra Pradesh, India using spot satellite datasets. *Environmental Monitoring and Assessment*. 123(1–3), 75–96.

Vadrevu, K.P. and Justice, C.O. 2011. Vegetation fires in the Asian region: Satellite observational needs and priorities. *Global Environmental Research*. 15(1), 65–76.

Vadrevu, K.P., Csiszar, I., Ellicott, E., Giglio, L., Badarinath, K.V.S., Vermote, E. and Justice, C. 2012. Hotspot analysis of vegetation fires and intensity in the Indian region. *IEEE Journal of Selected Topics in Applied Earth Observations and Remote Sensing*. 6(1), 224–238.

Vadrevu, K.P., Ohara, T. and Justice, C. (Eds.) 2018. *Land-Atmospheric Research Applications in South and Southeast Asia*. Springer, Cham.

Westerling, A.L., 2016. Increasing western US forest wildfire activity: Sensitivity to changes in the timing of spring. *Philosophical Transactions of the Royal Society B: Biological Sciences*. 371(1696), 20150178.

Williams, A.P., Seager, R., Macalady, A.K., Berkelhammer, M., Crimmins, M.A., Swetnam, T.W., Trugman, A.T., Buenning, N., Noone, D., McDowell, N.G. and Hryniw, N. 2015. Correlations between components of the water balance and burned area reveal new insights for predicting forest fire area in the southwest United States. *International Journal of Wildland Fire*. 24(1), 14–26.

16 Geochemical Evidence for Biomass Burning Signals on Tibetan Glaciers

Chao You
College of Environment and Ecology,
Chongqing University, China
Institute of Tibetan Plateau Research,
Chinese Academy of Sciences, China

Chao Xu
Institute of Atmospheric Physics, Chinese
Academy of Sciences, China

CONTENTS

INTRODUCTION

The Tibetan Plateau (TP) is a unique landform; it has an average elevation greater than 4000 m above sea level (a.s.l.) and is often called the Third Pole of the earth (Yao et al. 2012; 2015; Thompson et al. 2018; Yao et al. 2019). Also known as the roof of the world and the Asian water tower, over 100,000 km^2 of the TP and the surrounding high mountains are ice-covered; this ice cover is the largest ice mass outside the Arctic Circle and Antarctica (Yao et al. 2012; 2015; 2019). Due to the low-temperature environment and high accumulation rate, TP glaciers are ideal media for studying high-resolution climate and environmental changes over subtropical Eurasia (Xu et al. 2009b; Thompson et al. 2018).

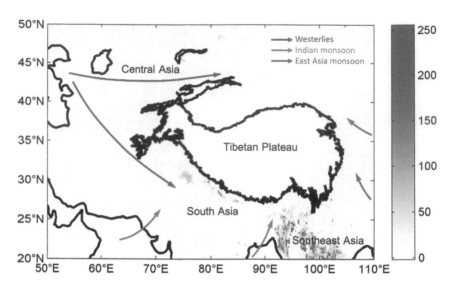

FIGURE 16.1 Biomass burning carbon emissions (g/m²/a) and prevailing atmospheric circulation for subtropical regions around the Tibetan Plateau.

The TP is located very close to South and Southeast Asia (Figure 16.1), which are considered to be some of the most intense biomass burning (BB) emission sources in the Northern Hemisphere (Marlon et al. 2016; Andela et al. 2017; van der Werf et al. 2017; Vadrevu et al. 2018; Vadrevu and Lasko 2018). BB emissions are considered to be a major source of the South Asian brown clouds that occur over the Indian peninsula (Kant et al. 2000; Gupta et al. 2001; Prasad et al. 2000; 2002; 2003; Ramanathan and Crutzen 2003; Badarinath et al. 2008a, b; Gustafsson et al. 2009), and which cause more than 100,000 deaths every year due to related cardiovascular and respiratory diseases (Gustafsson et al. 2009; Johnston et al. 2012).

Smoke plumes from intense BB events can be transported to the upper troposphere and even lower stratosphere by strong thermal convection processes (Vadrevu and Choi 2011; Vadrevu et al. 2012a,b; 2013; Xu et al. 2015), from which they may be transported to remote high-elevation regions (Xu et al. 2014; 2015). Both observational and modeling results have indicated that BB emissions can occasionally be transported to the high-elevation marginal regions of the TP, such as the Himalayas (Lu et al. 2012; Zhao et al. 2013; Xu et al. 2014; Cong et al. 2015; Xu et al. 2015; Vadrevu et al. 2015; Zheng et al. 2017; Xu et al. 2018; Vadrevu and Lasko 2018), and sometimes even reach the central TP (Xia et al. 2011; Wang et al. 2015c).

The important sources of BB in the region include slash-and-burn agriculture, crop residue burning, accidental fires, land-clearing purposes for growing plantations, and peatland burning (Prasad et al. 2001; 2008; Justice et al. 2015; Lasko et al. 2017; 2018a, b; Lasko and Vadrevu 2018; Vadrevu et al. 2006; Biswas et al. 2015a, b; Vadrevu et al. 2014a, b). Evidence from records of black carbon (BC) (Xu et al. 2006; 2009b; Li et al. 2016), dissolved organic matter (Xu et al. 2013), organic acids (Lee et al. 2003), and aromatic acids (Gao et al. 2015) has shown that Tibetan glaciers have been contaminated by BB emissions at different levels. Due to limited observations, however, BB signals on Tibetan glaciers are still poorly understood.

Furthermore, the abovementioned components have been reported alongside other sources besides BB emissions (Simoneit et al. 1999; Simoneit et al. 2004; You and Xu 2018). Monosaccharide anhydrides (MAs) including levoglucosan (LEV), mannosan (MAN), and galactosan (GAL) can only be generated from the pyrolysis of plant cellulose and hemicellulose that occurs during BB processes (Simoneit et al. 1999). LEV normally remains stable for several days under most atmospheric conditions (Simoneit et al. 1999; 2004; Hu et al. 2013) and can be transported long distances, thus ensuring its global distribution (Stohl et al. 2007; Hu et al. 2013; You and Xu 2018). LEV is usually recommended as a powerful tracer for indicating BB signals in environmental studies including atmospheric aerosols (Stohl et al. 2007; Cong et al. 2015), precipitation (Mullaugh et al. 2014; You et al. 2015), soil (Simoneit et al. 2004; Hopmans et al. 2013), lake and marine sediments (Elias et al. 2001; Lopes dos Santos et al. 2013; Callegaro et al. 2018; Schreuder et al. 2018), and polar ice sheets (Kehrwald et al. 2012; Zennaro et al. 2014; 2015; Battistel et al. 2018; Shi et al. 2019).

In this section, we report BB signals detected using the abovementioned proxies and mostly based on BC and LEV records in Tibetan glaciers. We summarize our understanding of LEV records on Tibetan glaciers in recent years. We first report the detected BB signals on Tibetan glaciers using LEV and other proxies. Then, we discuss the key factors that affect distributions of BB aerosols on Tibetan glaciers, based on observational data. Finally, we report a rapid increase in wildfires across the Himalayas and surrounding areas over recent years, using records of LEV concentrations from ice cores.

BIOMASS BURNING SIGNALS DETECTED IN TIBETAN GLACIERS

The possible sources of BB aerosols deposited on Tibetan glaciers are determined by the distribution of fire sources and the prevailing atmospheric circulations (You et al. 2016a, b; 2018; 2019). Climatic and environmental conditions over high-elevation Tibetan glaciers are impacted by the Indian summer monsoon in the southern TP during summer, and primarily by westerly winds during other seasons (Xu et al. 2009b; Yao et al. 2012; Thompson et al. 2018; Yao et al. 2019). However, physical blocking due to the very high mountains along the transport pathways (e.g., the Pamir Plateau and Himalaya) and formation of precipitation on the windward slope (Tian et al. 2007) can substantially block the moisture transport of both the monsoon and the westerlies. Therefore, BB emissions from West Asia and regions further away can be substantially affected by wet deposition throughout their long-range transport pathways. BB emissions from the southern Indian peninsula and southern Southeast Asia contribute little to the Tibetan glaciers because the prevailing winds impair the transport of smoke aerosols throughout the year (You et al. 2016a, b). Considering their low biomass loading capacity (Yao et al. 2015; 2019), fire emissions from the interior TP play a negligible role. As a result of these factors, central Asia and the northern Indian peninsula are considered to be the main sources of the LEV found in the Tibetan glaciers (Figure 16.1).

BC and LEV are two common tracers for interpreting BB signals in Tibetan glacier snow and ice (Xu et al. 2009b; Li et al. 2016; You et al. 2016a). BC data from Tibetan glaciers have shown a decreasing trend from east to west and from north to south (Xu et al. 2006), and some studies have demonstrated that BC can

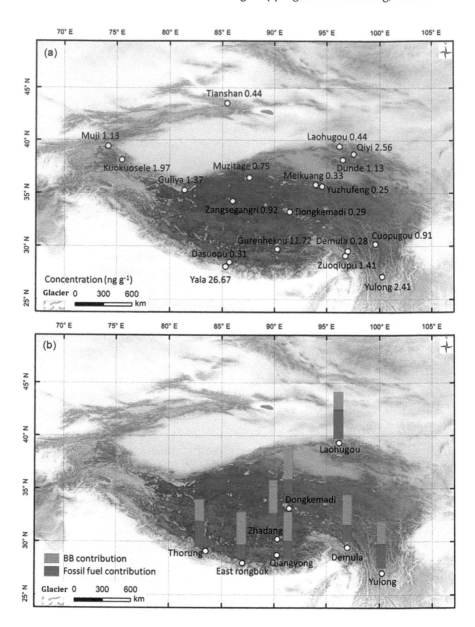

FIGURE 16.2 (a) Spatial distribution of LEV in Tibetan glaciers (data were from You et al 2016a, b and Li et al. 2018a); and (b) relative contributions of the combustion of BB and fossil fuels to BC in Tibetan glacier (data were from Li et al. 2016).

reach a concentration even higher than 1 mg/g in aged glacier snow samples from Tianshan in summertime (Xu et al. 2012; Yang et al. 2015), possibly corresponding to BB emissions over central Asia. Geochemical results have demonstrated that higher K+/BC ratios indicate a greater contribution of BC from BB (Xu et al. 2009b), especially during the pre-monsoon/spring time over southeastern Tibetan glaciers (Xu et al. 2009b; Wang et al. 2015a). In fact, isotopic evidence has indicated that BB can contribute more than half of the BC in glaciers in the southern TP (Figure 16.2; Li et al. 2016). However, BC records in glacier snow and ice layers have also shown very similar variations to fossil fuel usage in the windward regions throughout the past few decades, over both seasonal and annual timescales (Xu et al. 2009a, b; Wang et al. 2015a, b).

LEV concentrations in Tibetan glaciers range from below level of detection (LOD) to 32.90 ng/g (You et al. 2016a, b; 2017; Li et al. 2018b; You et al. 2018; 2019), with an average of less than 1.00 ng/g (You and Xu 2018). Regional differences in LEV concentrations were observed (Figure 16.2; You et al. 2016a). The average LEV concentration in glaciers on the northern TP was approximately 1.50 ng/g, while on the southern TP, it was approximately 2.65 ng/g. The lowest LEV concentrations were detected in the central TP glaciers, with an average of approximately 0.30 ng/g only. Similar spatial variations have previously been observed in BC concentrations in Tibetan glaciers (Xu et al. 2006; Li et al. 2016). LEV concentrations in central Tibetan glaciers are generally higher than those in Antarctic snow and ice samples (Battistel et al. 2018; Shi et al. 2019), but are close to the concentrations observed in samples from the Greenland ice sheet (Kehrwald et al. 2012; Zennaro et al. 2014; 2015). Tibetan glaciers, therefore, mainly act as receptors of BB emissions from surrounding regions. Large differences in LEV concentrations were observed between the northern and southern slopes of the Himalayas (You et al. 2016a). For example, LEV concentrations from the Yala glacier were about two orders of magnitude greater than those from other Tibetan glaciers (Figure 16.2a). Results demonstrate that the Himalayas can substantially block BB aerosols from the South Asian Brown Cloud from entering the TP. Although some studies have declared that internal Tibetan BB sources could be important contributions to BC on Tibetan glaciers (Li et al. 2016), our results support that central Tibetan glaciers are as clean as those representative of regional backgrounds (You et al. 2016a; Li et al. 2018b; You et al. 2019).

In addition, LEV has also been reported as a major organic component in cryoconites in Tibetan glaciers, with concentrations of about 2–3 orders of magnitude greater than those in snow and ice on Tibetan glaciers (Li et al. 2019).

LEVOGLUCOSAN AS A TRACER FOR INDICATING FIRE CHANGES

Although LEV is recommended as a tracer for indicating BB emissions in the polar ice sheets (Kehrwald et al. 2012; Zennaro et al. 2014; 2015; Shi et al. 2019), it has been sparsely evaluated in Tibetan glaciers. Usually, we obtain samples from the accumulation zone on a glacier in order to reveal climatic or environmental change information (Xu et al. 2009a; Thompson et al. 2018). Observations of geochemical processes on a glacier surface are, thus, crucial for understanding whether LEV can be used as a specific tracer for indicating fire changes. In previous studies, surface snow samples were collected at several elevations from 5149 to 5576 m a.s.l. on Zuoqiupu Glacier, which is affected by the Indian summer monsoon (You et al. 2016b), and from 4761 to 5530 m a.s.l. on Muji Glacier, which is dominated by westerlies (You et al. 2017), in order to address the distribution patterns of LEV at different altitudes. The highest LEV concentration was detected in the surface snow sample collected at an elevation of approximately 5412 m a.s.l. on Zuoqiupu Glacier and around 5291 m a.s.l. on Muji Glacier. The elevations of the LEV concentration maxima correspond roughly to those of the equilibrium line altitude of those two glaciers (Yao et al. 2012); LEV concentrations decreased sharply above and below these two elevations.

Previous work has detected BC (Xu et al. 2006; 2009a; Li et al. 2016), acetate (Lee et al. 2003), formate, p-hydroxybenzoic, vanillic, and dehydroabietic acids (Gao et al. 2015) in Tibetan glacier snow and ice. Some studies found that organic carbon and BC strongly correlated with K+ and LEV in aerosols at Mountain Everest, indicating that they mainly originated from BB (Cong et al. 2015). However, insoluble BC behaves differently to water-soluble LEV on glacier surfaces. LEV concentrations were analyzed in samples from the Zuoqiupu snow-pit and the Cuopugou ice core to investigate whether LEV could reflect the seasonal variations of BB signals on southeastern Tibetan glaciers. In the samples taken from the Zuoqiupu snow-pit, the LEV and BC tracers displayed similar variations; LEV concentrations during the pre-monsoon season were approximately three times those of the monsoon season and approximately two times those of the post-monsoon season (You et al. 2016b). These observations were consistent with the fire changes over the windward northern Indian Peninsula (Vadrevu et al. 2013).

LEV and BC concentrations, however, were found to be poorly correlated. These differences may indicate that LEV and BC deposited on southeastern Tibetan glaciers do not always share the same sources. Additionally, the different geochemical processes of LEV and BC during the transport process (e.g., different scavenging rates) may also contribute to the differences in LEV and BC concentrations found in TP glaciers.

Furthermore, it is possible that the post-depositional melt and refreeze processes redistribute LEV on Tibetan glacier surfaces. LEV is strongly water soluble (Simoneit et al. 2004), and may be easily leached during the melt–refreeze process. Soluble species have a greater affinity for meltwater than for ice and preferentially partition into the liquid phase. Consequently, they are removed at the beginning of the leaching process. When the meltwater reaches a depth at which the in situ temperature is lower than the freezing point, it refreezes and forms an ice layer. Void spaces are reduced during the freezing process and become closed after the formation of ice layers, potentially preventing any further infiltration of LEV (You et al. 2016b). This superimposed ice under the snowpack can then prevent further infiltration of meltwater and BC (Xu et al. 2012). The percolation of meltwater therefore acts to remove LEV from the glacier surface and concentrate it in layers close to superimposed ice below (Figure 16.3).

Although these results indicated that LEV can be redistributed by the melting and refreezing processes on the glacier surface, it does, however, remain as a proxy that can represent the primary characteristics of BB aerosols deposited in the accumulation areas of the SET glaciers, over seasonal timescales. For instance, LEV and BC both peaked at a depth of 90–135 cm in the Cuopugou ice core (Figure 16.3), possibly demonstrating the result of strong BB events. Furthermore, LEV and BC maxima both appeared near the dust layers (Figure 16.3), indicating significantly concentrated LEV and BC during the pre-monsoon season. These results have suggested that strong BB events may lead to a substantial increase in BC concentration

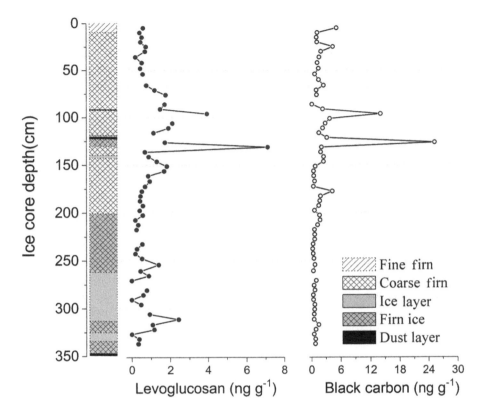

FIGURE 16.3 Visual stratigraphy of ice core characteristics and variation of LEV and BC concentrations in the Cuopugou ice core (You et al., 2016b).

on the SET glaciers, and may play an important role in glacier melting during the pre-monsoon season.

RECENT INCREASING WILDFIRE REGIMES DERIVED IN A CENTRAL TIBETAN ICE CORE

The IPCC AR5 predicts more frequent wildfires with climate warming in the near future (IPCC 2013); however, some recent works have indicated that human activities are responsible for a decline in the global burned area (Andela et al. 2017; Li et al. 2018a), at the same time as emphasizing that fire activity has increased in the densely populated subtropical Asian regions of India and China (Andela et al. 2017; Vadrevu et al. 2015; 2018; 2019; Vadrevu and Lasko 2018). These controversial issues draw our attention to understanding the recent fire changes for densely populated subtropical Asia regions. Could LEV in Tibetan glacier ice be a calibration tool for reconstructing past fire changes? Is fire change throughout subtropical Asia dominated by climate change, or controlled by human activities? We have tried to investigate these questions using LEV records derived from a central Tibetan ice core named the Zangsegangri (ZSGR) ice core (Figure 16.2).

Preliminary LEV concentration results show an increasing trend ($P<0.001$) from 1990 to 2012 (You et al. 2018; 2019). About 20% of the samples analyzed showed LEV concentrations lower than the LOD. The highest LEV concentration was 8.49 ng/g at a depth of 1.70 m, which was more than nine times the mean of 0.88 ng/g. Some samples had LEV concentrations higher than one standard deviation above the mean (>1.96 ng/g) and were considered to represent strong events; those samples that had LEV concentrations higher than two standard deviations above the mean (>3.04 ng/g) were considered extreme events. All of these strong and extreme LEV events are concentrated in the upper 2.0 m of the ice core, which corresponds to the period 2006–2012. This indicates that strong LEV events have occurred more frequently in recent years (Figures 16.4 and 16.5). In addition, annually resolved LEV concentration and flux were highly correlated with each other (You et al. 2018).

According to MODIS observations, central Asia experienced a decrease in fire activity over the period 2003–2012, while the northern Indian peninsula showed a significant increase in fire counts over this same period (You et al. 2018; 2019). Likewise, strong fire events were also becoming more frequent along the Himalayas and surrounds. Strong fire events with a fire radiative power (FRP) greater than 100 MW increased at a rate of approximately 21% per year from 2003 to 2012 over the northern Indian peninsula (You et al. 2018). Noticeably, the average of LEV concentrations corresponding to years after 2006 was about four times greater than that for years

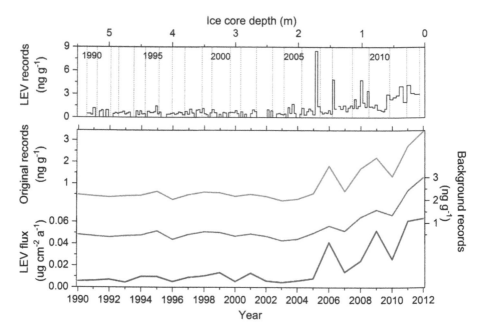

FIGURE 16.4 LEV records versus depth (dashed lines indicate the annual layer), annual LEV concentration (original and background), and flux (red) records from 1990 to 2012 in the ZSGR ice core (modified from You et al. 2018; 2019).

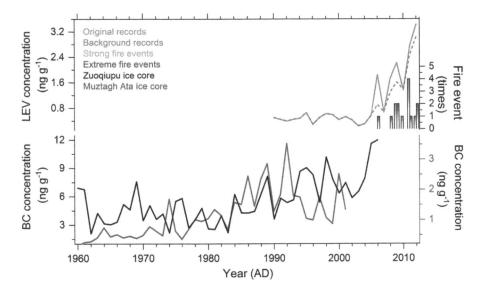

FIGURE 16.5 Variations in LEV-derived fire records and fire event times in the ZSGR ice core from 1990 to 2012 (You et al. 2018; 2019) and BC records in the Zuoqiupu (Xu et al. 2009a; Wang et al. 2015a) and Muztagh Ata ice cores (Wang et al. 2015b).

before 2006 (Figure 16.4) over this same area, with LEV maxima occurring more frequently (Figures 16.4 and 16.5). LEV records in the ZSGR ice core evidently reveal an increasing frequency of fires in the Himalayas and their surroundings during the first decade of the 21st century. Annually based LEV records in the ZSGR ice core indicate a recent increase in the number of fires in the TP and its surroundings from 1990 to 2012, in terms of both the long-term trend and fire events (Figures 16.4 and 16.5).

Andela et al. (2017) indicated that agricultural intensification might increase fire activity in densely populated India, where crop residue burning was suggested as the dominant fire type; however, there was no apparent increase in the agricultural acreage during the first decade of the 21st century over the Indian peninsula (Tamara and David 2014). Additionally, some studies have suggested that the average FRP for agricultural fires over this time was about 22 MW, only about half of the forest fires along the Himalayan regions (Vadrevu et al. 2012a,b; 2013). Those studies also indicated that smoke plumes from the burning of agricultural residue mainly reside below the planetary boundary layer. In comparison, smoke plumes from forest fires can be injected higher than 5 km and then transported at regional scales to the TP (Vadrevu et al. 2013). Strong and extreme fire events represented by samples with LEV concentrations higher than one standard deviation above the mean (1.96 ng/g) were all concentrated in the period from 2006 to 2012 (Figure 16.4), indicating that strong fire events have occurred more frequently in recent years. In addition, the annually resolved LEV background level displayed a sharply increasing trend (Figures 16.4 and 16.5), indicating that the increasing frequency of fire events has also strongly affected these background levels.

Variations of BC in found in Tibetan ice cores (Figure 16.5) also support this conclusion. The Muztagh Ata ice core was taken from the northwestern TP, an area dominated by westerly winds (Figure 16.2). Results from the Muztagh Ata ice core display a decreasing trend for BC since the mid-1990s (Wang et al. 2015b), most likely indicating a decrease in BB emissions from regions of central Asia. In contrast, the Zuoqiupu ice core from the southern TP (Figure 16.2) displayed increasing BC concentrations for this same time frame (Xu et al. 2009b; Wang et al. 2015a). Notably, the northern Indian Peninsula is the dominant source of the BC on southern Tibetan glaciers (Xu et al. 2009a, b; Lu et al. 2012; Xu et al. 2018) and contributes to about 80% of the BC deposited on the Zuoqiupu Glacier (Wang et al. 2015a).

The climate system has been considered as the principal driver of increasing wildfires in the Himalayas and surrounding areas over the past decade, despite the fact that, for fire activity worldwide over the past several decades, human activity has been deemed to be the major influencing factor (Marlon et al. 2016; Andela et al. 2017). Precipitation, rather than temperature, acts as the crucial component regulating terrestrial ecosystems across regions around the TP and controls the quantity of available biofuels (Yao et al. 2015; You et al., 2018). Decreasing precipitation has been reported in regions around the Himalayas since the 1990s (Yao et al. 2012), especially in the central and eastern Himalayas (80–95°E). These decreasing trends have been more significant during the pre-monsoon and winter seasons. Similar decreasing precipitation patterns were also observed in the Gangetic Plain in northern Indian peninsula, in areas such as Haryana and Bihar, from 2000 to 2012 (http://www.tropmet.res.in/Home). Decreasing precipitation can lead to prolonged dry seasons in the Himalayas and surrounding areas, which, for recent years, has ultimately resulted in an increased frequency of wildfires. On the other hand, it appears that decreasing Indian summer monsoon rainfall does not influence the available biofuels in the humid regions, because precipitation is the dominant factor controlling available biomass fuels across the semiarid Indus Plain. Notably, significant increases in precipitation were observed on the Indus Plain from 2000 to 2012 (You et al. 2018), which may have yielded increased biofuels. The northern Indian peninsula has therefore suffered increasing wildfires over the past decade.

CONCLUSIONS

In this chapter, BB signals are reported in snow and ice from different Tibetan glaciers, primarily using LEV and BC records. Although many factors can influence the spatial and temporal distribution regimes of LEV concentration on Tibetan glaciers, LEV deposited in the accumulation zone remains a proxy for representing the primary characteristics of wildfire changes, at least on seasonal-to-annual timescales. Continuous LEV records in Tibetan ice cores can capture the signals of discrete fire events as well as the long-term trend and can provide additional fire event information for reconstructing past fire histories. LEV records in a central Tibetan ice core reveal a rapid increase in wildfires across the Himalayas and surrounding areas at the beginning of the 21st century, for which climate change is concluded to be the dominant cause.

ACKNOWLEDGMENTS

This work was supported by the Youth Innovation Promotion Association CAS (2020071 of Dr. YOU Chao), Chongqing University, and the National Natural Science Foundation of China (41701078 and 41805127).

REFERENCES

Andela, N., Morton, D.C., Giglio, L., Chen, Y., Van Der Werf, G.R., Kasibhatla, P.S., DeFries, R.S., Collatz, G.J., Hantson, S., Kloster, S. and Bachelet, D. 2017. A human-driven decline in global burned area. *Science.* 356(6345), 1356–1362.

Badarinath, K.V.S., Kharol, S.K., Krishna Prasad, V., Kaskaoutis, D.G. and Kambezidis, H.D. 2008a. Variation in aerosol properties over Hyderabad, India during intense cyclonic conditions. *International Journal of Remote Sensing.* 29(15), 4575–4597.

Badarinath, K.V.S., Kharol, S.K., Prasad, V.K., Sharma, A.R., Reddi, E.U.B., Kambezidis, H.D. and Kaskaoutis, D.G. 2008b. Influence of natural and anthropogenic activities on UV Index variations–A study over tropical urban region using ground based observations and satellite data. *Journal of Atmospheric Chemistry.* 59(3), 219–236.

Battistel, D., Kehrwald, N.M., Zennaro, P., Pellegrino, G., Barbaro, E., Zangrando, R., Pedeli, X.X., Varin, C., Spolaor, A., Vallelonga, P.T. and Gambaro, A., 2018. High-latitude Southern Hemisphere fire history during the mid to late Holocene (6000–750 BP). *Climate of the Past.* 14(6).

Biswas, S., Vadrevu, K.P., Lwin, Z.M., Lasko, K. and Justice, C.O. 2015a. Factors controlling vegetation fires in protected and non-protected areas of Myanmar. *PLoS One.* 10(4), e0124346.

Biswas, S., Lasko, K.D. and Vadrevu, K.P. 2015b. Fire disturbance in tropical forests of Myanmar—Analysis using MODIS satellite datasets. *IEEE Journal of Selected Topics in Applied Earth Observations and Remote Sensing.* 8(5), 2273–2281.

Callegaro, A., Battistel, D., Kehrwald, N.M., Matsubara Pereira, F., Kirchgeorg, T., del Carmen Villoslada Hidalgo, M., Bird, B.W. and Barbante, C. 2018. Fire, vegetation, and Holocene climate in a southeastern Tibetan lake: A multi-biomarker reconstruction from Paru Co. *Climate of the Past.* 14(10), 1543–1563.

Cong, Z., Kang, S., Kawamura, K., Liu, B., Wan, X., Wang, Z., Gao, S. and Fu, P. 2015. Carbonaceous aerosols on the south edge of the Tibetan Plateau: Concentrations, seasonality and sources. *Atmospheric Chemistry and Physics.* 15(3), 1573–1584.

dos Santos, R.A.L., De Deckker, P., Hopmans, E.C., Magee, J.W., Mets, A., Damsté, J.S.S. and Schouten, S. 2013. Abrupt vegetation change after the Late Quaternary megafaunal extinction in southeastern Australia. *Nature Geoscience.* 6(8), 627–631.

Elias, V.O., Simoneit, B.R.T., Cordeiro, R.C. and Turcq, B. 2001. Evaluating levoglucosan as an indicator of biomass burning in Carajas, Amazonia: A comparison to the charcoal record. *Geochimica et Cosmochimica Acta.* 65(2), 267–272.

Gao, S., Liu, D., Kang, S., Kawamura, K., Wu, G., Zhang, G. and Cong, Z. 2015. A new isolation method for biomass-burning tracers in snow: Measurements of p-hydroxybenzoic, vanillic, and dehydroabietic acids. *Atmospheric Environment.* 122, 142–147.

Gustafsson, Ö., Kruså, M., Zencak, Z., Sheesley, R.J., Granat, L., Engström, E., Praveen, P.S., Rao, P.S.P., Leck, C. and Rodhe, H. 2009. Brown clouds over South Asia: Biomass or fossil fuel combustion? *Science.* 323(5913), 495–498.

Gupta, P.K., Prasad, V.K., Sharma, C., Sarkar, A.K., Kant, Y., Badarinath, K.V.S. and Mitra, A.P. 2001. CH_4 emissions from biomass burning of shifting cultivation areas of tropical deciduous forests–Experimental results from ground-based measurements. *Chemosphere-Global Change Science.* 3(2), 133–143.

Hopmans, E.C., dos Santos, R.A.L., Mets, A., Sinninghe Damsté, J.S., Schouten, S. 2013. A novel method for the rapid analysis of levoglucosan in soils and sediments. *Organic Geochemistry*. 58, 86–88.

Hu, Q.H., Xie, Z.Q., Wang, X.M., Kang, H., Zhang, P. 2013. Levoglucosan indicates high levels of biomass burning aerosols over oceans from the Arctic to Antarctic. *Scientific Reports*. 3, 3119.

IPCC. 2013. Climate change 2013: The physical science basis. *Contribution of Working Group I to the Fifth Assessment Report of the Intergovernmental Panel on Climate Change*. Cambridge, United Kingdom and New York, NY, USA. Cambridge University Press.

Johnston, F.H., Henderson, S.B., Chen, Y., Randerson, J.T., Marlier, M., DeFries, R.S., Kinney, P., Bowman, D.M. and Brauer, M. 2012. Estimated global mortality attributable to smoke from landscape fires. *Environmental Health Perspectives*. 120(5), 695–701.

Justice, C., Gutman, G. and Vadrevu, K.P. 2015. NASA land cover and land use change (LCLUC): An interdisciplinary research program. *Journal of Environmental Management*. 148(15), 4–9.

Kant, Y., Ghosh, A.B., Sharma, M.C., Gupta, P.K., Prasad, V.K., Badarinath, K.V.S. and Mitra, A.P. 2000. Studies on aerosol optical depth in biomass burning areas using satellite and ground-based observations. *Infrared Physics & Technology*. 41(1), 21–28.

Kehrwald, N., Zangrando, R., Gabrielli, P., Jaffrezo, J.L., Boutron, C., Barbante, C. and Gambaro, A. 2012. Levoglucosan as a specific marker of fire events in Greenland snow. *Tellus B: Chemical and Physical Meteorology*. 64(1), 18196.

Lasko, K. and Vadrevu, K.P. 2018. Improved rice residue burning emissions estimates: Accounting for practice-specific emission factors in air pollution assessments of Vietnam. *Environmental Pollution*. 236(5), 795–806.

Lasko, K., Vadrevu, K.P., Tran, V.T., Ellicott, E., Nguyen, T.T., Bui, H.Q. and Justice, C. 2017. Satellites may underestimate rice residue and associated burning emissions in Vietnam. *Environmental Research Letters*. 12(8), 085006.

Lasko, K., Vadrevu, K.P., Tran, V.T. and Justice, C. 2018a. Mapping double and single crop paddy rice with Sentinel-1A at varying spatial scales and polarizations in Hanoi, Vietnam. *IEEE Journal of Selected Topics in Applied Earth Observations and Remote Sensing*. 11(2), 498–512.

Lasko, K., Vadrevu, K.P. and Nguyen, T.T.N. 2018b. Analysis of air pollution over Hanoi, Vietnam using multi-satellite and MERRA reanalysis datasets. *PLoS One*. 13(5), 0196629.

Lee, X.Q., Qin, D.H., Jiang, G.B., Duan, K.Q., Zhou, H. 2003. Atmospheric pollution of a remote area of Tianshan Mountain: Ice core record. *Journal of Geophysical Research: Atmospheres*. 108(D14). doi: 10.1029/2002JD002181.

Li, C., Bosch, C., Kang, S., Andersson, A., Chen, P., Zhang, Q., Cong, Z., Chen, B., Qin, D. and Gustafsson, Ö. 2016. Sources of black carbon to the Himalayan–Tibetan Plateau glaciers. *Nature Communications*. 7(1), 1–7.

Li, F., Lawrence, D.M. and Bond-Lamberty, B. 2018a. Human impacts on 20th century fire dynamics and implications for global carbon and water trajectories. *Global Planet Change*. 162, 18–27.

Li, Q., Wang, N., Barbante, C., Kang, S., Yao, P., Wan, X., Barbaro, E., Hidalgo, M.D.C.V., Gambaro, A., Li, C. and Niu, H. 2018b. Levels and spatial distributions of levoglucosan and dissolved organic carbon in snowpits over the Tibetan Plateau glaciers. *Science of The Total Environment*. 612, 1340–1347.

Li, Q., Wang, N., Barbante, C., Kang, S., Callegaro, A., Battistel, D., Argiriadis, E., Wan, X., Yao, P., Pu, T. and Wu, X. 2019. Biomass burning source identification through molecular markers in cryoconites over the Tibetan Plateau. *Environmental Pollution*. 244, 209–217.

Lu, Z.F., Streets, D.G., Zhang, Q. and Wang, S.W. 2012. A novel back-trajectory analysis of the origin of black carbon transported to the Himalayas and Tibetan Plateau during 1996–2010. *Geophysical Research Letters.* 39, 871–886. doi: 10.1029 /2011GL049903.

Marlon, J.R., Kelly, R., Daniau, A.L., Vannière, B., Power, M.J., Bartlein, P., Higuera, P., Blarquez, O., Brewer, S., Brücher, T. and Feurdean, A. 2016. Reconstructions of biomass burning from sediment charcoal records to improve data-model comparisons. *Biogeosciences (BG).* 13, 3225–3244.

Mullaugh, K.M., Byrd, J.N., Avery Jr, G.B., Mead, R.N., Willey, J.D. and Kieber, R.J. 2014. Characterization of carbohydrates in rainwater from the Southeastern North Carolina. *Chemosphere.* 107, 51–57.

Prasad, V.K., Gupta, P.K., Sharma, C., Sarkar, A.K., Kant, Y., Badarinath, K.V.S., Rajagopal, T. and Mitra, A.P. 2000. NO_x emissions from biomass burning of shifting cultivation areas from tropical deciduous forests of India–estimates from ground-based measurements. *Atmospheric Environment.* 34(20), 3271–3280.

Prasad, V.K., Kant, Y. and Badarinath, K.V.S. 2001. CENTURY ecosystem model application for quantifying vegetation dynamics in shifting cultivation areas: A case study from Rampa Forests, Eastern Ghats (India). *Ecological Research.* 16(3), 497–507.

Prasad, V.K., Kant, Y., Gupta, P.K., Elvidge, C. and Badarinath, K.V.S. 2002. Biomass burning and related trace gas emissions from tropical dry deciduous forests of India: A study using DMSP-OLS data and ground-based measurements. *International Journal of Remote Sensing.* 23(14), 2837–2851.

Prasad, V.K., Lata, M. and Badarinath, K.V.S. 2003. Trace gas emissions from biomass burning from northeast region in India—estimates from satellite remote sensing data and GIS. *Environmentalist.* 23(3), 229–236.

Prasad, V.K., Badarinath, K.V.S. and Eaturu, A. 2008. Biophysical and anthropogenic controls of forest fires in the Deccan Plateau, India. *Journal of Environmental Management.* 86(1), 1–13.

Ramanathan, V. and Crutzen, P.J. 2003. New directions: Atmospheric brown "Clouds". *Atmospheric Environment.* 37(28), 4033–4035.

Schreuder, L.T., Hopmans, E.C., Stuut, J.-B.W., Sinninghe Damsté, J.S. and Schouten, S. 2018. Transport and deposition of the fire biomarker levoglucosan across the tropical North Atlantic Ocean. *Geochimica et Cosmochimica Acta.* 227, 171–185.

Shi, G., Wang, X.C., Li, Y., Trengove, R., Hu, Z., Mi, M., Li, X., Yu, J., Hunter, B. and He, T. 2019. Organic tracers from biomass burning in snow from the coast to the ice sheet summit of East Antarctica. *Atmospheric Environment.* 201, 231–241.

Simoneit, B.R., Kobayashi, M., Mochida, M., Kawamura, K., Lee, M., Lim, H.J., Turpin, B.J. and Komazaki, Y. 2004. Composition and major sources of organic compounds of aerosol particulate matter sampled during the ACE-Asia campaign. *Journal of Geophysical Research: Atmospheres,* 109(D19).

Simoneit, B.R., Schauer, J.J., Nolte, C.G., Oros, D.R., Elias, V.O., Fraser, M.P., Rogge, W.F. and Cass, G.R. 1999. Levoglucosan, a tracer for cellulose in biomass burning and atmospheric particles. *Atmospheric Environment.* 33(2), 173–182.

Stohl, A., Berg, T., Burkhart, J.F., Fjæraa, A.M., Forster, C., Herber, A., Hov, Ø., Lunder, C., McMillan, W.W., Oltmans, S. and Shiobara, M. 2007. Arctic smoke—record high air pollution levels in the European Arctic due to agricultural fires in Eastern Europe in spring 2006. *Atmospheric Chemistry and Physics.* 7(2), 511–534.

Tamara, B.-A. and David, M. 2014. Decomposing global crop yield variability. *Environmental Research Letters.* 9(11), 114011.

Thompson, L.G., Yao, T., Davis, M.E., Mosley-Thompson, E., Wu, G., Porter, S.E., Xu, B., Lin, P.N., Wang, N., Beaudon, E. and Duan, K. 2018. Ice core records of climate variability on the Third Pole with emphasis on the Guliya ice cap, western Kunlun Mountains. *Quaternary Science Reviews.* 188, 1–14.

Tian, L., Yao, T., MacClune, K., White, J.W.C., Schilla, A., Vaughn, B., Vachon, R. and Ichiyanagi, K. 2007. Stable isotopic variations in west China: A consideration of moisture sources. *Journal of Geophysical Research: Atmospheres.* 112(D10). doi: 10.1029 /2006JD007718

Vadrevu, K.P. and Choi, Y. 2011. Wavelet analysis of airborne CO_2 measurements and related meteorological parameters over heterogeneous landscapes. *Atmospheric Research.* 102(1–2), 77–90.

Vadrevu, K.P. and Lasko, K. 2018. Intercomparison of MODIS AQUA and VIIRS I-Band fires and emissions in an agricultural landscape—Implications for air pollution research. *Remote Sensing.* 10(7), 978. doi:10.3390/rs10070978.

Vadrevu, K.P., Eaturu, A. and Badarinath, K.V.S. 2006. Spatial distribution of forest fires and controlling factors in Andhra Pradesh, India using spot satellite datasets. *Environmental Monitoring and Assessment.* 123(1–3), pp. 75–96.

Vadrevu, K.P., Csiszar, I., Ellicott, E., Giglio, L., Badarinath, K.V.S., Vermote, E. and Justice, C. 2012a. Hotspot analysis of vegetation fires and intensity in the Indian region. *IEEE Journal of Selected Topics in Applied Earth Observations and Remote Sensing.* 6(1), 224–238.

Vadrevu, K.P., Ellicott, E., Giglio, L., et al. 2012b. Vegetation fires in the Himalayan region – Aerosol load, black carbon emissions and smoke plume heights. *Atmospheric Environment.* 47(0), 241–251.

Vadrevu, K.P., Giglio, L. and Justice, C. 2013. Satellite based analysis of fire–carbon monoxide relationships from forest and agricultural residue burning (2003–2011). *Atmospheric Environment.* 64, 179–191.

Vadrevu, K.P., Ohara, T. and Justice, C. 2014a. Air pollution in Asia. *Environmental Pollution.* 12, 233–235.

Vadrevu, K.P., Lasko, K., Giglio, L. and Justice, C. 2014b. Analysis of Southeast Asian pollution episode during June 2013 using satellite remote sensing datasets. *Environmental Pollution.* 12, 245–256.

Vadrevu, K.P., Lasko, K., Giglio, L. and Justice, C. 2015. Vegetation fires, absorbing aerosols and smoke plume characteristics in diverse biomass burning regions of Asia. *Environmental Research Letters.* 10(10), 105003.

Vadrevu, K.P., Ohara, T. and Justice, C. eds. 2018. *Land-Atmospheric Research Applications in South and Southeast Asia.* Springer, Cham.

Vadrevu, K.P., Lasko, K., Giglio, L., Schroeder, W., Biswas, S. and Justice, C. 2019. Trends in vegetation fires in South and Southeast Asian countries. *Scientific Reports.* 9(1), 7422. doi:10.1038/s41598-019-43940-x.

Van Der Werf, G.R., Randerson, J.T., Giglio, L., Van Leeuwen, T.T., Chen, Y., Rogers, B.M., Mu, M., Van Marle, M.J., Morton, D.C., Collatz, G.J. and Yokelson, R.J. 2017. Global fire emissions estimates during 1997–2016. *Earth System Science Data.* 9(2), 697–720.

Wang, M., Xu, B., Cao, J., Tie, X., Wang, H., Zhang, R., Qian, Y., Rasch, P.J., Zhao, S., Wu, G. and Zhao, H. 2015a. Carbonaceous aerosols recorded in a southeastern Tibetan glacier: Analysis of temporal variations and model estimates of sources and radiative forcing. *Atmospheric Chemistry and Physics.* 15(3), 1191–1204.

Wang, M., Xu, B., Kaspari, S.D., Gleixner, G., Schwab, V.F., Zhao, H., Wang, H. and Yao, P. 2015b. Century-long record of black carbon in an ice core from the Eastern Pamirs: Estimated contributions from biomass burning. *Atmospheric Environment.* 115, 79–88.

Wang, Q.Y., Huang, R.J., Cao, J.J., Tie, X.X., Ni, H.Y., Zhou, Y.Q., Han, Y.M., Hu, T.F., Zhu, C.S., Feng, T. and Li, N. 2015c. Black carbon aerosol in winter northeastern Qinghai–Tibetan Plateau, China: The source, mixing state and optical property. *Atmospheric Chemistry and Physics.* 15(22), 13059–13069.

Xia, X., Zong, X., Cong, Z., Chen, H., Kang, S. and Wang, P. 2011. Baseline continental aerosol over the central Tibetan plateau and a case study of aerosol transport from South Asia. *Atmospheric Environment*. 45(39), 7370–7378.

Xu, B., Yao, T., Liu, X., Wang, N. 2006. Elemental and organic carbon measurements with a two-step heating-gas chromatography system in snow samples from the Tibetan Plateau. *Annals of Glaciology*. 43, 257–262.

Xu, B., Cao, J., Hansen, J., Yao, T., Joswia, D.R., Wang, N., Wu, G., Wang, M., Zhao, H., Yang, W. and Liu, X. 2009a. Black soot and the survival of Tibetan glaciers. *Proceedings of the National Academy of Sciences*. 106(52), 22114–22118.

Xu, B.Q., Wang, M., Joswiak, D.R., Cao, J.J., Yao, T.D., Wu, G.J., Yang, W. and Zhao, H.B. 2009b. Deposition of anthropogenic aerosols in a southeastern Tibetan glacier. *Journal of Geophysical Research: Atmospheres*. 43 (D17), 257–262.

Xu, B., Cao, J., Joswiak, D.R., Liu, X., Zhao, H. and He, J. 2012. Post-depositional enrichment of black soot in snow-pack and accelerated melting of Tibetan glaciers. *Environmental Research Letters*. 7(1), 014022.

Xu, J., Zhang, Q., Li, X., Ge, X., Xiao, C., Ren, J. and Qin, D. 2013. Dissolved organic matter and inorganic ions in a central Himalayan glacier insights into chemical composition and atmospheric sources. *Environmental Science & Technology*. 47(12), 6181–6188.

Xu, C., Ma, Y.M., Panday, A., Cong, Z.Y., Yang, K., Zhu, Z.K., Wang, J.M., Amatya, P.M. and Zhao, L., 2014. Similarities and differences of aerosol optical properties between southern and northern sides of the Himalayas. *Atmospheric Chemistry and Physics*. 14(6), 3133–3149.

Xu, C., Ma, Y.M., You, C. and Zhu, Z.K. 2015. The regional distribution characteristics of aerosol optical depth over the Tibetan Plateau. *Atmospheric Chemistry and Physics*. 15(20), 12065.

Xu, R., Tie, X., Li, G., Zhao, S., Cao, J., Feng, T. and Long, X. 2018. Effect of biomass burning on black carbon (BC) in South Asia and Tibetan Plateau: The analysis of WRF-Chem modeling. *Science of The Total Environment*. 645, 901–912.

Yang, S., Xu, B., Cao, J., Zender, C.S. and Wang, M. 2015. Climate effect of black carbon aerosol in a Tibetan Plateau glacier. *Atmospheric Environment*. 111, 71–78.

Yao, T., Thompson, L., Yang, W., Yu, W., Gao, Y., Guo, X., Yang, X., Duan, K., Zhao, H., Xu, B. and Pu, J. 2012. Different glacier status with atmospheric circulations in Tibetan Plateau and surroundings. *Nature Climate Change*. 2(9), 663–667.

Yao, T., Wu, F., Ding, L., Sun, J., Zhu, L., Piao, S., Deng, T., Ni, X., Zheng, H. and Ouyang, H. 2015. Multispherical interactions and their effects on the Tibetan Plateau's earth system: A review of the recent researches. *National Science Review*. 2(4), 468–488.

Yao, T., Xue, Y., Chen, D., Chen, F., Thompson, L., Cui, P., Koike, T., Lau, W.K.M., Lettenmaier, D., Mosbrugger, V. and Zhang, R. 2019. Recent third pole's rapid warming accompanies cryospheric melt and water cycle intensification and interactions between monsoon and environment: Multidisciplinary approach with observations, modeling, and analysis. *Bulletin of the American Meteorological Society*. 100(3), 423–444.

You, C. and Xu, C. 2018. Review of levoglucosan in glacier snow and ice studies: Recent progress and future perspectives. *Science of The Total Environment*. 616–617, 1533–1539.

You, C., Gao, S.P. and Xu, C. 2015. Biomass burning emissions contaminate winter snowfalls in urban Beijing: A case study in 2012. *Atmospheric Pollution Research*. 6(3), 376–381.

You, C., Xu, C., Xu, B., Zhao, H. and Song, L. 2016a. Levoglucosan evidence for biomass burning records over Tibetan glaciers. *Environmental Pollution*. 216, 173–181.

You, C., Yao, T., Xu, B., Xu, C., Zhao, H. and Song, L. 2016b. Effects of sources, transport, and postdepositional processes on levoglucosan records in southeastern Tibetan glaciers. *Journal of Geophysical Research: Atmospheres*. 121(14), 8701–8711.

You, C., Yao, T.D., Xu, C. and Song, L.L. 2017. Levoglucosan on Tibetan glaciers under different atmospheric circulations. *Atmospheric Environment*. 152, 1–5.

You, C., Yao, T.D. and Xu, C. 2018. Recent increases in wildfires in the Himalayas and surrounding regions detected in Central Tibetan ice core records. *Journal of Geophysical Research: Atmospheres.* 123(6), 3285–3291.

You, C., Yao, T. and Xu, C. 2019. Environmental significance of levoglucosan records in a central Tibetan ice core. *Science Bulletin.* 64(2), 122–127.

Zennaro, P., Kehrwald, N., McConnell, J.R., Schüpbach, S., Maselli, O.J., Marlon, J., Vallelonga, P., Leuenberger, D., Zangrando, R., Spolaor, A. and Borrotti, M. 2014. Fire in ice: Two millennia of boreal forest fire history from the Greenland NEEM ice core. *Climate of the Past.* 10(5), 1905–1924.

Zennaro, P., Kehrwald, N., Marlon, J., Ruddiman, W.F., Brücher, T., Agostinelli, C., Dahl-Jensen, D., Zangrando, R., Gambaro, A. and Barbante, C. 2015. Europe on fire three thousand years ago: Arson or climate? *Geophysical Research Letters.* 42(12), 5023–2033.

Zhao, Z., Cao, J., Shen, Z., Xu, B., Zhu, C., Chen, L.W.A., Su, X., Liu, S., Han, Y., Wang, G. and Ho, K. 2013. Aerosol particles at a high-altitude site on the Southeast Tibetan Plateau, China: Implications for pollution transport from South Asia. *Journal of Geophysical Research: Atmospheres.* 118(19), 11–360.

Zheng, J., Hu, M., Du, Z., Shang, D., Gong, Z., Qin, Y., Fang, J., Gu, F., Li, M., Peng, J. and Li, J., 2017. Influence of biomass burning from South Asia at a high-altitude mountain receptor site in China. *Atmospheric Chemistry and Physics.* 17(11), 6853.

Index

Note: Page numbers in **bold** and *italics* refer to **tables** and *figures*.

299

Milton Keynes UK
Ingram Content Group UK Ltd.
UKHW022037141024
449569UK00014B/640

9 781032 013510